New Advances into Nanostructured Oxides

New Advances into Nanostructured Oxides

Editor

Roberto Nisticò

MDPI • Basel • Beijing • Wuhan • Barcelona • Belgrade • Manchester • Tokyo • Cluj • Tianjin

Editor
Roberto Nisticò
Department of Materials Science
University of Milano-Bicocca
Milano
Italy

Editorial Office
MDPI
St. Alban-Anlage 66
4052 Basel, Switzerland

This is a reprint of articles from the Special Issue published online in the open access journal *Inorganics* (ISSN 2304-6740) (available at: www.mdpi.com/journal/inorganics/special_issues/nanostructured_oxides).

For citation purposes, cite each article independently as indicated on the article page online and as indicated below:

LastName, A.A.; LastName, B.B.; LastName, C.C. Article Title. *Journal Name* **Year**, *Volume Number*, Page Range.

ISBN 978-3-0365-7523-0 (Hbk)
ISBN 978-3-0365-7522-3 (PDF)

© 2023 by the authors. Articles in this book are Open Access and distributed under the Creative Commons Attribution (CC BY) license, which allows users to download, copy and build upon published articles, as long as the author and publisher are properly credited, which ensures maximum dissemination and a wider impact of our publications.
The book as a whole is distributed by MDPI under the terms and conditions of the Creative Commons license CC BY-NC-ND.

Contents

About the Editor vii

Roberto Nisticò
New Advances into Nanostructured Oxides
Reprinted from: *Inorganics* 2023, 11, 130, doi:10.3390/inorganics11030130 1

Davide Palma, Francesca Deganello, Leonarda Francesca Liotta, Valeria La Parola, Alessandra Bianco Prevot and Mery Malandrino et al.
Main Issues in the Synthesis and Testing of Thermocatalytic Ce-Doped SrFeO$_3$ Perovskites for Wastewater Pollutant Removal
Reprinted from: *Inorganics* 2023, 11, 85, doi:10.3390/inorganics11020085 7

Arianna Actis, Francesca Sacchi, Christos Takidis, Maria Cristina Paganini and Erik Cerrato
Changes in Structural, Morphological and Optical Features of Differently Synthetized C$_3$N$_4$-ZnO Heterostructures: An Experimental Approach
Reprinted from: *Inorganics* 2022, 10, 119, doi:10.3390/inorganics10080119 25

Maria I. Chebanenko, Sofia M. Tikhanova, Vladimir N. Nevedomskiy and Vadim I. Popkov
Synthesis and Structure of ZnO-Decorated Graphitic Carbon Nitride (g-C$_3$N$_4$) with Improved Photocatalytic Activity under Visible Light
Reprinted from: *Inorganics* 2022, 10, 249, doi:10.3390/inorganics10120249 39

Ridha Ben Said, Seyfeddine Rahali, Mohamed Ali Ben Aissa, Abuzar Albadri and Abueliz Modwi
Uptake of BF Dye from the Aqueous Phase by CaO-g-C$_3$N$_4$ Nanosorbent: Construction, Descriptions, and Recyclability
Reprinted from: *Inorganics* 2023, 11, 44, doi:10.3390/inorganics11010044 51

Michel F. G. Pereira, Mayane M. Nascimento, Pedro Henrique N. Cardoso, Carlos Yure B. Oliveira, Ginetton F. Tavares and Evando S. Araújo
Preparation, Microstructural Characterization and Photocatalysis Tests of V^{5+}-Doped TiO$_2$/WO$_3$ Nanocomposites Supported on Electrospun Membranes
Reprinted from: *Inorganics* 2022, 10, 143, doi:10.3390/inorganics10090143 67

Marcos E. Peralta, Alejandro Koffman-Frischknecht, M. Sergio Moreno, Daniel O. Mártire and Luciano Carlos
Application of Biobased Substances in the Synthesis of Nanostructured Magnetic Core-Shell Materials
Reprinted from: *Inorganics* 2023, 11, 46, doi:10.3390/inorganics11010046 85

Sander Dekyvere, Mohamed Elhousseini Hilal, Somboon Chaemchuen, Serge Zhuiykov and Francis Verpoort
Sacrificial Zinc Oxide Strategy-Enhanced Mesoporosity in MIL-53-Derived Iron–Carbon Composite for Methylene Blue Adsorption
Reprinted from: *Inorganics* 2022, 10, 59, doi:10.3390/inorganics10050059 99

Mohamed F. Nawar, Alaa F. El-Daoushy, Ahmed Ashry, Mohamed A. Soliman and Andreas Türler
Evaluating the Sorption Affinity of Low Specific Activity ^{99}Mo on Different Metal Oxide Nanoparticles
Reprinted from: *Inorganics* 2022, 10, 154, doi:10.3390/inorganics10100154 111

Chun-Wei Chang, Chen-Han Tsou, Bai-Hung Huang, Kuo-Sheng Hung, Yung-Chieh Cho and Takashi Saito et al.
Fabrication of a Potential Electrodeposited Nanocomposite for Dental Applications
Reprinted from: *Inorganics* **2022**, *10*, 165, doi:10.3390/inorganics10100165 . . . . **125**

Silvia Sfameni, Mariam Hadhri, Giulia Rando, Dario Drommi, Giuseppe Rosace and Valentina Trovato et al.
Inorganic Finishing for Textile Fabrics: Recent Advances in Wear-Resistant, UV Protection and Antimicrobial Treatments
Reprinted from: *Inorganics* **2023**, *11*, 19, doi:10.3390/inorganics11010019 . . . . **135**

Francesca Scarpelli, Nicolas Godbert, Alessandra Crispini and Iolinda Aiello
Nanostructured Iridium Oxide: State of the Art
Reprinted from: *Inorganics* **2022**, *10*, 115, doi:10.3390/inorganics10080115 . . . . **167**

Roberto Nisticò
A Comprehensive Study on the Applications of Clays into Advanced Technologies, with a Particular Attention on Biomedicine and Environmental Remediation
Reprinted from: *Inorganics* **2022**, *10*, 40, doi:10.3390/inorganics10030040 . . . . **189**

About the Editor

Roberto Nisticò

Roberto Nisticò (RN) is Fixed-Time Assistant Professor (RTD-b), i.e., tenure track Associate Professor, in General and Inorganic Chemistry at the University of Milano-Bicocca (Department of Materials Science, Italy). Up until today, RN published more than 70 papers on renowned international journals (mostly as first and/or corresponding author) and two book chapters. Additionally, RN is member of the Editorial Board of the journal *Inorganics* (Inorganic Materials Section). The research activity of RN mainly focuses on several aspects at the interface between inorganic chemistry, nanotechnology and materials science, always looking for novel and appealing solutions for a sustainable future. The principal field of interest is the development of magnet-responsive nanomaterials (i.e., iron oxides, ferrites) and other inorganic systems for environmental remediation of contaminated wastewater, (photo)catalysis, and for energetic applications. Other fields of interest are functional coatings and surface functionalization of nanoparticles, mesoporous oxides, metallic nanomaterials, and inorganic fillers for new generations of nanocomposites.

Editorial

New Advances into Nanostructured Oxides

Roberto Nisticò

Department of Materials Science, INSTM, University of Milano-Bicocca, Via R. Cozzi 55, 20125 Milano, Italy; roberto.nistico@unimib.it; Tel.: +39-02-6448-5111

Inorganic nanostructured (metal) oxides are a large class of inorganic materials extensively investigated for their unique and outstanding properties that allow for their use within a multitude of technological fields of emerging interest, such as (photo)catalysis, environmental remediation processes, energy storage, controlled transport and/or release of drugs and chemicals, biomedicine, sensing, development of smart materials, stimuli-responsive materials, and nanocomposites [1–6]. The large applicability of nanostructured (metal) oxides is primarily due to the possibility of easily exerting a high level of control in the design of their morphology, in particular in terms of particle dimensions, shapes, and surface porosities, by means of proper synthetic routes and innovative templating processes [7–9]. However, exerting morphological control might be not always enough. In fact, even surface residual functionalities and surface reactivity play key roles in the determination of a nanomaterial's final properties [10]. Hence, for this reason, nanostructured (metal) oxides often require surface functionalization with specific chemical moieties (thus, eventually forming inorganic–organic hybrid systems) to further extend their field of application.

Nowadays, the research on this class of inorganic materials is still very dynamic, as documented by the recent achievements in the field, and it is expected to further increase in the coming years, with particular attention on the continuous search for more ecofriendly synthesis and templating approaches alternative to the traditional inorganic chemistry routes [11]. This Special Issue, entitled "New Advances into Nanostructured Oxides", is specifically dedicated to the recent advancements in the design of the most promising nanostructured (metal) oxides and their technological application. Altogether, this Special Issue collects nine original research articles and three review papers, covering diverse nanostructured (metal) oxides and different technological applications.

Palma et al. [12] reported the effects of the experimental parameters over the solution combustion synthesis of Ce-doped Sr ferrates with the perovskite structure. The authors demonstrated that the full gelification of the *sol* is crucial to obtain single-phase perovskite materials, avoiding secondary phase formation and cerium oxide segregation, whereas the presence of secondary phases might induce the formation of Sr carbonate at the perovskite's surface that negatively affects the material's catalytic activity. Interestingly, the authors also evaluated that the presence of segregated cerium oxide has a positive effect on the availability of oxygen vacancies in low temperature conditions, thus favoring the interaction with polar substrates and bettering the material's catalytic activity. Concerning the thermocatalytic activity of these nanostructured materials, tests were performed against two different target substrates, namely: Orange II dye and Bisphenol A at alkaline pH (to avoid metal leaching from perovskites). Experimental results indicated that the best performances are obtained for samples containing the highest amount of segregated cerium oxide, probably due to the formation of a higher number of oxygen vacancies, which are available during the catalytic process.

The formation of heterojunctions between C_3N_4 and nanostructured (metal) oxides is a very promising technological solution to exploit sunlight (i.e., visible frequencies) in a multitude of photocatalytic processes, such as fuel production (i.e., CO_2 photo-reduction, water photo-splitting), and photo-degradation of contaminants in wastewater. In this context, Actis et al. [13] systematically investigated different synthesis techniques used to prepare bare C_3N_4 and combined C_3N_4/ZnO mixed systems. In particular, the authors compared different synthetic protocols, namely: (i) the formation of combined C_3N_4/ZnO mixed systems from supramolecular precursors, (ii) the direct growth of C_3N_4 on pre-formed ZnO nanoparticles, and (iii) the ultrasonic/mechanical solid-state mixing of both pre-formed C_3N_4 and ZnO nanopowders. Experimental data pointed out a trend existing between these three different synthetic processes. In detail, in the case of the synthesis from supramolecular precursors, a tighter association between the two phases forming the heterojunction has been recognized (with the appearance of adsorption bands in the visible region) coupled with low crystallinity. An opposite trend was registered in the case of ultrasonic/mechanical solid-state mixing. Crystalline materials with a modified morphology are obtained this way, with the consequent formation of a high-crystalline heterojunction characterized by lower visible light harvesting, deriving from the minor cooperation between the two components at the interface. Lastly, the approach based on C_3N_4 growth on pre-formed ZnO shows an intermediate condition in terms of both crystallinity and optical activity under visible frequencies. Chebanenko et al. [14] reported the production of ZnO-decorated graphitic carbon nitride (g-C_3N_4) nanocomposites synthesized following a three-step process, namely: (i) heat treatment of urea, (ii) ultrasonic exfoliation of the colloidal solution by introducing different ratios of ZnO precursor, and (iii) final thermal treatment. The authors reported a very interesting photocatalytic system by placing g-C_3N_4 in close contact with ZnO nanoparticles, since the formation of a heterojunction between these two components allows for the achievement of both excitation by visible light and a better charge separation with the formation of O-containing radicals. Nanocomposites were tested in the photocatalytic abatement of Methylene Blue dye as a target molecule under visible light. Experimental results indicated that the best performances were obtained for the nanocomposite containing 7.5 wt.% of ZnO. Moreover, Said et al. [15] reported the synthesis of a mesoporous g-C_3N_4/CaO nanocomposite using a step-by-step ultrasonication technique. Furthermore, such a composite was successfully tested in the capture of Basic Fuchsin dye by means of a sorption mechanism. After the optimization of the sorption conditions, it was demonstrated that: (i) dye sorption is pH-independent, (ii) interaction between nanocomposite and dye follows the Freundlich model, (iii) sorption capacity increases with contact time (reaching an equilibrium after ca. 30 min), and (iv) maximum sorption capacity at the optimized condition is quantified as 813 mg/g. Furthermore, the authors experimentally demonstrated that the adsorption mechanism involves both hydrogen bonds and π-π stacking interactions.

Pereira et al. [16] reported the preparation of V-doped titania/WO_3 (1:1 molar ratio) nanocomposites with different dopant concentrations and dispersed them in high-surface-area fibrous polymeric membranes obtained by means of electrospinning deposition. Evidence demonstrated that: (i) V^{5+} ions doped the atomic structure of titania (anatase phase), (ii) the titania crystallite sizes slightly increased by increasing the dopant concentration, and (iii) both W^{6+} and V^{5+} occupies Ti^{4+} vacancies in the anatase crystal lattice, allowing for a greater degree of local vibrations, which favors the occurrence of photocatalytic processes. Furthermore, the introduction of V into the TiO_2/WO_3 system caused a significant decrease in the titania bandgap as it generates an n-type extrinsic semiconductor. Electrospun fibrous membranes containing the nanocomposites were successfully tested in the photocatalytic abatement of Rhodamine B dye from the water environment under visible light. Experimental results indicated that the best performances (92% dye degradation after 2 h of visible light irradiation) are reached by the nanocomposite containing 5 wt.% of vanadium oxide, as it shows an improved photoresponse and charge separation efficiency, coupled with the presence of high active site density.

Peralta et al. [17] reported novel ecofriendly routes for the synthesis of nanostructured core–shell magnetic particles based on the polyol method and using biowaste-derived bio-based substances (BBSs) as both stabilizers and structure-directing agents. Subsequently, the covering of such BBS-stabilized magnetite nanoparticles with either mesoporous silica shell or titanium dioxide was performed by means of a sol–gel method carried out in alkaline environment. Experimental results evidenced that the size/shape and the aggregation of the magnetite nanoparticles strongly depend on the starting BBS concentration (the higher the concentration of BBS, the smaller the sizes of the magnetite nanoparticles). Additionally, the presence of BBS is also crucial concerning the growth of both the mesoporous silica shell and the homogeneous titania shell, probably due to interactions involving the carboxylic groups of BBS. Such core–shell systems were successfully tested as nanocarriers of ibuprofen and adsorbing agents against Methylene Blue dye from aqueous solution, and as photocatalysts in the degradation of Methylene Blue dye. In particular, the mesoporous silica core–shell systems show an ibuprofen loading capacity of 13% and a very fast adsorption capacity of Methylene Blue dye, achieving 76% of dye removal within the first 15 min, whereas the titania core–shell systems show a photocatalytic degradation efficiency against Methylene Blue dye, analogous to bare titania. This last result is very surprising (and potentially a breakthrough), as the direct contact between magnetite and titania typically brings an unfavorable heterojunction, which accelerates the recombination of the electron–hole pairs and weakens the photocatalytic activity of the catalyst.

Dekyvere et al. [18] reported a novel strategy for the fabrication of nano zero-valent-iron (nZVI) carbon-based mesoporous materials derived from iron-containing metal-organic frameworks (MOFs) using zinc oxide nanorods as sacrificial nuclei. Here, the breakthrough idea is the utilization of the nanostructured zinc oxide under the shape of nanorods (i.e., aspect ratio 10) as sacrificial consumable nuclei for shaping the final highly porous materials (with specific surface area in the 185–270 m^2/g range). The mesoporous materials thus obtained were tested as adsorbing agents against Methylene Blue dye, reaching an impressive sorption of 78% after only 2 min and close to total sorption (99%) after 6 min. The best-performing system also showed excellent recyclability, with 99% of abatement after 60 min of contact time for 10 consecutive cycles. As highlighted by the authors, the high number of mesopores produced by the sacrificial zinc oxide rods significantly enhanced the adsorption capability of the nanomaterial by increasing the contact area with the adsorbate. Furthermore, this innovative strategy opens the possibility of exerting further control over the pore size when forming the final porous material by changing the morphology of the starting nanostructured sacrificial nuclei.

Newar et al. [19], instead, tested the use of different commercial (metal) oxide nanoparticles as sorbing systems against low specific activity ^{99}Mo radiotracer solution. The commercial (metal) oxides investigated were: silica, titania, zirconium oxide, cerium oxide, tin oxide, and mixed systems, such as an aluminosilicate, cerium aluminum oxide, aluminum titanium oxide, and cerium zirconium oxide. The authors evaluated the adsorption behavior under different experimental conditions (pH, initial concentration of Mo, contact time, and temperature). Experimental results showed that: (i) the optimum adsorption pH for all systems is pH 2–4, (ii) the Freundlich isotherm model fitted the experimental data (mainly physisorption), (iii) the maximum adsorption capacity at optimized conditions was registered by zirconium oxide (73 mg/g), followed by cerium oxide and titania. This study clearly demonstrates that the investigated nanostructured (metal) oxides showed higher static sorption capacities than conventionally used alumina (2–20 mg/g) and potentially opens up the possibility of further improving the sorption capacity of these systems by varying both morphologies and porosities of these nanostructured (metal) oxides.

Chang et al. [20] evaluated the possibility of using a metal matrix nanocomposite reinforced with titania nanoparticles towards a potential endodontic instrument using a co-deposition approach with pulse electroplating for dental applications. In this study, the authors registered that the electrodeposition of the nanocomposite produced a smooth sidewall and surface, thus maintaining structural integrity and revealing a promoted formability, whereas cytotoxicity tests revealed high cell viability, high cell proliferation behavior and excellent biocompatibility.

Sfameni et al. [21] contributed to this Special Issue with an interesting overview dealing with the surface modification of textile fabrics to confer specific implemented and new properties by means of inorganic sol–gel approaches. In this review, different examples and methodologies based on sol–gel inorganic coatings for textile finishing are evaluated, considering a wide range of inorganic precursors and functional additives to fabricate textiles showing different/implemented properties. Moreover, with respect to conventional textile finishing processes, synthetic routes based on functional nanoparticles and nanosols represent a more ecofriendly and safe approach than the most commonly employed formulations containing harmful substances. In the present review, sol modification techniques and functional applications for textile materials are classified into three sections, properly discussed with examples taken from the literature. Inorganic protocols are organized into methods for improving wear resistance of textile fabrics, for UV protection textile finishing, and for antimicrobial textile finishing, with a further classification based on the active nanoparticles involved.

Scarpelli et al. [22] reviewed the recent advancements in the synthesis of nanostructured iridium oxide. Iridium oxide is a very peculiar (metal) oxide as it is characterized by showing metallic-type conductivity and also displaying a low surface work function (which is the energy required for moving an electron from the Fermi level to the local vacuum level), high chemical stability, and good stability under the influence of high electric fields. Iridium oxide can be synthesized as various nanostructures, which significantly affect its surface work function. In this regard, the authors provided a very interesting overview of the nanostructuring of iridium oxide, with a special emphasis on the different strategies to drive the synthesis toward specific nanostructures. In this context, the authors organized the text into four sections, dedicated to the description of the different synthesis methods to obtain spherical nanoparticles, one-dimensional nanostructures (e.g., nanotubes, nanorods, nanowires, and nanofibers), unusual shapes (e.g., urchin-like, and nanoneedles), and thin films.

Nisticò [23] reviewed the recent advances in the exploitation of naturally occurring clays (phyllosilicates), including more non-conventional applications. Within the entire review, the author analyzed clays' peculiar properties, such as their ability to exchange (capture) ions, their layered structure, surface area and reactivity, and their biocompatibility, pointing out the deep correlation existing between the field of application and the structure–property relationships involved. After providing an introduction mainly focused on the economic analysis of the global trade in clays, the review was organized into two mains sections. Section I is dedicated to the classification of clays based on their structural and chemical composition, together with a schematic summary of the main relevant structure-induced properties, which are strongly correlated to the nature and quantity of chemical species at the interlayer, exchange capacity, interlayer thickness, surface area, and hydration/gel-forming capacity. Section II, instead, is dedicated to the analysis of the unconventional advanced uses of clays in technological fields of emerging interest, providing a further organization into biomedical applications (ranging from the development of scaffolds and drug delivery systems to cancer diagnosis and therapy), environmental applications (wastewater treatments and membrane technology), additive manufacturing, and sol–gel processes.

Personally, the hope is that readers will enjoy this Special Issue, now edited as a book. Many examples of interesting and challenging research are collected in this "scientific compilation". In some cases, the focus was on the nanostructuring of (metal) oxides, in others, their exploitation towards target advanced applications, but the common thread remains the continuous progress of this important class of inorganic materials.

Acknowledgments: I personally would like to thank all authors, reviewers, and the entire editorial staff of *Inorganics* who assisted me in the realization of the present Special Issue.

Conflicts of Interest: The author declares no conflict of interest.

References

1. Fernandez-Garcia, M.; Martinez-Arias, A.; Hanson, J.C.; Rodriguez, J.A. Nanostructured oxides in chemistry: Characterization and properties. *Chem. Rev.* **2004**, *104*, 4063–4104. [CrossRef] [PubMed]
2. Ciriminna, R.; Fidalgo, A.; Pandarus, V.; Beland, F.; Ilharco, L.M.; Pagliaro, M. The sol-gel route to advanced silica-based materials and recent applications. *Chem. Rev.* **2013**, *113*, 6592–6620. [CrossRef] [PubMed]
3. Aricò, A.S.; Bruce, P.; Scrosati, B.; Tarascon, J.-M.; van Schalkwijk, W. Nanostructured materials for advanced energy conversion and storage devices. *Nat. Mater.* **2005**, *4*, 366–377. [CrossRef]
4. Cheng, Y.-J.; Hu, J.-J.; Qin, S.-Y.; Zhang, A.-Q.; Zhang, X.-Z. Recent advances in functional mesoporous silica-based nanoplatforms for combinational photo-chemotherapy of cancer. *Biomaterials* **2020**, *232*, 119738. [CrossRef] [PubMed]
5. Polliotto, V.; Pomilla, F.R.; Maurino, V.; Marcì, G.; Bianco Prevot, A.; Nisticò, R.; Magnacca, G.; Paganini, M.C.; Ponce Robles, L.; Perez, L.; et al. Different approaches for the solar photocatalytic removal of micro-contaminants from aqueous environment: Titania vs. hybrid magnetic iron oxides. *Catal. Today* **2019**, *328*, 164–171. [CrossRef]
6. Jadhav, S.A.; Nisticò, R.; Magnacca, G.; Scalarone, D. Packed hybrid silica nanoparticles as sorbents with thermo-switchable surface chemistry and pore size for fast extraction of environmental pollutants. *RSC Adv.* **2018**, *8*, 1246–1254. [CrossRef]
7. Kaur, A.; Bajaj, B.; Kaushik, A.; Saini, A.; Sud, A. A review on template assisted synthesis of multi-functional metal oxide nanostructures: Status and prospects. *Mater. Sci. Eng. B* **2022**, *286*, 116005. [CrossRef]
8. Pal, N.; Bhaumik, A. Soft templating strategies for the synthesis of mesoporous materials: Inorganic, organic–inorganic hybrid and purely organic solids. *Adv. Colloid Interface Sci.* **2013**, *189–190*, 21–41. [CrossRef]
9. Lu, A.-H.; Schuth, F. Nanocasting: A versatile strategy for creating nanostructured porous materials. *Adv. Mater.* **2006**, *18*, 1793–1805. [CrossRef]
10. Wieszczycka, K.; Staszak, K.; Wozniak-Budych, M.J.; Litowczenko, J.; Maciejewska, B.M.; Jurga, S. Surface functionalization—The way for advanced applications of smart materials. *Coord. Chem. Rev.* **2021**, *436*, 213846. [CrossRef]
11. Gregory, D.H. Innovative inorganic synthesis. *Inorganics* **2014**, *2*, 552–555. [CrossRef]
12. Palma, D.; Deganello, F.; Liotta, L.F.; La Parola, V.; Bianco Prevot, A.; Malandrino, M.; Laurenti, E.; Boffa, V.; Magnacca, G. Main issues in the synthesis and testing of thermocatalytic Ce-doped SrFeO3 perovskites for wastewater pollutant removal. *Inorganics* **2023**, *11*, 85. [CrossRef]
13. Actis, A.; Sacchi, F.; Takidis, C.; Paganini, M.C.; Cerrato, E. Changes in structural, morphological and optical features of differently synthetized C3N4-ZnO heterostructures: An experimental approach. *Inorganics* **2022**, *10*, 119. [CrossRef]
14. Chebanenko, M.I.; Tikhanova, S.M.; Nevedomskiy, V.N.; Popkov, V.I. Synthesis and structure of ZnO-decorated graphitic carbon nitride (g-C3N4) with improved photocatalytic activity under visible light. *Inorganics* **2022**, *10*, 249. [CrossRef]
15. Said, R.B.; Rahali, S.; Ben Aissa, M.A.; Albadri, A.; Modwi, A. Uptake of BF dye from the aqueous phase by CaO-g-C3N4 nanosorbent: Construction, descriptions, and recyclability. *Inorganics* **2023**, *11*, 44. [CrossRef]
16. Pereira, M.F.G.; Nascimento, M.M.; Cardoso, P.H.N.; Oliveira, C.Y.B.; Tavares, G.F.; Araujo, E.S. Preparation, Microstructural characterization and photocatalysis tests of V^{5+}-doped TiO2/WO3 nanocomposites supported on electrospun membranes. *Inorganics* **2022**, *10*, 143. [CrossRef]
17. Peralta, M.E.; Koffman-Frischknecht, A.; Moreno, M.S.; Martire, D.O.; Carlos, L. Application of biobased substances in the synthesis of nanostructured magnetic core-shell materials. *Inorganics* **2023**, *11*, 46. [CrossRef]
18. Dekyvere, S.; Hilal, M.E.; Chaemchuen, S.; Zhuiykov, S.; Verpoort, F. Sacrificial zinc oxide strategy-enhanced mesoporosity in MIL-53-derived iron-carbon composite for methylene blue adsorption. *Inorganics* **2022**, *10*, 59. [CrossRef]
19. Newar, M.F.; El-Daoushy, A.F.; Ashry, A.; Soliman, M.A.; Turler, A. Evaluating the sorption affinity of low specific activity [99]Mo on different metal oxide nanoparticles. *Inorganics* **2022**, *10*, 154. [CrossRef]
20. Chang, C.-W.; Tsou, C.-H.; Huang, B.-H.; Hung, K.-S.; Cho, Y.-C.; Saito, T.; Tsai, C.-H.; Hsieh, C.-C.; Liu, C.-M.; Lan, W.-C. Fabrication of a potential electrodeposited nanocomposite for dental applications. *Inorganics* **2022**, *10*, 165. [CrossRef]
21. Sfameni, S.; Hadhri, M.; Rando, G.; Drommi, D.; Rosace, G.; Trovato, V.; Plutino, M.R. Inorganic finishing for textile fabrics: Recent advances in wear-resistant, UV protection and antimicrobial treatments. *Inorganics* **2023**, *11*, 19. [CrossRef]
22. Scarpelli, F.; Godbert, N.; Crispini, A.; Aiello, I. Nanostructured iridium oxide: State of the art. *Inorganics* **2022**, *10*, 115. [CrossRef]
23. Nisticò, R. A comprehensive study on the applications of clays into advanced technologies, with a particular attention on biomedicine and environmental remediation. *Inorganics* **2022**, *10*, 40. [CrossRef]

Disclaimer/Publisher's Note: The statements, opinions and data contained in all publications are solely those of the individual author(s) and contributor(s) and not of MDPI and/or the editor(s). MDPI and/or the editor(s) disclaim responsibility for any injury to people or property resulting from any ideas, methods, instructions or products referred to in the content.

Article

Main Issues in the Synthesis and Testing of Thermocatalytic Ce-Doped SrFeO$_3$ Perovskites for Wastewater Pollutant Removal

Davide Palma [1], Francesca Deganello [2], Leonarda Francesca Liotta [2], Valeria La Parola [2], Alessandra Bianco Prevot [1,*], Mery Malandrino [1], Enzo Laurenti [1], Vittorio Boffa [3] and Giuliana Magnacca [1,4]

1. Dipartimento di Chimica, Università di Torino, Via P. Giuria 7, 10125 Torino, Italy
2. Istituto per lo Studio dei Materiali Nanostrutturati (ISMN)—Consiglio Nazionale delle Ricerche, 90146 Palermo, Italy
3. Department of Chemistry and Biochemistry, Aalborgh University, 9220 Aalborg, Denmark
4. NIS and INSTM Reference Centre, Via P. Giuria 7, 10125 Torino, Italy
* Correspondence: alessandra.biancoprevot@unito.it

Abstract: The effect of the synthesis and processing parameters on the thermocatalytic performance of Ce-doped SrFeO$_3$ inorganic perovskites was investigated to improve the reproducibility and reliability of the synthetic methodology and of the testing procedure. A structural, surface and redox characterization was performed to check the extent of variability in the chemical–physical properties of the prepared materials, revealing that a strict control of the synthesis parameters is indeed crucial to optimize the thermocatalytic properties of Ce-doped SrFeO$_3$ inorganic perovskites. The thermocatalytic tests, aimed to degrade organic pollutants in water, were performed using Orange II and Bisphenol A as target compounds, in view of a later technological application. The main issues in the synthesis and testing of Ce-doped SrFeO$_3$ perovskite thermocatalysts are highlighted and described, giving specific instructions for the resolution of each of them. A limited number of prepared materials showed an efficient thermocatalytic effect, indicating that a full gelification of the sol, an overstoichiometric reducer-to-oxidizer ratio, a nominal cerium content of 15 mol%, slightly higher than its solubility limit (i.e., 14 mol%), a pH of 6 and a thermal treatment at 800 °C/2 h are the best synthesis conditions to obtain an effective Ce-doped SrFeO$_3$ perovskite. Regarding the testing conditions, the best procedure is to follow the degradation reaction without any preconditioning with the pollutant at room temperature. The severe leaching of the active perovskite phase during tests conducted at acidic pH is discussed. Briefly, we suggest confining the application of these materials to a limited pH range. Variability between thermocatalysts prepared in two different laboratories was also checked. The issues discussed and the proposed solutions overcome some of the obstacles to achieving a successful scale up of the synthesis process. Our results were favorable in comparison to those in the literature, and our approach can be successfully extended to other perovskite catalysts.

Keywords: advanced oxidation processes; experimental protocols; heterogeneous catalysis; inorganic thermal activation; organic pollutants; perovskites; physico-chemical characterization; solution combustion synthesis; strontium ferrates; water treatment

1. Introduction

The global increase in the use of freshwater, resulting from population growth, economic development, and changes in consumption patterns, combined with the increasingly evident effects of climate change, has led to increased incidence of water stress and to situations of scarce water resources, even in non-intrinsically arid areas. The problem is not only quantitative but also qualitative, because in many areas of the planet, water use is not followed by a treatment that restores its initial quality, and even where treatments are applied, they are not always completely effective [1]. This has led to a progressive deterioration of water resources with heavy repercussions on the sustainability of our society. It is therefore necessary to rethink water resource management, i.e., to make urban wastewater and industrial wastewater treatment more efficient and extend them worldwide, aiming for a transition from a linear model (capture, use, discharge) to a circular model that facilitates the "regeneration" and reuse of water, minimizing withdrawal from primary sources.

In this context, a crucial aspect is represented by the removal of anthropogenic contaminants of emerging concern (CECs) which are resistant to biological wastewater treatment and are detected at very low concentrations in surface water bodies (and in some cases, in groundwater) [2,3]. Despite the negative impact of these substances on ecosystems and human health, there is still no comprehensive European regulation; nonetheless, there is an increasing focus on their monitoring in anticipation of regulatory compliance. The effective and sustainable removal of CECs will contribute to the "circular" management of water, reducing the impact of climate change on water availability.

Among the possible approaches for CECs removal, the application of Advanced Oxidation Processes (AOPs) has gained interest among researchers [4,5]. AOPs rely on the generation of highly reactive species (above all HO• radicals) which are able to initiate the degradation of pollutants, in many cases degrading through complete oxidation to CO_2, H_2O and inorganic ions [6,7]. Over the years, great attention has been devoted to the application of photocatalysis for CEC degradation, and a variety of novel materials have been prepared and tested [8]. The spectrum of photocatalysts is indeed wide, spanning from semiconductor oxides to composite materials, often aiming to optimize light harvesting [9–11]. One of the main drawbacks of AOPs, and this is especially true for photocatalysis, is that the generation of reactive species is mostly light assisted, thus introducing additional energy costs when artificial light is employed or infrastructural aspects in case of harvesting solar light. Other AOPs operating in dark conditions, on the other hand, typically require the addition of reagents (e.g., ozonation, Fenton process, peroxone).

Recently, the capability has been shown of perovskite-type mixed oxides to be thermally activated and to promote the degradation of organic compounds [12,13], as well as their capability of activating peroxymonosulfate (PMS) in AOPs, where both free and non-free radical pathways may occur [14]. Inorganic perovskite-type materials are a versatile and multifunctional class of mixed oxides with the general formula ABO_3, where A sites are occupied by larger cations, B sites are occupied by smaller cations, and O sites can be occupied by oxygen or by other anions [15]. Several papers rely on the use of $Sr_{0.85}Ce_{0.15}FeO_{3-\delta}$ perovskite, which has been successfully tested (i) for the degradation of model organic pollutants such as azo-dye Orange II [16], acid orange 8 (AO8) [17], Bisphenol A [18], and acetamiprid (AAP), (ii) for the abatement of oil residues in water [19], (iii) in combination with graphene oxide as the active layer deposited over commercial flat-sheet polyethersulfone nanofiltration membranes, yielding an improved catalytic activity for the abatement of bisphenol A [20], (iv) in a novel strategy for water purification that involves the integration of membrane filtration and thermocatalytic pollutants degradation, using an alumina tubular support coated with an Al_2O_3-doped NF silica layer for the filtration step, aiming to simultaneous degrade micropollutants and mitigate fouling [21], and (v) in an integrated process based on membrane distillation and thermocatalytic oxidation, simultaneously using the thermal energy to drive the permeation of pure water through a hydrophobic membrane and to activate the perovskite [18].

Strontium ferrates-based materials belong to a class of perovskite-type compounds in which iron is mainly stabilized as Fe^{4+} [22]. When doped with Cerium at the Sr site, the cubic structure is stabilized and the redox couple Fe^{4+}/Fe^{3+} plays a fundamental role in the functional properties of the strontium ferrates. Solution Combustion Synthesis has been frequently used for the synthesis of Ce-doped $SrFeO_3$ powders due to its versatility and efficiency in terms of energy, time, and effort [23]. However, the large number of synthesis parameters that can be changed in this synthesis can either be considered as an advantage or as an obstacle to the preparation of reproducible and efficient materials, given that a preliminary investigation of the synthesis-structure-properties relationships is missing. In this work, detailed information about the main issues that might be faced in the solution combustion synthesis and testing of Ce-doped SrFeO$_3$ perovskite thermocatalysts is given. The effect of the specific synthesis and processing parameters on the structure, surface, redox behavior, and texture was investigated in order to select the best conditions to obtain an optimized powder. Moreover, the testing procedure was also studied and applied using different samples in order to get reliable information on their thermocatalytic behavior, highlighting the paramount importance of the reproducibility of the synthesis and procedures in the interpretation of the scientific results and the need to create reliable protocols for the synthesis and testing of the materials.

Assessing the synthesis-structure-properties relationships of such materials could open a new technology-oriented field, aiming at preparing reproducible and scalable materials, suitable for being activated at relatively mild temperatures and, in the perspective of application at real scales, exploiting the residual heat which is often available in an industrial context.

2. Results and Discussion

2.1. Synthesis and Characterization

The reproducibility of a solution combustion synthesis technique largely depends on the understanding of which preparation parameters influence the properties of the final material. Therefore, a deep knowledge on the synthesis-properties relationships is mandatory. Additionally, a reliable scale up process requires a high degree of reproducibility to maintain the same material performance observed at the lab scale. In addition, an accurate characterization protocol and a knowledge of the variability in the characterization results helps in determining the real differences among the materials. Therefore, the effect of the synthesis and processing parameters was investigated to ensure high reproducibility of the powder properties. A complementary structural and redox characterization was also performed to check the extent of variability in the characterization results.

2.1.1. Effect of the Synthesis Parameters

A series of powders with the same (or slightly different) nominal composition, $Sr_{0.86}Ce_{0.14}FeO_{3-\delta}$, but prepared using different synthesis parameters, were compared accordingly to some of their chemical–physical properties, with the aim of classifying the importance of the experimental conditions in an effort to obtain reproducible and effective catalysts. In the first comparison, two powders prepared with the same nominal composition, i.e., $Sr_{0.86}Ce_{0.14}FeO_{3-\delta}$ at pH = 7, but using two different degrees of gelification, NPW9-1000 and NPW4-1000, are shown. The use of pH = 6–7 ensures a good interaction between citric acid and metal cations, leading to a complete incorporation of cerium dopant into the perovskite structure, as well as minimizing the addition of ammonia solution into the combustion mixture. In NPW4-1000, the gel was combusted just after its formation from the sol, whereas in NPW9-1000, the gel was left to dry for another 10 min under magnetic stirring, i.e., until the magnetic bar could no longer move due to the full densification of the gel (high gelification degree). Another powder, NPW12-1000, was prepared at pH = 7 using a higher reducer-to-oxidizer ratio, (1.6). The result was a fully dried gel with a slightly different nominal composition of $Sr_{0.85}Ce_{0.15}FeO_{3-\delta}$, where the nominal Ce content exceeded by 1 mol% the maximum solubility of cerium in SrFeO$_3$. NPW4-1000, NPW9-1000, and NPW12-1000 were thermally treated at 1000 °C for 5 h and characterized for their structural (bulk and surface) and redox properties. NPW4-1000 was thermally treated using a temperature ramp of 2 °C/min instead of 10 °C/min, in contrast to the other two samples. However, the effect of the temperature ramp was strongly dependent on the gelification degree, albeit less important than the effect of gelification degree, as will be discussed in Section 2.1.2. The bulk structural characterization results of these three samples are reported in Figure 1.

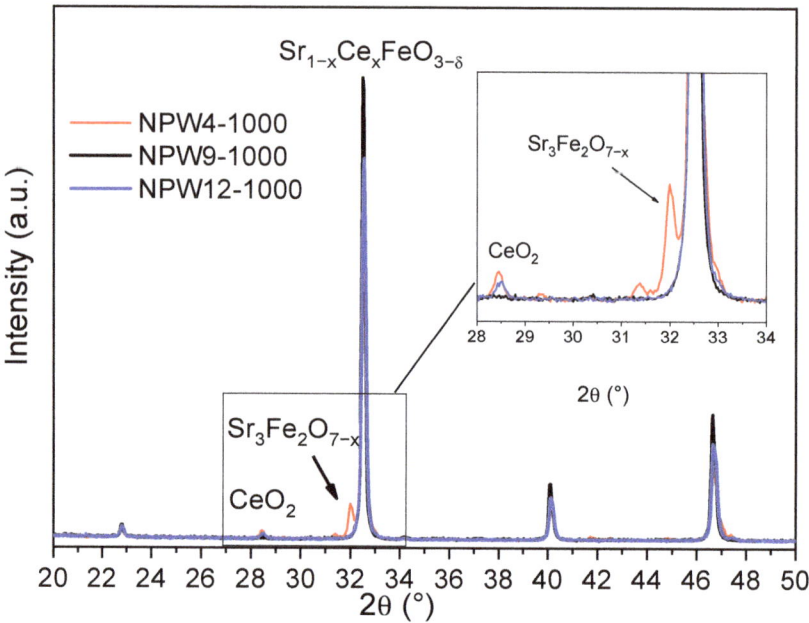

Figure 1. XRD patterns of NPW4-1000, NPW9-1000, and NPW12-1000.

The single-phase perovskite material could be obtained only for the powder prepared with fully dried gels, NPW9-1000, whereas some secondary phases were clearly visible in the sample where the gel was not completely dried, NPW4-1000 (Figure 1). These secondary phases were mainly layered tetragonal perovskites of the type $Sr_3Fe_2O_{7-\delta}$, together with minor percentages of CeO_2, formed to balance the perovskite composition after Sr and Fe segregation as layered tetragonal perovskite phases. In NPW12-1000, only the main perovskite phase was present, together with <1 wt% of segregated CeO_2, and no layered tetragonal perovskites of the type $Sr_3Fe_2O_{7-\delta}$ were formed. Despite the absence of layered tetragonal phases, segregated cerium oxide was expected for this sample, since the maximum Ce solubility in $SrFeO_3$ is lower than 15 mol% [16]. It is worth noting that the XRD pattern of NPW10-1000 (not shown), prepared using identical conditions to those of NPW9-1000, was identical to that of NPW9-1000, demonstrating the importance of controlling the gelification degree. Therefore, full gelification of the sol is a very important synthesis parameter for a good reproducibility of the structural properties, avoiding $Sr_3Fe_2O_{7-\delta}$ layered perovskite and, consequently, cerium oxide segregation.

The same three samples were characterized for their surface structure by X-ray photoelectron spectroscopy (XPS). Figure 2B shows the Sr3d, Fe2p, and O1s regions of the two samples, while Table 1 shows the binding energy and the relative abundance of the different components. The presence of cerium is evidenced by the analysis, but its low intensity and the complexity of the Ce3d peak make the region analysis unmeaningful. The Fe2p region shows a similar profile for the three samples (see Figure 2B) with the typical complex profile of the Fe2p region with the Fe2p3/-Fe2p1/2 separation of 13.3 eV and the presence of a small shake up feature.

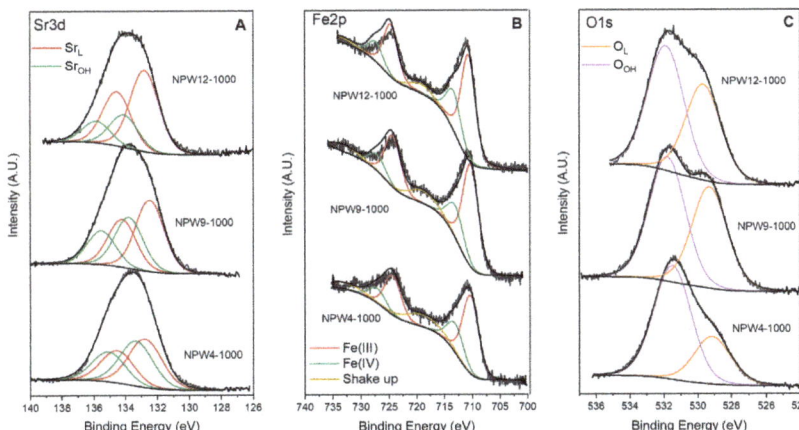

Figure 2. (**A**) Sr 3d 5/2, (**B**) Fe2p3/2 and (**C**) O1s XPS regions for NPW4-1000, NPW9-1000, and NPW12-1000.

Table 1. Binding energy and relative abundance of the different components of the Sr 3d 5/2, Fe2p3/2 and O1s XPS regions for NPW4-1000, NPW9-1000, and NPW12-1000 powders.

Sample	Sr 3d 5/2	Fe2p3/2	O1s
NPW4-1000	132.8 (52%)	710.3 (76%)	529.1 (30%)
	133.3 (48%)	713.0 (24%)	531.5 (70%)
NPW9-1000	132.5 (58%)	710.3 (73%)	529.3 (45%)
	133.8 (42%)	713.1 (27%)	531.8 (55%)
NPW12-1000	132.8 (71%)	710.7 (73%)	529.7 (43%)
	134.1 (29%)	713.3 (27%)	531.9 (57%)

The main peaks can be deconvoluted into two components at ca. 711 and 713 eV, which are attributed to Fe(III) and Fe(IV), respectively [24,25]. The relative percentage of the two species does not change along the series. The Sr3d region shows a peak, which is a convolution of Sr3d5/2 and Sr3d3/2 (DE = 1.8 eV), as evidenced in Figure 2A. The shape of the peak points to the presence of two doublets with the Sr3d5/2 centered at ca. 132.5 eV and 134 eV, which, according to the more common interpretation, may be attributed to Sr in the perovskite structure (Sr_L) and $SrCO_3$ (strontium carbonate), respectively [26–28]. $SrCO_3$ is often formed in $SrFeO_3$-based compounds due to the high affinity of Sr for atmospheric CO_2, especially in the presence of humidity [29]. The O1s region is characterized by two components (Figure 2C): O_L, at 529.6 eV, is attributed to oxygen in the perovskites lattice, while and O_{OH}, at 531.5 eV, is attributed to adsorbed oxygen species and/or carbonates [25,30]. Differences in the ratio between the two oxygen components may be indicative of differences in the oxygen vacancies present in the materials, even though due to the presence of carbonates, the correlation is not straightforward. NPW4-1000 has the smallest O_L/O_{OH} ratio. However, this does not indicate a higher number of oxygen vacancies present in the structure, but a higher percentage of $SrCO_3$ species at the surface (Figure 2 and Table 1). Therefore, it seems that the presence of secondary phases in the bulk, as detected by XRD, favors the formation of carbonate on the surface, and this could negatively affect the thermocatalytic activity, as recently observed in the literature [29]. Looking at the XPS data, NPW9-1000 and NPW12-1000 showed similar O_L/O_{OH} ratios, indicating that the oxygen surface distribution remained identical, although less surface carbonate formation occurred.

The redox properties of the three samples were investigated using TPR, TPO, and TPD techniques. The characterization results are shown in Figure 3. The TPR profile of the single-phase powder, NPW9, is characterized by two main peaks: the first one centered at 455 °C and the second one above ~750 °C, with a maximum at around 1000 °C (Figure 3A). According with our previous investigation of Ce-doped strontium ferrates [31], the low temperature peak is ascribed to the reduction of Fe^{4+} species, typically present in the perovskite, to Fe^{3+}, while in the region 550–700 °C, the reduction of Ce^{4+} may occur, and above 700–750 °C, reduction of Fe^{3+} to Fe^{2+} and eventually to Fe^0 takes place. Similar reduction profiles were registered for the other two samples, NPW4-1000 and NPW12-1000, with both containing minor percentages of segregated phases as $Sr_3Fe_2O_{7-\delta}$ layered perovskites. In both cases, a shift of the first main reduction peak at low temperature occurred; this effect was more important for NPW12-1000. Therefore, the presence of the segregated layered perovskite phases in NPW4-1000 had a positive effect on the oxygen vacancies availability at lower temperatures, although the most important improvement was for NPW12-1000, due to the use of an over-stoichiometric reducers-to-oxidizers (Φ) ratio (1.6) and the presence of segregated ceria phase. The TPO curves, registered on the samples reduced up to ~1000 °C (Figure 3B), show very similar profiles with two main peaks: the first at around 200 °C and the second one at 380–400 °C. The curves for NPW9-1000 and NPW12-1000 are overlapping, while the reoxidation for the reduced phases formed in NPW4-1000 seems to occur at slightly higher temperatures than for the previous two, indicating that the reduced Fe-containing phases formed upon the previous reduction treatment (TPR) were slightly more stable in NPW4-1000 than in the other two samples. The TPD curves registered after the TPO (Figure 3C) for all the three samples showed desorption peaks at around 335–360 °C. According to the literature, such peaks correspond to the desorption of suprafacial α-species [32]. The sample NPW4-1000 was characterized by the highest desorption peak, suggesting the presence of the highest amount of active surface oxygen species that can be easily released at relatively low temperature (starting from 250 °C, with a maximum at 335 °C) and, therefore, that could better participate in a thermal oxidation reaction.

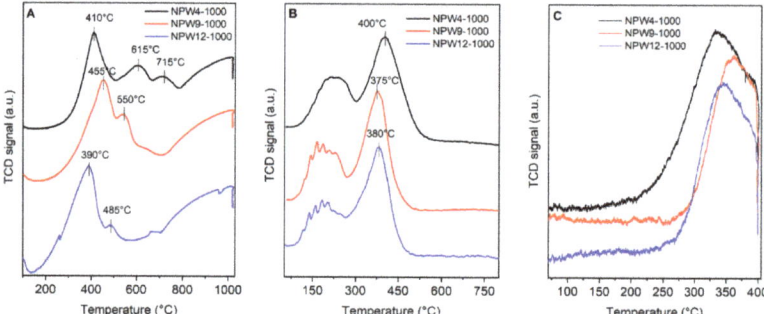

Figure 3. TPR (**A**), TPO (**B**) and TPD (**C**) profiles of NPW4-1000, NPW9-1000, and NPW12-1000 samples.

Looking at the comparison of the redox properties of NPW4-1000, NPW9-1000, and NPW12-1000, the presence of segregated perovskite layered phases could favor the interaction with polar substrates, thereby inducing a better catalytic activity of the material. In order to investigate this aspect, another powder was considered: NPW6-1000 was prepared under conditions similar to those applied for NPW4-1000, with changes in pH and gelification degree. These modifications (namely, a pH value of 6 and a medium degree of gelification) should guarantee the formation of layered perovskite and segregated ceria, as confirmed by XRD measurements, in addition to the increase of perovskite fraction and maximization of absorbance response. Similarly, NPW6-1000 contains a smaller amount of layered perovskite phase with respect to NPW4-1000 (see Table 2). This sample was analyzed by FTIR spectroscopy and the related spectra, reported in Figure 4, show the scattering profile typical of a nanopowder, with signals specifically related to the perovskite structure at 600 cm^{-1} and to the surface carbonate groups produced by interactions with atmospheric CO_2 at 1100 cm^{-1}. The high frequency range (3700–2500 cm^{-1}) is typically related to the presence of OH, NH, and CH vibrational features. A strong and large signal is visible in the spectra of the material, indicating that such species are present, even if a more specific attribution of the signal was not possible, given the wideness of the band. To verify that these groups were actually exposed at the sample surface, a contact with D_2O vapor pressure was performed in order to exchange the hydrogen atoms in the OH, NH, and CH groups with deuterium, with consequent modification of the spectra profile. No modification of the signals was observed, indicating that their concentration at the surface of the material was very low or that the hydrogenated groups were not reachable by D_2O molecules and, therefore, these groups, if present, were in the bulk of the material or under its surface.

Table 2. Details of the synthesis condition and relevant chemical–physical features of the investigated Ce-doped $SrFeO_3$ samples.

Sample/ Ce mol%	Synthesis Conditions (Reducers-to-Oxidizers Ratio, pH, Gelification Degree and Laboratory)	Processing Conditions (calcination Temperature and Heating Ramp)	BET Specific Surface Area (m^2/g)	Phase Composition (from Rietveld Refinement)		
				$Sr_{0.86}Ce_{0.14}FeO_3$ Amount (% wt)	CeO_2 Amount (% wt)	$Sr_3Fe_2O_{7-\delta}$ Phase Amount (% wt)
NPW9_1000/ 14 mol%	1 7 High Lab B	1000 °C/5 h, 10 °C/min	nd	100	0	0
NPW4-1000/ 14 mol%	1 7 Low Lab B	1000 °C/5 h, 2 °C/min	~1	88.4	1.6	10
NPW3-1000/ 14 mol%	1 7 High Lab B	1000 °C/5 h, 2 °C/min	~1	99	1	0
NPW6-1000/ 14 mol%	1 6 Medium Lab B	1000 °C/5 h, 2 °C/min	~1	93.4	1.6	5
NPW10-1000/ 14 mol%	1 7 High Lab B	1000 °C/5 h, 10 °C/min	~1	100	0	0
NPW12-1000/ 15 mol%	1.63 7 High Lab B	1000 °C/5 h, 10 °C/min	~1	99	1	0
NPW13-1000_A/15 mol%	1.63 6 High LabA	1000 °C/5 h, 5 °C/min	~1	98	2	0
NPW13-1000_B/15 mol%	1.63 6 High Lab B	1000 °C/5 h, 10 °C/min	~1	98	2	0
NPW13-NC/ 15 mol%	1.63 6 High Lab B	No thermal treatment (as-combusted powder)	~6	94	0	4 ($SrCO_3$) 2 (Fe_3O_4)
NPW13-800/ 15 mol%	1.63 6 High Lab B	800 °C/2 h, 10 °C/min	~4	98.8	1.2	traces ($SrCO_3$)

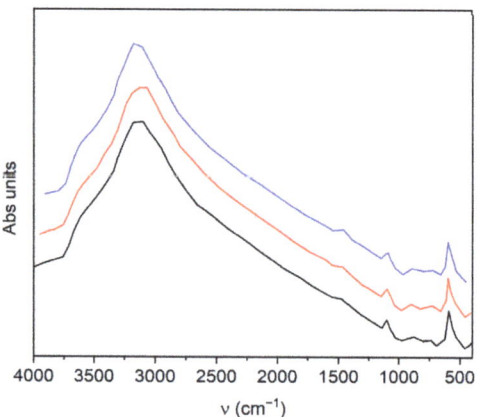

Figure 4. FTIR absorbance spectra of NPW6-1000 in air (black curve), after 1 min outgas (red curve) and after 30 min outgas (blue curve). The curves were shifted along the y-axis to better display the spectra details.

For the identification of surface active sites, the powders, after pre-activation in a vacuum at 30, 150, 300, 400, 500, 600, and 700 °C, were contacted with the CO, NO, and CO_2 gaseous probes in order to study the surface Lewis acidity of the material given by cationic species, which is prone to establishing specific interactions with the probe molecules (for instance, it is known that CO and NO show a very important affinity with Fe^{n+}, whereas CO can establish important interactions with Ce^{4+}). In no case did the probe molecules show visible interactions with the material, suggesting a limited concentration of the adsorbing surface sites and confirming again that the exposed surface of the material was very limited. Additionally, gaseous O_2 was used as a probe to check the redox behavior of the material, considering that activation in vacuo at high temperature favors the reduction of the perovskite, inducing a consequent interaction with molecular oxygen. The interaction was studied by microgravimetry. The results of the test performed on NPW6-1000 sample outgassed at RT, 150, 300, and 400 °C are reported in Figure 5 and indicate a limited interaction of the material with O2, increasing with the temperature of the preliminary outgassing of the sample. The uptake on the sample treated at 300 and 400 °C was similar, indicating that the activation of the sample at 400 °C did not significantly change the O_2 interaction properties of the material.

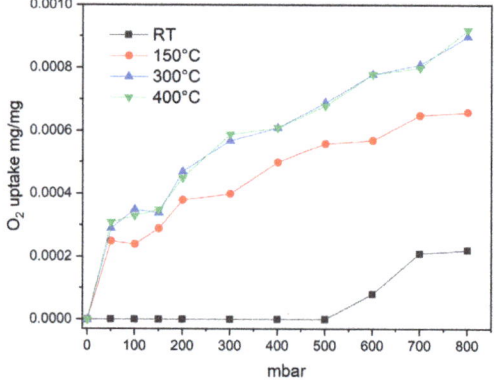

Figure 5. Microgravimetric O_2 uptake with NPW6-1000 activated in vacuo at RT (nearly 30 °C), 150, 300, and 400 °C.

2.1.2. Effect of the Processing Parameters

The same batch of as-burned powder was divided in two portions and treated at 1000 °C/5 h with a different temperature ramp: 2 °C/min or 10 °C/min. The XRD patterns registered for the samples treated with different temperature ramps were slightly different for samples prepared from gels with low and medium gelification degrees, although the effect was null for samples prepared using a complete gelification. It can be hypothesized that the not fully densified gel still contained some water, whose evaporation process velocity may have affected the phase composition of the final powder. As a consequence, the applied temperature ramp value was not a parameter controlling the process if the gelification degree was high enough. From the N_2 adsorption results, it was found that all the powders showed a specific surface area of about 1 m^2/g. A washing treatment carried out by sonicating the powder in bi-distilled water for two hours, according to a procedure previously used for other perovskites prepared by the same methodology [33], did not significantly change the textural feature of the materials. At the same time, a check of the washing medium was performed to evaluate the release of unreacted precursors during sonication. Again, the total organic carbon measurements did not evidence substantial release, confirming the N_2 adsorption results. To increase the specific surface area of the materials and, consequently, the activity, the effect of the calcination temperature was studied on a sample that was prepared under optimized conditions of gelification degree, pH, and chemical composition. The comparison was done between the as-burned sample, NPW13-NC, the sample treated at 800 °C/2 h, NPW13-800, and the sample treated at 1000 °C/5 h, NPW13-1000_B. NPW13_NC shows a specific surface area of about 6 m^2/g, i.e., slightly higher than that of NPW13-800 (about 4 m^2/g), which is slightly higher than that of NPW13-1000 (about 1 m^2/g). This trend may reflect both a modification of the crystal size during the calcination process (increasing from 71 nm to 150 nm with the increase of calcination temperature) or the presence of unburned precursors that may have contributed to the increase of the area of non-calcined or 800 °C-calcined samples. [23]. Beside the slight change in the surface area, the perovskite was present as the main phase at all the firing temperatures, although some $SrCO_3$ was clearly visible in the non-calcined sample and in the 800 °C-calcined sample (traces), whereas segregated CeO_2 was visible in both the calcined samples, although the CeO_2 peaks of the 800 °C-calcined sample were very large (Figure 6 and Table 2).

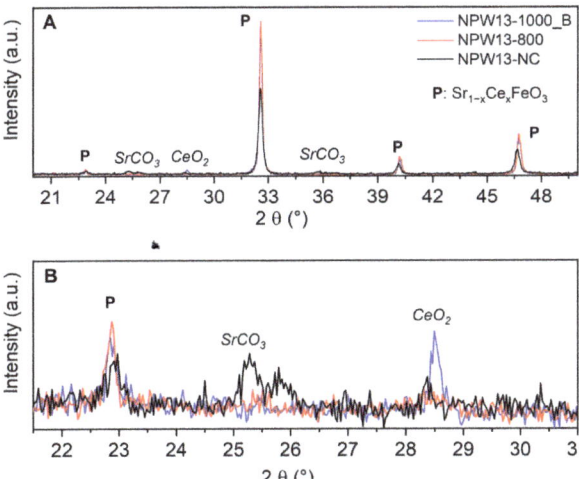

Figure 6. (**A**) XRD patterns of NPW13-NC, NPW13-800, and NPW13-1000_B; (**B**) an inset of the 21.5–31° 2θ region showing the secondary phases.

2.2. Testing Procedures

2.2.1. Effect of pH on Perovskite Stability and Metal Leaching

The results of the stability tests performed on the sample NPW13-1000_A are reported in Figure 7. NPW13-1000-A was chosen for this test as it is one of the most active thermocatalysts studied (see Section 2.2.2), for which the definition of the stability pH range could be interesting for the future applications. Strontium was released in the entire examined range of pH, but the most dramatic situation was observed at pH ≈ 2, where Sr, Ce, and Fe were all released in solution. Interestingly, the ratio between the amount of Sr and Ce released in solution at that pH was almost the same as that calculated for the perovskite phase (Sr/Ce = 0.86/0.14), whereas the amount of Fe released was much lower than expected, probably because it forms insoluble species.

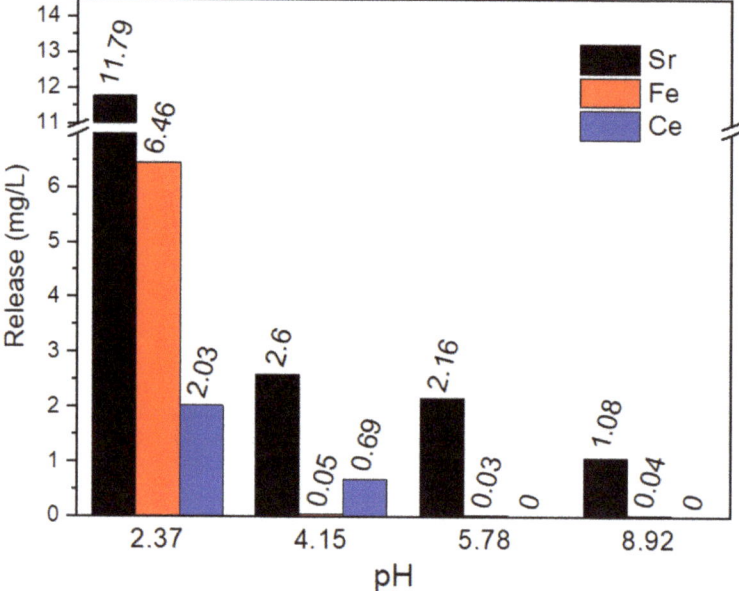

Figure 7. Sr, Fe, and Ce concentrations in solution released from perovskite at different pH values, for the NPW13-1000_A sample.

Apart from a small amount of strontium released at pH > 5, the results indicate a good material stability and suggest a possible use on pollutant removal from freshwater sources, whose pH generally ranges between 6 and 8. Otherwise, these materials seem to be not suitable for applications requiring acidic media (pH lower than 4), e.g., for the treatment of waters contaminated by inorganic contaminants.

2.2.2. Effect of the Experimental Conditions Used for the Thermocatalytic Tests

The thermocatalytic performance of the Ce-doped $SrFeO_3$ perovskite was studied on a sample prepared under optimized conditions of gelification degree, pH, and chemical composition, and subjected to different thermal treatments. To this end, Orange II (OII) and bisphenol A (BPA) were selected as model contaminants. All experiments were performed at the pH obtained after the solubilization of the target pollutant and the suspension of the perovskite. The obtained pH values typically ranged between pH 9 and 10. The strong solution basification driven by these perovskites was likely due to protonation/deprotonation equilibria at the surface of the catalyst, as confirmed by the behavior that was already reported in the literature for similar materials [16]. Control experiments at room temperature, as well as adsorption tests, were also performed, followed by degradation experiments after raising the temperature to 70 °C. Using Orange II (OII), no degradation was observed after 200 min of contact at 70 °C with NPW6-1000, NPW10-1000, and NPW12-1000, while a different trend was observed when using the samples prepared under optimized conditions, as reported in Figure 8.

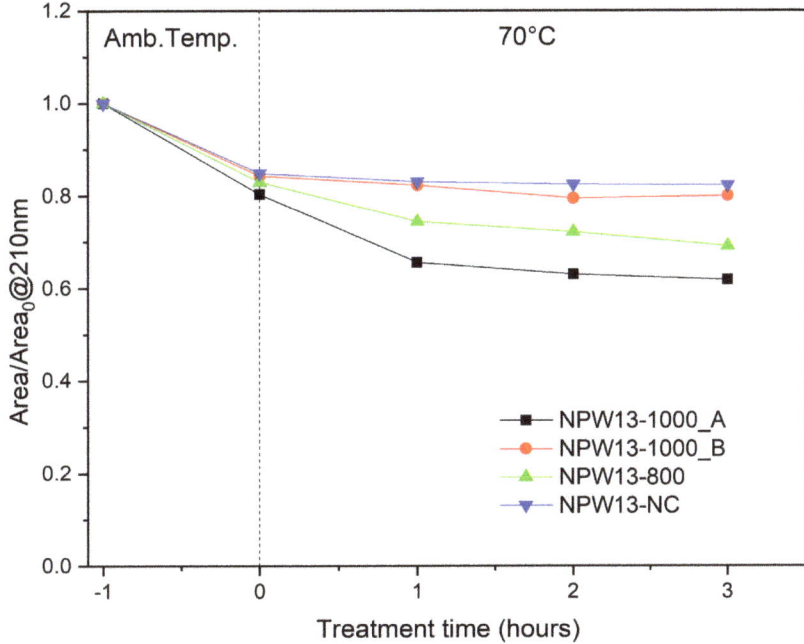

Figure 8. Orange II degradation vs. contact time in the presence of NPW13-NC and NPW13 calcined at different temperatures and prepared by Lab B, in comparison with NPW13-1000_A.

As can be noticed after 1 h contact at room T with OII, all the powders showed a comparable decrease in OII concentration; this can be reasonably ascribed to adsorption. Afterwards, the most efficient material was NPW13-1000_A (prepared in the laboratory A using the same synthetic procedure adopted for sample NPW13-1000_B in laboratory B), followed by NPW13-800, whereas NPW13-NC, and NPW13-1000_B were negligibly able to promote OII degradation.

A similar behavior was observed when testing the material with BPA, as can be observed in Figure 9. Figure 9 reports a quite odd shape of the curves, with an unexpected change at about 2 h of reaction time. The action of the thermocatalysts was stimulated at 60–70 °C; these conditions could facilitate the desorption of initially adsorbed compounds, resulting in the odd trend observed in Figure 9 after around 2 h of reaction time. Once the system had stabilized, because the abatement approaches the steady step, a normal trend returned and the curves became similar to what one would expect for all the samples except for NPW13-NC. This sample was not submitted to a post-synthesis calcination and the relative trend observed confirms that the calcination of the system makes it possible to clean the material of possible unburned precursors affecting the reaction and standardizes the product for its subsequent application.

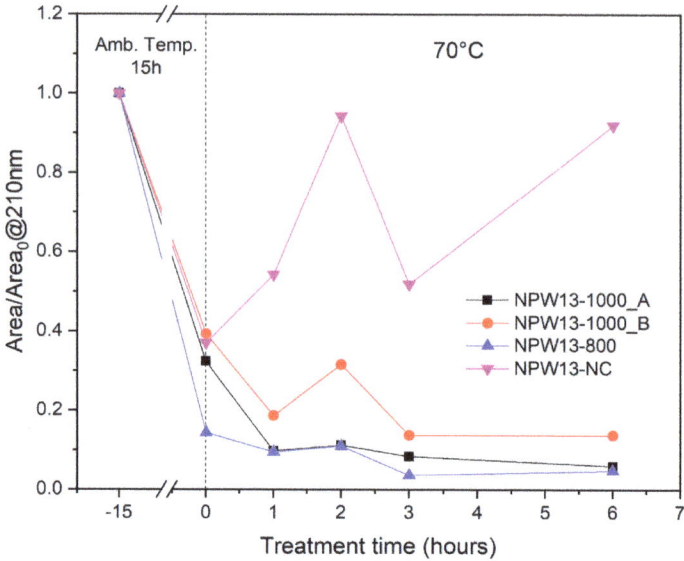

Figure 9. Degradation of BPA 5 mg/L in the presence of NPW13-NC and NPW13 calcined at different temperatures and prepared by Lab B, in comparison with NPW13-1000_A.

However, with BPA, the contact time at room T was prolonged to 15 h and the percentage of BPA decrease was higher than the one observed for OII that was kept in contact at room T with the perovskite for only 1 h before raising the T. It is reasonable to consider that more adsorption took place after 15 h at ambient temperature, also considering the absence of secondary peaks in the HPLC profiles. However, we cannot exclude the possibility that the material had already catalyzed the BPA degradation during the preconditioning, in agreement with the observed reactivity of the material at room temperature (see next discussion on Figure 10A). In contrast, at 70 °C, the further BPA decrease corresponded to the appearance of a new peak in the HPLC profile, thus indicating that BPA had undergone degradation.

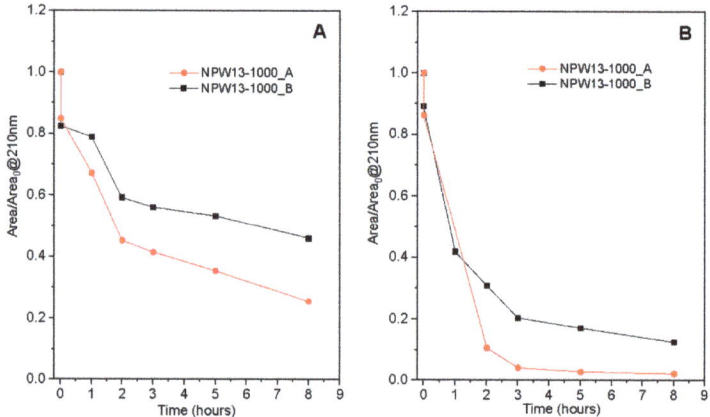

Figure 10. Difference in the performances of BPA degradation at ambient T and at 70 °C of two NPW13-1000 powders prepared following the same procedure but in two different laboratories (**A**,**B**).

Further experiments were performed using BPA as a target compound in the presence of NPW13-1000_A or NPW13-1000_B in order to compare the thermocatalytic performance of these two nominally identical materials prepared in different laboratories. BPA was again added at an initial concentration of 5 mg/L, and two series of experiments, at room temperature and at 70 °C, were performed. In both scenarios, the sample prepared in laboratory A had better performances, removing after 8 h of treatment 70% of BPA at room temperature (Figure 10A) and reaching almost complete BPA removal at 70 °C (Figure 10B).

Both batches could promote the thermal degradation of BPA at room temperature and at 70 °C, although there was a surprising and not negligible difference in their efficiency. The possible causes for this, beside the temperature ramp of 5 °C/min, may be the use of glass beaker by laboratory A and a stainless-steel beaker by laboratory B (which better retains the heat during combustion), to the beaker geometry, to the laboratory temperature during combustion, and to the technical characteristics of the muffle furnace.

To get insights on the possible species involved in the thermocatalytic process and how the presence of segregated phases can influence those species, the formation of catalyzed ROS in two samples, NPW3-1000 e NPW6-1000, containing different amount of segregated layered perovskite phases, was studied by EPR spectroscopy. In analogy with a previous study on the properties of $Sr_{0.85}Ce_{0.15}FeO_{3-\delta}$ perovskite [16], our investigations were performed at different temperatures and in the presence of a specific spin-trap (DMPO for hydroxyl radicals and TMP for singlet oxygen). The experiments in the presence of DMPO seemed to indicate, for both samples, the generation of a similar amount of hydroxyl radicals when heated at 70–75 °C, regardless of the presence of secondary phases. As shown in Figure 11, the correspondent EPR signal intensity was time-dependent, and a maximum was reached after about two hours of heating for both the samples, in agreement with the results obtained for NPW13-1000_A and close to the results obtained for NPW13-1000_B (Figure 10B). In the inset of Figure 11, the spectra are shown at 25 and 50 °C after 2 h of heating in the presence of NPW6-1000 in order to demonstrate the absence of appreciable concentrations of DMPO-OH• adduct in these conditions. Similar experiments performed in the presence of TMP did not lead to the appearance of any EPR signal, which was indicative of a lack of singlet oxygen formation in these conditions.

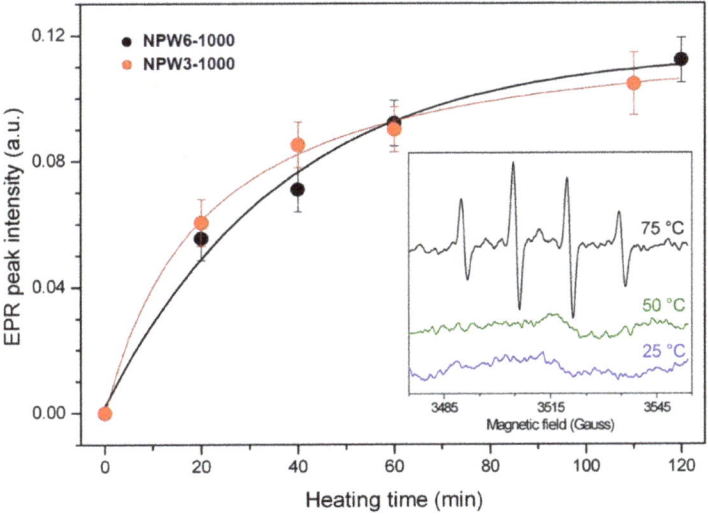

Figure 11. Intensity of DMPO-OH·EPR spectra of NPW3-1000 (red circles) and NPW6-1000 (black circles) at different heating times (inset: EPR spectra obtained with NPW6-1000 after 120 min heating at different temperatures).

3. Materials and Methods

3.1. Perovskite Oxides Synthesis

Ce-doped $SrFeO_3$ powders were obtained by solution combustion synthesis using different synthesis and processing parameters. The general methodology is described elsewhere [16]. Briefly, iron, strontium, and cerium nitrates were dissolved in distilled water in the presence of citric acid as a reducer/complexing agent/microstructural template. pH was regulated by adding ammonia solution (30 vol%) and the reducer-to-oxidizer ratio was regulated by adding ammonium nitrate as an additional oxidant, maintaining a citric acid-to-metal cation ratio of 2. The combustion mixture was left to evaporate until a gel was formed and the combustion process was initiated on the gel by setting the hotplate temperature to about 300 °C. The varied parameters were pH, gelification degree, calcination temperature/ramp, reducers-to-oxidizers ratio, and Ce nominal amount in mol%. The applied reducer was citric acid, whereas oxidizers were metal nitrate precursors and ammonium nitrate as an additional oxidant.

Gelification degree was evaluated qualitatively as follows:

(i) low: the gel is still fluid at 80 °C and densifies only after it is left at room temperature;

(ii) medium: the gel is only partially densified at 80 °C and the magnetic stirrer can still rotate and mix the gel;

(iii) high: the gel is fully densified already at 80 °C and the magnetic stirrer is blocked and cannot mix the gel anymore.

Most of the samples were calcined at 1000 °C for 5 h after the synthesis to remove any unreacted precursor and any residual carbon formed from incomplete citric acid combustion, to allow the perovskite to fully form, and to allow the Sr from $SrCO_3$ to be fully incorporated in the perovskite structure. In addition, one sample was left uncalcined, while another sample was calcined at 800 °C/2 h.

Referring to Table 2, the applied synthesis conditions allowed us to obtain materials containing (i) $Sr_{0.86}Ce_{0.14}FeO_3$ as the unique phase (NPW9-1000 and NPW10-1000), (ii) $Sr_{0.86}Ce_{0.14}FeO_3$ as the main phase together with low percentage of CeO_2 as segregated oxide (NPW12-1000, NPW3-1000, NPW13-1000_B), or (iii) $Sr_{0.86}Ce_{0.14}FeO_3$ as the main phase together with low percentage of CeO_2 as segregated oxide and of another secondary phase in variable amounts (NPW4-1000, NPW6-1000). Additionally, the effect of the post-synthesis calcination treatment was considered when comparing the features of the sample prepared under optimized conditions with those of the as burned sample (no calcination) (NPW13-NC) and after calcination (ramp 10 °C/min) for 2 h at 800 °C (NPW13-800) or for 5 h at 1000 °C (NPW13-1000_B). The optimized conditions used for NPW13-1000_B were coincident with the ones applied in a previous paper [16]. Finally, a sample was prepared by some of the authors in another laboratory (called Laboratory A) following the same procedure used for NPW13-1000_B (prepared in the Lab B) but using a temperature ramp of 5 °C/min, and was named NPW13-1000_A (prepared in the Lab A).

3.2. Perovskite Oxides Characterization

X-ray diffraction (XRD) measurements were carried out on a Bruker-Siemens D5000 X-ray powder diffractometer equipped with a Kristalloflex 760 X-ray generator and with a curved graphite monochromator using Cu Kα radiation (40 kV/30 mA). The 2θ step size was 0.03, the integration time was 20 s per step, and the 2θ scan ranged from 10° to 90°. The powder diffraction patterns were analyzed by Rietveld refinement using the GSAS II software [34]. The compound $Sr_{0.9}Ce_{0.1}FeO_3$ (ICDD PDF4+ inorganic database—PDF Card n° 04-014-0169) was chosen as a starting model for the Rietveld Refinement, setting the Sr and the Ce occupancies to 0.86 and 0.14, respectively. A Chebyschev polynomial function with eight polynomial coefficients was chosen for the background, and the Pseudo Voigt function was used for the peak profile fitting. In the structure refinement lattice constants, Debye Waller factors, microstrain, and crystal size were considered as variable parameters. Crystal size was obtained directly from the GSAS II software output based on the Scherrer equation, while microstrain is a unitless number obtained directly from the GSAS II software output and describing a range of lattice constants through the equation $10^{-6} \cdot (\Delta\text{-d})/\text{d}$. From the fitting results, the structural parameters of the investigated compounds and the phase composition and the relative cell edge lengths were obtained. The agreement between fitted and observed intensities, the Rw factor, and the χ^2 were acceptable [35].

To remove and quantify the unburned carbon, the powder was suspended in MilliQ™ water (500 mg L^{-1}) and stirred for 2 h.

The specific surface area was determined by studying the gas-volumetric N_2 uptake at N_2 boiling point ($-196\ °C$) using an ASAP2020 gas-volumetric apparatus by Micromeritics. Sample analyses were performed after outgassing the materials in a vacuum (residual pressure ~10^{-2} mbar) for several hours at 100 °C to achieve good surface cleaning. The BET model was applied for surface area determination.

The stability of the sample NPW13-1000_A was tested at different pH values keeping 20 mL of the powder suspensions (750 mg/L) under constant stirring for 24 h. Suspensions at pH 2, 4, and 6 were obtained via acidifying with HCl, and a test was also performed at the natural pH of the powder (pH > 8). After 24 h of contact, the suspensions were filtered and the solutions, properly acidified, and submitted to inductively coupled plasma–optical emission spectroscopy (ICP-OES Optima 7000 DV Perkin Elmer) analysis to verify the release of Sr, Fe, and Ce. This instrument was equipped with a PEEK Mira Mist nebulizer, a cyclonic spray chamber, and an Echelle monochromator. The wavelengths were 238.204, 413.764, and 407.771 nm for Fe, Ce, and Sr, respectively. Each concentration value was averaged on three instrumental measurements. A dilution (1:10) was necessary as a final step to determine Sr concentrations in all the solutions.

X-ray photoelectron spectroscopy (XPS) analyses of the powders were performed with a VG Microtech ESCA 3000 Multilab (VG Scientific, Sussex, UK), using Al Kα source (1486.6 eV) running at 14 kV and 15 mA, and in CAE analyzer mode. For the individual peak energy regions, a pass energy of 20 eV across the hemispheres was used. The constant charging of the samples was removed by referencing all the energies to the C 1s peak energy (set at 285.1 eV) arising from adventitious carbon. Analyses of the peaks were performed using the CASA XPS software (version 2.3.17, Casa Software Ltd.Wilmslow, Cheshire, UK, 2009). Gaussian (70%)-Lorentzian (30%), defined in Casa XPS as GL(30) profiles were used for each component of the main peaks after a Shirley type baseline subtraction. The binding energy values are quoted with a precision of ±0.15 eV and the atomic percentages with a precision of ±10%.

The redox properties of the perovskites were investigated by performing Temperature Programmed Reduction (TPR) and Temperature Programmed Oxidation (TPO) experiments. TPR were performed by flowing a mixture of H_2/Ar (5%, 30 mL/min) over the sample (~0.1 g) and increasing the temperature in the range between rt and ~1000 °C (heating rate 10 °C/min). Experiments were carried out with a Micromeritics Autochem 2910 instrument equipped with a thermal conductivity detector (TCD). All the powders were pre-treated in O_2/He (5%, 30 mL/min) at 400 °C for 1 h to remove any impurities and then cooled down under Ar atmosphere before starting the TPR. After reduction up to 1050 °C, the materials were cooled to room temperature under Ar and then subjected to a TPO experiment to reoxidize all the reduced species. TPO were performed as described for TPR but while flowing a mixture of O_2/He (5%, 30 mL/min) and increasing the temperature to 1000 °C. After TPO, the oxidized samples were cooled to rt under the same oxidizing atmosphere with the aim of keeping all the oxygen vacancies filled. Finally, a Temperature Programmed Desorption (TPD) experiment was carried out by heating the samples from rt to 400 °C under He flow to desorb the "suprafacial" chemisorbed oxygen species.

3.3. Thermocatalytic Tests and EPR Measurements

For the thermocatalytic tests, aqueous suspensions of perovskite materials (750 mg L^{-1}) and substrate, Orange II (sodium salt, >85%) or Bisphenol A (>95%), 5 mg L^{-1}, were prepared with Milli-Q™ ultrapure water, placed in ultrasonic bath for 30 s, and then stirred for 30 min to homogenize the suspension before treatment. The thermal treatments were performed using 50 mL of suspension in closed Pyrex® cells under continuous stirring in a thermostatic bath maintained at 70 °C; control dark tests at room temperature were also conducted. After different times, the samples were taken and filtered on a PTFE membrane using syringe Minisart® and Syringe Filters 0.45 µm. The compound disappearance was followed by HPLC-UV using a Merk Hitachi L-6200 system coupled with a UV-Vis L-4200 detector; the column used was a RP-18 (5 µm) LIChrospher® 100 Merck (125 × 4 mm). Compound detection was performed at the wavelength of maximum absorbance of each compound (484 nm for OII and 225 nm for BPA) under isocratic conditions using acidic ultrapure water (0.1% H3PO4) and methanol as eluents mixed in a ratio 88/12 for OII and 60/40 for BPA. Study of the reactive species was conducted by Electron Paramagnetic Resonance (EPR) in the presence of 5,5-dimethyl-1-pyrroline-N-oxide (DMPO) or 2,2,6,6-tetramethyl-4-piperidone hydrochloride (4- oxo-TMP) [36,37]. EPR spectra were recorded at rt with a X-band Bruker-EMX spectrometer equipped with a cylindrical cavity operating at 100 kHz field modulation. The acquisition parameters were as follows: frequency—9.86 GHz, microwave power—5 mW, modulation amplitude—2 G, and conversion time—30.68 msec. As previously reported [38,39], the experiments were carried out by adding the spin trap in the cell at the desired temperature; then, after the treatment, a portion of the sample was withdrawn in a capillary quartz tube and the EPR spectrum was immediately acquired.

4. Conclusions

The obtained results indicate how crucial the control of the synthetic parameters is in obtaining a material with the desired efficiency and featuring reproducible properties when different batches of the same materials are synthesized. With regards to the thermocatalytic efficiency of these materials with Orange II or BPA, the most promising ones seem to be those containing the highest amount of segregated CeO_2, i.e., NPW13-1000_A and NPW13-1000_B, probably due to the higher oxygen vacancy availability induced by the CeO_2 segregation. On the other hand, the segregation of tetragonal layered $Sr_3Fe_2O_{7-\delta}$ phases has to be avoided, since it favors $SrCO_3$ formation on the surface, with the risk of blocking the thermocatalytic reaction without causing an evident improvement in the oxygen vacancy availability. Furthermore, the surprising and not negligible difference in the thermocatalytic efficiency of two materials prepared with the same procedure but in two different laboratories evidences the sensitivity of the synthesis to the experimental set-up and instrumentation used. Despite the low surface area values of the investigated samples, a surface area increase from 1 to 5 m^2/g had a positive effect on the thermocatalytic performance. An in-depth study of the degradation pathways of contaminants and a better comprehension of the adsorption phenomena at the surface of the catalyst is still to be done. The overall picture obtained by this work confirms the potential application of perovskite materials in thermocatalytic AOPs, but, at the same time, focuses the attention on the need to fully identify the key synthetic parameters to be tuned in order to obtain materials with fully reproducible characteristics and performance.

Author Contributions: Conceptualization, A.B.P. and F.D.; validation, D.P.; investigation, D.P., E.L., G.M., V.B., L.F.L., V.L.P. and M.M.; resources, A.B.P., G.M. and F.D.; writing—original draft preparation, A.B.P., D.P. and F.D.; writing—review and editing, D.P. All authors have read and agreed to the published version of the manuscript.

Funding: This research was funded under ERA-NET AquaticPollutants Joint Transnational Call, grant number 869178 and under Eurostars joint programme, grant number E!113844, NanoPerWater project. The authors would like to thank European Commission and FCT (Portugal), IFD (Denmark), MUR (Italy) for funding in the frame of the collaborative international consortium NanoTheC-Aba financed under the 2020 AquaticPollutants Joint call of the AquaticPollutants ERA-NET Cofund (GA N° 869178). This ERA-NET is an integral part of the activities developed by the Water, Oceans and AMR JPIs.

Informed Consent Statement: Not applicable.

Data Availability Statement: Data available on request due to privacy restrictions.

Acknowledgments: The authors like to thank Sara Giovine for her contribution to the experimental work.

Conflicts of Interest: The authors declare no conflict of interest.

References

1. Jones, E.R.; van Vliet, M.T.H.; Qadir, M.; Bierkens, M.F.P. Country-level and gridded estimates of wastewater production, collection, treatment and reuse. *Earth Syst. Sci. Data* **2021**, *13*, 237–254. [CrossRef]
2. Biel-Maeso, M.; Corada-Fernández, C.; Lara-Martín, P.A. Removal of personal care products (PCPs) in wastewater and sludge treatment and their occurrence in receiving soils. *Water Res.* **2019**, *150*, 129–139. [CrossRef] [PubMed]
3. Clara, M.; Strenn, B.; Gans, O.; Martinez, E.; Kreuzinger, N.; Kroiss, H. Removal of selected pharmaceuticals, fragrances and endocrine disrupting compounds in a membrane bioreactor and conventional wastewater treatment plants. *Water Res.* **2005**, *39*, 4797–4807. [CrossRef] [PubMed]
4. Rivera-Utrilla, J.; Sánchez-Polo, M.; Ferro-García, M.Á.; Prados-Joya, G.; Ocampo-Pérez, R. Pharmaceuticals as emerging contaminants and their removal from water. A review. *Chemosphere* **2013**, *93*, 1268–1287. [CrossRef] [PubMed]
5. Rodriguez-Narvaez, O.M.; Peralta-Hernandez, J.M.; Goonetilleke, A.; Bandala, E.R. Treatment technologies for emerging contaminants in water: A review. *Chem. Eng. J.* **2017**, *323*, 361–380. [CrossRef]
6. Miklos, D.B.; Remy, C.; Jekel, M.; Linden, K.G.; Drewes, J.E.; Hübner, U. Evaluation of advanced oxidation processes for water and wastewater treatment—A critical review. *Water Res.* **2018**, *139*, 118–131. [CrossRef]
7. Li, X.; Wang, B.; Cao, Y.; Zhao, S.; Wang, H.; Feng, X.; Zhou, J.; Ma, X. Water Contaminant Elimination Based on Metal–Organic Frameworks and Perspective on Their Industrial Applications. *ACS Sustain. Chem. Eng.* **2019**, *7*, 4548–4563. [CrossRef]
8. Anucha, C.B.; Altin, I.; Bacaksiz, E.; Stathopoulos, V.N. Titanium dioxide (TiO_2)-based photocatalyst materials activity enhancement for contaminants of emerging concern (CECs) degradation: In the light of modification strategies. *Chem. Eng. J. Adv.* **2022**, *10*, 100262. [CrossRef]
9. Castanheira, B.; Otubo, L.; Oliveira, C.L.P.; Montes, R.; Quintana, J.B.; Rodil, R.; Brochsztain, S.; Vilar, V.J.P.; Teixeira, A.C.S.C. Functionalized mesoporous silicas SBA-15 for heterogeneous photocatalysis towards CECs removal from secondary urban wastewater. *Chemosphere* **2022**, *287*, 132023. [CrossRef]
10. Senobari, S.; Nezamzadeh-Ejhieh, A. A novel ternary nano-composite with a high photocatalyitic activity: Characterization, effect of calcination temperature and designing the experiments. *J. Photochem. Photobiol. Chem.* **2020**, *394*, 112455. [CrossRef]
11. Noruozi, A.; Nezamzadeh-Ejhieh, A. Preparation, characterization, and investigation of the catalytic property of α-Fe_2O_3-ZnO nanoparticles in the photodegradation and mineralization of methylene blue. *Chem. Phys. Lett.* **2020**, *752*, 137587. [CrossRef]
12. Chen, H.; Motuzas, J.; Martens, W.; Diniz da Costa, J.C. Degradation of azo dye Orange II under dark ambient conditions by calcium strontium copper perovskite. *Appl. Catal. B Environ.* **2018**, *221*, 691–700. [CrossRef]
13. Mahmoudi, F.; Saravanakumar, K.; Maheskumar, V.; Njaramba, L.K.; Yoon, Y.; Park, C.M. Application of perovskite oxides and their composites for degrading organic pollutants from wastewater using advanced oxidation processes: Review of the recent progress. *J. Hazard. Mater.* **2022**, *436*, 129074. [CrossRef] [PubMed]
14. Yang, L.; Jiao, Y.; Xu, X.; Pan, Y.; Su, C.; Duan, X.; Sun, H.; Liu, S.; Wang, S.; Shao, Z. Superstructures with Atomic-Level Arranged Perovskite and Oxide Layers for Advanced Oxidation with an Enhanced Non-Free Radical Pathway. *ACS Sustain. Chem. Eng.* **2022**, *10*, 1899–1909. [CrossRef]
15. Žužić, A.; Ressler, A.; Macan, J. Perovskite oxides as active materials in novel alternatives to well-known technologies: A review. *Ceram. Int.* **2022**, *48*, 27240–27261. [CrossRef]
16. Tummino, M.L.; Laurenti, E.; Deganello, F.; Bianco Prevot, A.; Magnacca, G. Revisiting the catalytic activity of a doped $SrFeO_3$ for water pollutants removal: Effect of light and temperature. *Appl. Catal. B Environ.* **2017**, *207*, 174–181. [CrossRef]
17. Leiw, M.Y.; Guai, G.H.; Wang, X.; Tse, M.S.; Ng, C.M.; Tan, O.K. Dark ambient degradation of Bisphenol A and Acid Orange 8 as organic pollutants by perovskite $SrFeO_3-\delta$ metal oxide. *J. Hazard. Mater.* **2013**, *260*, 1–8. [CrossRef] [PubMed]
18. Janowska, K.; Boffa, V.; Jørgensen, M.K.; Quist-Jensen, C.A.; Hubac, F.; Deganello, F.; Coelho, F.E.B.; Magnacca, G. Thermocatalytic membrane distillation for clean water production. *npj Clean Water* **2020**, *3*, 34. [CrossRef]
19. Østergaard, M.B.; Strunck, A.B.; Jørgensen, M.K.; Boffa, V. Abatement of oil residues from produced water using a thermocatalytic packed bed reactor. *J. Environ. Chem. Eng.* **2021**, *9*, 106749. [CrossRef]
20. Bortot Coelho, F.E.; Nurisso, F.; Boffa, V.; Ma, X.; Rasse-Suriani, F.; Roslev, P.; Magnacca, G.; Candelario, V.; Deganello, F.; Parola, V. A thermocatalytic perovskite-graphene oxide nanofiltration membrane for water depollution. *J. Water Process Eng.* **2022**, *49*, 102941. [CrossRef]
21. Janowska, K.; Ma, X.; Boffa, V.; Jørgensen, M.K.; Candelario, V.M. Combined Nanofiltration and Thermocatalysis for the Simultaneous Degradation of Micropollutants, Fouling Mitigation and Water Purification. *Membranes* **2021**, *11*, 639. [CrossRef] [PubMed]
22. Tummino, M.L. $SrFeO_3$ peculiarities and exploitation in decontamination processes and environmentally-friendly energy applications. *Curr. Res. Green Sustain. Chem.* **2022**, *5*, 100339. [CrossRef]

23. Deganello, F.; Tyagi, A.K. Solution combustion synthesis, energy and environment: Best parameters for better materials. *Prog. Cryst. Growth Charact. Mater.* **2018**, *64*, 23–61. [CrossRef]
24. Wu, M.; Chen, S.; Xiang, W. Oxygen vacancy induced performance enhancement of toluene catalytic oxidation using $LaFeO_3$ perovskite oxides. *Chem. Eng. J.* **2020**, *387*, 124101. [CrossRef]
25. Hu, H.; Zhang, Q.; Wang, C.; Chen, M.; Wang, Q. Facile synthesis of $CaMn_{1-x}Fe_xO_3$ to incorporate Fe(IV) at high ratio in perovskite structure for efficient in situ adsorption-oxidation of As(III). *Chem. Eng. J.* **2022**, *435*, 134894. [CrossRef]
26. Diodati, S.; Nodari, L.; Natile, M.M.; Russo, U.; Tondello, E.; Lutterotti, L.; Gross, S. Highly crystalline strontium ferrites $SrFeO_{3-\delta}$: An easy and effective wet-chemistry synthesis. *Dalton Trans.* **2012**, *41*, 5517–5525. [CrossRef] [PubMed]
27. Kuyyalil, J.; Newby, D.; Laverock, J.; Yu, Y.; Cetin, D.; Basu, S.N.; Ludwig, K.; Smith, K.E. Vacancy assisted SrO formation on $La_{0.8}Sr_{0.2}Co_{0.2}Fe_{0.8}O_{3-\delta}$ surfaces—A synchrotron photoemission study. *Surf. Sci.* **2015**, *642*, 33–38. [CrossRef]
28. Bukhtiyarova, M.V.; Ivanova, A.S.; Slavinskaya, E.M.; Plyasova, L.M.; Rogov, V.A.; Kaichev, V.V.; Noskov, A.S. Catalytic combustion of methane on substituted strontium ferrites. *Fuel* **2011**, *90*, 1245–1256. [CrossRef]
29. Østergaard, M.B.; Strunck, A.B.; Boffa, V.; Jørgensen, M.K. Kinetics of Strontium Carbonate Formation on a Ce-Doped $SrFeO_3$ Perovskite. *Catalysts* **2022**, *12*, 265. [CrossRef]
30. Paswan, S.K.; Kumari, S.; Kar, M.; Singh, A.; Pathak, H.; Borah, J.P.; Kumar, L. Optimization of structure-property relationships in nickel ferrite nanoparticles annealed at different temperature. *J. Phys. Chem. Solids* **2021**, *151*, 109928. [CrossRef]
31. Deganello, F.; Liotta, L.F.; Longo, A.; Casaletto, M.P.; Scopelliti, M. Cerium effect on the phase structure, phase stability and redox properties of Ce-doped strontium ferrates. *J. Solid State Chem.* **2006**, *179*, 3406–3419. [CrossRef]
32. Fino, D.; Russo, N.; Saracco, G.; Specchia, V. The role of suprafacial oxygen in some perovskites for the catalytic combustion of soot. *J. Catal.* **2003**, *217*, 367–375. [CrossRef]
33. Deganello, F.; Tummino, M.L.; Calabrese, C.; Testa, M.L.; Avetta, P.; Fabbri, D.; Prevot, A.B.; Montoneri, E.; Magnacca, G. A new, sustainable $LaFeO_3$ material prepared from biowaste-sourced soluble substances. *N. J. Chem.* **2015**, *39*, 877–885. [CrossRef]
34. Toby, B.H.; Von Dreele, R.B. GSAS-II: The genesis of a modern open-source all purpose crystallography software package. *J. Appl. Crystallogr.* **2013**, *46*, 544–549. [CrossRef]
35. Toby, B.H. R factors in Rietveld analysis: How good is good enough? *Powder Diffr.* **2006**, *21*, 67–70. [CrossRef]
36. Dikalov, S.; Jiang, J.; Mason, R.P. Characterization of the high-resolution ESR spectra of superoxide radical adducts of 5-(diethoxyphosphoryl)-5-methyl-1-pyrroline N-oxide (DEPMPO) and 5,5-dimethyl-1-pyrroline N-oxide (DMPO). Analysis of conformational exchange. *Free Radic. Res.* **2005**, *39*, 825–836. [CrossRef] [PubMed]
37. Moan, J.; Wold, E. Detection of singlet oxygen production by ESR. *Nature* **1979**, *279*, 450–451. [CrossRef]
38. Bianco Prevot, A.; Avetta, P.; Fabbri, D.; Laurenti, E.; Marchis, T.; Perrone, D.G.; Montoneri, E.; Boffa, V. Waste-derived bioorganic substances for light-induced generation of reactive oxygenated species. *ChemSusChem* **2011**, *4*, 85–90. [CrossRef]
39. Avetta, P.; Bella, F.; Bianco Prevot, A.; Laurenti, E.; Montoneri, E.; Arques, A.; Carlos, L. Waste cleaning waste: Photodegradation of monochlorophenols in the presence of waste-derived photosensitizer. *ACS Sustain. Chem. Eng.* **2013**, *1*, 1545–1550. [CrossRef]

Disclaimer/Publisher's Note: The statements, opinions and data contained in all publications are solely those of the individual author(s) and contributor(s) and not of MDPI and/or the editor(s). MDPI and/or the editor(s) disclaim responsibility for any injury to people or property resulting from any ideas, methods, instructions or products referred to in the content.

Article

Changes in Structural, Morphological and Optical Features of Differently Synthetized C_3N_4-ZnO Heterostructures: An Experimental Approach

Arianna Actis [1], Francesca Sacchi [1], Christos Takidis [1], Maria Cristina Paganini [1,*] and Erik Cerrato [2]

1. Department of Chemistry and NIS, University of Torino, Via P. Giuria 7, 10125 Torino, Italy
2. NRIM Istituto Nazionale di Ricerca Metrologica, I-10135 Torino, Italy
* Correspondence: mariacristina.paganini@unito.it

Abstract: C_3N_4 is an innovative material that has had huge success as a photocatalyst in recent years. More recently, it has been coupled to robust metal oxides to obtain more stable materials. This work is focused on the different synthesis techniques used to prepare bare C_3N_4 and combined C_3N_4/ZnO mixed systems. Different precursors, such as pure melamine and cyanuric acid-based supramolecular complexes, were employed for the preparation of the C_3N_4 material. Moreover, different solvents were also used, demonstrating that the use of water leads to the formation of a more stable heterojunction. Structural (XRD), morphological (FESEM) and optical (UV-vis) measurements underlined the role of the precursors used in the preparation of the materials. A clear trend can be extrapolated from this experimental approach involving different intimate contacts between the two C_3N_4 and ZnO phases, strictly connected to the particular preparation method adopted. The use of the supramolecular complexes for the preparation of C_3N_4 leads to a tighter association between the two phases at the heterojunction, resulting in much higher visible light harvesting (connected to lower band gap values).

Keywords: C_3N_4; ZnO; heterojunction; photocatalyst

Citation: Actis, A.; Sacchi, F.; Takidis, C.; Paganini, M.C.; Cerrato, E. Changes in Structural, Morphological and Optical Features of Differently Synthetized C_3N_4-ZnO Heterostructures: An Experimental Approach. *Inorganics* **2022**, *10*, 119. https://doi.org/10.3390/inorganics10080119

Academic Editors: Roberto Nisticò and Antonino Gulino

Received: 14 July 2022
Accepted: 11 August 2022
Published: 16 August 2022

Publisher's Note: MDPI stays neutral with regard to jurisdictional claims in published maps and institutional affiliations.

Copyright: © 2022 by the authors. Licensee MDPI, Basel, Switzerland. This article is an open access article distributed under the terms and conditions of the Creative Commons Attribution (CC BY) license (https://creativecommons.org/licenses/by/4.0/).

1. Introduction

The search for new photoactive materials working in the visible frequency range of the solar spectrum is one of the most interesting and urgent challenges that the scientific community has to face. Among all of the possible candidates, C_3N_4-based materials have played a prominent role since the beginning of the century. The classic photocatalytic process is based on the exploitation of light energy, possibly falling into visible solar frequencies, to promote chemical transformations [1–3]. Photocatalytic reactions are employed for many purposes: from the decontamination of pollutants in wastewater [4,5] to the generation of fuels from the water photo-splitting reaction (producing H_2) [6,7] and the photo-reduction of CO_2 to generate useful chemical compounds (i.e., usually methanol) [8,9]. All of these processes aim to reduce the environmental impact caused by humans through the contamination and consumption of the planet's resources [10,11]. Recent studies demonstrated that the most suitable systems are semiconductor metal oxides thanks to their robustness, stability and availability. Their intrinsic band structure allows them to induce the promotion of electrons into the conduction band (CB), leaving holes in the valence band (VB) when an appropriate amount of energy, equal to or higher than the band gap width, is applied to the material [2]. Once at the semiconductor surface, the photo-generated charge carriers, i.e., electrons and holes, can facilitate reductive (H_2 photo-production and CO_2 photo-reduction) and oxidative (photo-degradation) redox reactions, respectively. The band gap of the materials, their surface area and their crystallinity are paramount parameters in determining photoactivity. In many cases, the most stable and widely used materials (i.e., TiO_2) present a high band gap, much higher than visible light wavelengths. To overcome this inconvenience, two main strategies are followed: the doping of materials and the formation of heterojunctions. The doping procedure has been widely described in the past, while the use of heterojunctions is still under debate. The formation of heterojunctions at material interfaces occurs in the band alignment; the best performance of this technology has turned out to be for interfaced systems in which the band gaps of the semiconductors are staggered, with the subsequent improvement of charge carrier separation [12]. Enhanced photocatalytic performance arises from the fact that photo-generated charge carriers are stabilized on two different semiconductors, achieving limited electron–hole pair recombination. The formation of a direct Z-scheme or S-scheme heterojunction flanking two semiconductors without electron mediators was proposed by Prof. Yu quite recently [13]. Among all of the materials proposed, g-C_3N_4, derived by the direct heating of a cyanamide precursor in air at different temperatures, has risen in importance thanks to the high production of H_2 from the water photo-splitting process upon visible light irradiation. However, despite the undoubted promising properties characterizing the promising photocatalyst exhibited by g-C_3N_4, some drawbacks have limited its practical application: the high electron–hole recombination rate, low quantum and separation efficiencies and the small specific surface area, resulting in large nanoparticles, dramatically affect its photocatalytic performance. For this reason, g-C_3N_4 has been coupled with other semiconductors, such as TiO_2, ZnO and WO_3, forming the direct heterojunction Z-scheme, obtaining very promising results [14–18].

In this preliminary work, we explored different synthesis processes for the preparation of mixed systems of C_3N_4/ZnO. The goal was to tune a facile, cheap and green synthesis process for the preparation of a mixed photocatalyst for oxidation and reduction reactions. We used different precursors and different approaches. We characterized the obtained materials via structural, optical and morphological analyses.

2. Results and Discussion

2.1. Structural Characterization of Supramolecular Precursors

Figure 1 reports the X ray Powder Diffraction (XRPD) patterns of supramolecular complexes (CMW and CME) and pure melamine (M). The supramolecular complexes were prepared by mixing melamine and cyanuric acid in either a water (CMW) or ethanol (CME) environment. The samples were dried in the oven at 70 °C for two days. Melamine was used as received. Indeed, before moving on to the discussion about the structural, morphological and optical features characterizing and differentiating the synthetized samples, it is worth evaluating the structural properties of the prepared precursors for subsequent C_3N_4 production. Accordingly, both of the new molecular arrangements exhibited by the supramolecular structural patterns show reflections at 10.67°, 18.48° and 21.41° 2θ values, indexed as (100), (100) and (200), respectively, in previous papers [19–22]. These low-angle reflections are considered evidence of the in-plane hexagonal channel structure within this supramolecular structure. Moreover, Ref. [23] attributed the higher-angle reflection present at 2θ values of 27.9° characterized by higher intensity to a d-spacing of about 0.320 nm, compatible with a graphite-like stacking of individual 2D sheets. The same evidence should be revealed by the reflection at 29.75°, in particular, for the (002) family plane, as proposed by Shalom et al. [19]. Finally, it is possible to note that the intensity of the two patterns differs a bit when changing the reaction environment, with the supramolecular structure grown in water (CMW) slightly more intense than the one in ethanol (CME): this could also be an indication of the different morphologies of the two synthetized precursors due to an altered growth path dictated by the solvent.

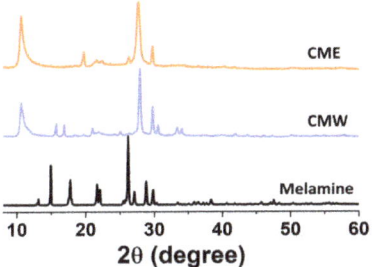

Figure 1. XRPD (X ray Powder Diffraction) pattern of the supramolecular structures synthetized starting from a mixture of melamine and cyanuric acid in water (CMW, light-blue line) and ethanol (CME, yellow line) and XRPD pattern of pure melamine.

2.2. Characterization of Bare C_3N_4

As described in Section 3.1.2, bare C_3N_4 batches were synthetized both by the direct thermal condensation of melamine (at 550 °C for 4 h in air, C_3N_4-M) and by thermal polycondensation of the previously prepared melamine–cyanuric acid supramolecular complexes, CMW (C_3N_4-CMW) and CME (C_3N_4-CME). Figure 2a reveals that the C_3N_4-M sample displays the typical XRD fingerprint of carbon nitride materials, namely, the presence of an intense reflection around a 27° 2θ angle, corresponding to an inter-planar distance of almost 0.326 nm and considered a peculiar feature associated with the (002) plane of two-dimensional aromatic systems in graphite-like structures [24,25]. However, it should be considered that the typical lamellar structure of C_3N_4 is significantly denser than graphene layers in crystalline graphite; this higher density of aromatic system stacking can be attributed to easier electron localization, which would result in a stronger interaction between the superimposed planes [26]. This attribution is confirmed by the pronounced broadening of this reflection, suggesting that overlapping only involves a limited number of layers [27]. Analyzing the pattern, an additional weaker and broader reflection at approximately 13.2° is also noticeable and associated with the (100) plane, indicating the structural periodicity within the individual two-dimensional layers: in particular, this reveals the alternating of tri-s-triazine or s-heptazine repetition units.

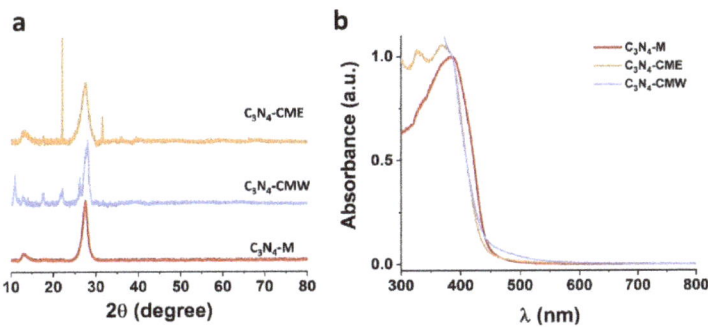

Figure 2. (a) XRPD pattern and (b) DRS-UV spectra of bare C_3N_4-M samples (red line), C_3N_4-CMW (light-blue line) and C_3N_4-CME (yellow line).

The described reflections are also visible in the patterns of C_3N_4-CMW and C_3N_4-CME samples, where additional signals are present, evidencing impurities in the final material arising from the melamine–cyanuric acid supramolecular complex. Thus, the sample synthesized by melamine thermal condensation shows a higher purity. However, all of the materials are characterized by the same degree of crystallinity. In this regard, C_3N_4-CME and C$_3$N$_4$-CMW seem to have a much broader reflection at higher angles (27°) than C$_3$N$_4$-M, suggesting a porous structure in these materials induced by the supramolecular precursor. As reported in Table 1, the surface areas of the prepared materials C$_3$N$_4$-CME, C$_3$N$_4$-CMW and C$_3$N$_4$-M obtained with the BET method are respectively 32, 33 and 11 m^2/g.

Table 1. Summary of the synthetic routes, precursors and thermal treatments employed for the sample preparation. M: melamine (C$_3$H$_6$N$_6$); C: cyanuric acid (C$_3$H$_3$N$_3$O$_3$); EtOH: ethanol (CH$_3$CH$_2$OH). The energy gaps were evaluated from the Tauc plot (see Section 3.2); the instrumental error is specified as ±0.05 eV.

Sample	Abbreviation	Synthetic procedure	Precursors	Supramol. Complexes	Hydroth. Step	Calcin. Step	Energy Gap (eV)	Surf. Area BET m^2/g
ZnO	ZnO_H	Hydrothermal	Zn(NO$_3$)$_2$		175 °C for 16 h		3.27	<10
C$_3$N$_4$	C$_3$N$_4$ –M	Thermal condensation	M			550 °C for 4 h	2.87	~11
C$_3$N$_4$	C$_3$N$_4$ - CME	Thermal polycondensation	M, C, EtOH (M-C molar ratio 1:1)	X		550 °C for 4 h	3.01	~32
C$_3$N$_4$	C$_3$N$_4$ - CMW	Thermal polycondensation	M, C, H$_2$O (M-C molar ratio 1:1)	X		550 °C for 4 h	3.04	~33
ZnO-C$_3$N$_4$	ZnO-CME	Thermal polycondensation	Zn(NO$_3$)$_2$, M, C, EtOH (M-C molar ratio 1:1)	X		550 °C for 4 h	3.20	<10
ZnO-C$_3$N$_4$	ZnO-CMW	Thermal polycondensation	Zn(NO$_3$)$_2$, M, C, H$_2$O (M-C molar ratio 1:1)	X		550 °C for 4 h	3.19	<10
ZnO-C$_3$N$_4$	ZnO-CMW_H	Thermal polycondensation + Hydrothermal	Zn(NO$_3$)$_2$, M, C, H$_2$O (M-C molar ratio 1:1)	X	175 °C for 16 h (after first calcination step)	550 °C for 4 h		<10
ZnO-C$_3$N$_4$	ZnO-M	Thermal condensation	Zn(NO$_3$)$_2$, M, EtOH			550 °C for 4 h	3.19	~12
ZnO_H-C$_3$N$_4$	DEP 31	Deposition	ZnO_H, M, H$_2$O (ZnO_H-C molar ratio 3:1)			550 °C for 4 h	3.21	<10
ZnO_H-C$_3$N$_4$	DEP 21	Deposition	ZnO_H, M, H$_2$O (ZnO_H-M molar ratio 2:1)			550 °C for 4 h	3.21	<10
ZnO_H-C$_3$N$_4$	DEP 21_H	Deposition	ZnO_H, M, H$_2$O (ZnO_H-M molar ratio 2:1)		175 °C for 16 h (after first calcination step)	550 °C for 4 h	3.21	<10
ZnO_H-C$_3$N$_4$	DEP 11	Deposition	ZnO_H, M, H$_2$O (ZnO_H-M molar ratio 1:1)			550 °C for 4 h	3.21	<10
ZnO_H-C$_3$N$_4$	DEP 11_H	Deposition	ZnO_H, M, H$_2$O (ZnO_H-M molar ratio 1:1)		175 °C for 16 h (after first calcination step)	550 °C for 4 h	3.21	<10
ZnO_H - C$_3$N$_4$	US 31	Ultrasonic/ mechanical mixture	ZnO-H, C$_3$N$_4$–M, H$_2$O (ZnO_H-C$_3$N$_4$–M molar ratio 3:1)				3.20	<10
ZnO_H-C$_3$N$_4$	US 21	Ultrasonic/ mechanical mixture	ZnO-H, C$_3$N$_4$–M, H$_2$O (ZnO_H- C$_3$N$_4$–M molar ratio 2:1)				3.20	<10

In terms of optical spectroscopy, whose spectra are reported in Figure 2b, it can be observed that a qualitative evaluation of the various absorbances does not evidence any significant variation from one sample to the other. The electronic transition from the valence band, mainly constituted by nitrogen p$_z$ orbitals, to the conduction band, made up of carbon p$_z$ orbitals, can be highlighted; this excitation is predominantly promoted by the wavelength around 460 nm and corresponds to an energy value of almost 2.8 eV, in line with previous works [28–32]. However, considering the band gap values obtained by the Tauc plot application, it is possible to see some differences: indeed, we obtained 2.87 eV for C$_3$N$_4$-M and 3 eV for C$_3$N$_4$-CMW and C$_3$N$_4$-CME (Table 1). Considering this result, as evidenced by X-ray analysis, the higher porosity of the suggested structure for the material synthetized by the supramolecular complex, and thus the more extended tri-s-triazine units conjugate system of the C$_3$N$_4$, may be the origin of the higher band gap value. Finally, the slightly longer absorption extension in the visible range for C_3N_4-CMW may be due to the higher presence of impurities in the unreacted supramolecular structure.

2.3. Characterization of C_3N_4-ZnO Heterojunction Synthetized by Prepared Supramolecular Complexes

As described in Section 3.1.3, different synthetic approaches were adopted for the preparation of the double-phase C_3N_4-ZnO heterostructure; the aim was to understand the different structural and optical features that a given synthetic route can lead to in the final material. All of the mixed ZnO-C_3N_4 samples obtained with different synthesis processes present the surface area of pure ZnO_H, about 10 m^2/g, which means that the presence of C_3N_4 does not affect the structure of the bare oxide.

Figure 3a reports the XRPD patterns of the C_3N_4-ZnO heterojunction formed by the simultaneous growth of the phases in the final calcination step (550 °C for 4 h) in air; as described in Table 1, pristine ZnO (used as a benchmark in this study) was produced via hydrothermal synthesis; the heterostructured samples labeled ZnO-CMW and ZnO-CME were obtained using the prepared supramolecular complexes as precursors for the carbon nitride phase, while the sample labeled ZnO-M used C_3N_4 as a precursor in the thermal condensation of melamine. As is observable, except for ZnO-CME, the prepared materials' XRPD patterns are dominated by the typical reflections of ZnO corresponding to the (100), (002), (101), (102), (110), (103), (200), (112), (201), (004) and (104) planes, attributable to the typical wurtzite hexagonal crystal phase (JCPDS 36-1451) [33]; the reflection sharpness confirms the purity and high crystallinity of this phase. Conversely, for the ZnO-CME sample, the ZnO phase reflections are not so intense: only the reflection around the 32° 2θ angle is of appreciable intensity, indicating that the preferential orientation for the nanostructure produced in this sample is probably caused by the nature of the supramolecular complex. However, ZnO-CME and ZnO-CMW are characterized by additional reflections due to impurities caused by unreacted supramolecular reagents, as similarly described in Section 2.2 for the bare C_3N_4. It should be noted that no appreciable signals ascribable to C_3N_4 were observed due to the very low intensity of these reflections.

In an attempt to avoid the observed impurities induced by the unreacted supramolecular complexes, we introduced a hydrothermal step (175 °C for 16 h) before the final calcination treatment in air for the ZnO-CMW sample, now named ZnO-CMW_H. Correspondingly, in the XRPD pattern of hydrothermally treated ZnO-CMW_H, the reflections for the impurities disappeared, and only those related to the ZnO phase are identifiable. Accordingly, hydrothermal treatment by means of the high pressure employed is able to stabilize the organic complex, promoting the complexation of melamine and cyanuric acid in an ordered arrangement of the two monomers, now held together by a greater number of hydrogen bonds and allowing for the net stabilization of the structure [34].

An additional general feature can be derived by analyzing these patterns: the different relative ratio intensities of the three main reflections of ZnO phases (between 2θ values of 30° and 40°) in the sample grown simultaneously to the carbon nitrate phase compared to pristine ZnO are a clear indication of a certain preferential growth direction for ZnO in the presence of the extra phase, where the nature of the precursor of the latter also induces a guiding effect, tailoring the nanoparticle geometry and aspect ratio.

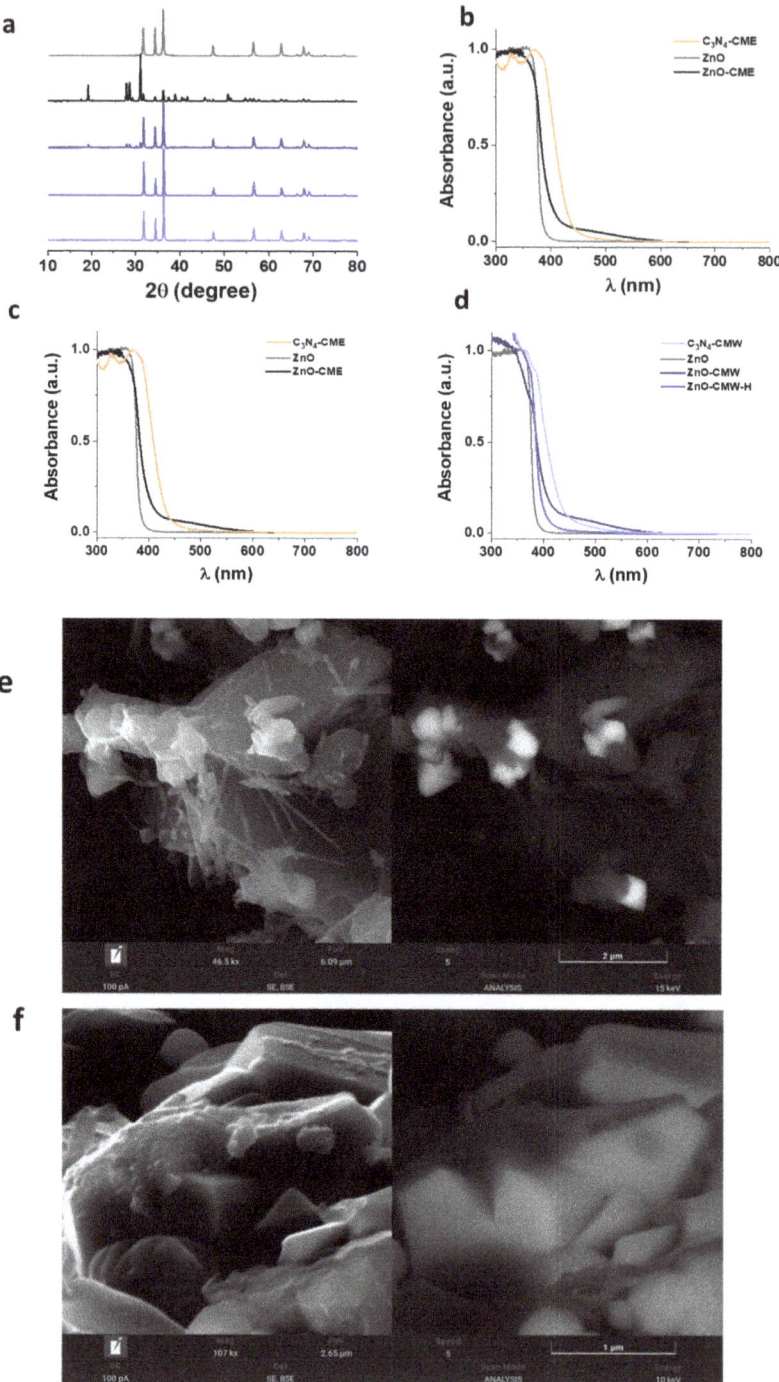

Figure 3. (**a**) XRPD pattern, (**b**–**d**) DRS-UV spectra and FESEM images (secondary electrons on the left and backscattered electrons on the right) of (**e**) ZnO-CME and (**f**) ZnO-CMW heterojunctions.

The absence of C_3N_4 crystallographic peaks in the combined samples is due to different reasons: first, the very low mass of C and N nuclei, and second, the formation of needles or, in some cases, of thin sheets that do not allow the formation of a periodic structure detectable by XRD. Moreover, in support of this suggestion, FESEM images reported in Figure 3e,f highlight the morphological differences in the ZnO-CMW and ZnO-CME samples for both the ZnO and C_3N_4 phases of the mixed materials: while ZnO-CME is composed of large hexagonal base shaped ZnO nanorods decorated by C_3N_4 nanoneedles, ZnO-CMW is characterized by narrower nanorods, with a smaller hexagonal base shape, embedded in the C_3N_4 matrix, thus providing a higher interaction area between the two phases. Moreover, the morphological differences between the two samples may suggest that water allows stronger mixing with the reagent and more intimate contact between the two phases.

However, in contrast to other papers [15,35,36], the crystallinity of the ZnO phase is not degraded in the final heterojunction; inter alia, the reflections for ZnO in the mixed materials remain as high and sharp as those for pristine ZnO. Furthermore, FESEM images also agree with this assumption, as already discussed.

While XRPD analysis does not evidence the presence of C_3N_4, DRS UV-vis spectroscopy shows features attributable to the carbon nitride phase, as observable in Figure 3b–d. Generally, all of the mixed C_3N_4-ZnO materials exhibit an intermediate behavior between the two constituting phases: this is deducible by the energy gap values calculated with the Tauc plot and reported in Table 1, displaying a visible red-shift relative to ZnO optical absorption. The visible photon absorption improvement in the case of the heterojunctions is attributable to the presence of the nitride, suggesting the strong cooperation of the two phases at the interface. Moreover, the absorption tail in the mixed samples around 500 nm, which is not present in the bare C_3N_4 samples, may be attributed to the higher defectiveness of this phase in the mixed system as a result of simultaneous crystallization with the ZnO phase during C_3N_4 condensation, as already discussed in previous research [15,35,37–39]: this seems to be particularly important for the more defective materials, such as ZnO-CMW, ZnO-CME and ZnO-M. This effect decreases for the thermally treated ZnO-CMW_H, probably due to the higher sintering process of the two phases and the loss of many intimate contacts among them. Then, it can be confirmed that the simultaneous growth of the two phases from different precursors and their contemporary existence are beneficial in terms of the wider range of photons absorbed compared to the bare materials in both the UV and visible ranges, revealing intimate cooperation during light absorption, determined by a successful charge carrier transfer at the interfaces. Photoactivity measurements such as EPR (electron paramagnetic resonance) experiments under light irradiation will be necessary to confirm this behavior [15]. Furthermore, the formation of the heterojunction will be demonstrated via photocurrent experiments [40–42].

2.4. Characterization of C_3N_4-ZnO Heterojunction Synthetized by Deposition Method

The C_3N_4-ZnO heterojunction was also prepared by applying a deposition method, as described in Section 3.1.5. We attempted to let melamine condensate and polymerize in the presence of the already-formed ZnO nanoparticles (ZnO_H), thus trying to realize C_3N_4 film growth on the ZnO surface. Referring to the results presented in the previous section and, in particular, to the FESEM images, we observed that the water medium could potentially allow establishing more extended surface contact between the two components of the heterojunction (as for ZnO-CMW). Therefore, deposition of C_3N_4 was performed by mixing melamine and ZnO in H_2O; this series of samples is named DEP followed by the indication of the ZnO/melamine molar ratio, thus obtaining DEP 31, DEP 21 and DEP 11 (see Table 1). Evaluating the XRPD analysis results for this set of samples in Figure 4a, one can realize that only DEP 31 shows the typical fingerprints of ZnO, as described in Section 2.3; indeed, the DEP 21 and DEP 11 patterns are studded by many reflections that are attributable to neither the ZnO phase nor the C_3N_4 phase. For this reason, such samples were subjected to a second calcination step, in which the impurities were completely eliminated, as appreciable in the extension of Figure 4a. Then, it emerges that, depending on the relative concentration of the melamine precursors, the final material shows a different purity degree, requiring longer thermal treatment: it could be that, in the presence of too high an amount of organic precursor, the surface sites of ZnO nanoparticles from which the C_3N_4 phase nucleates are completely occupied, leaving the excess precursor unreacted. This consideration is also supported by FESEM images of sample DEP 31, visible in Figure 4c: the ZnO nanorods, hardly recognizable by the hexagonal base shape, are submerged in an environment of C_3N_4 lamellar nanoparticles, appearing much more crystalline when compared to the ones derived from the supramolecular complexes (see FESEM images in Figure 3e,f).

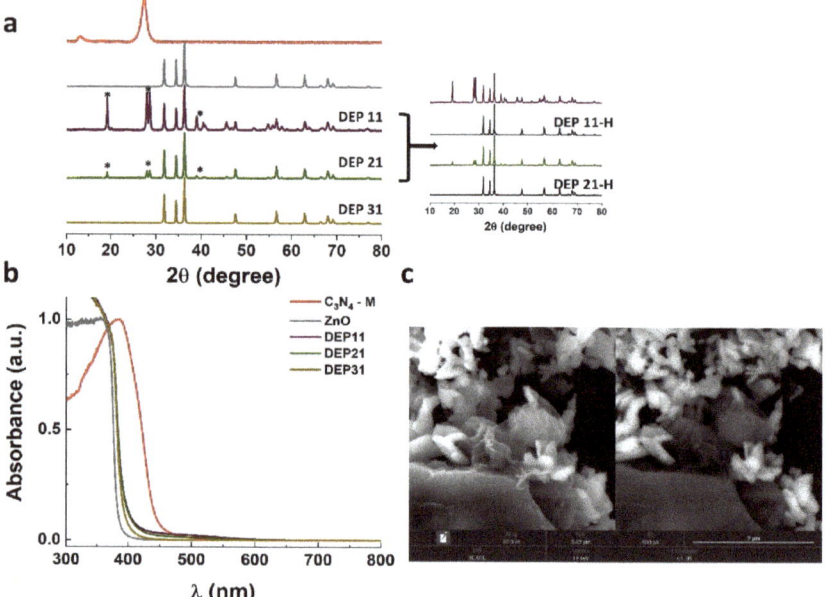

Figure 4. (a) XRPD pattern (in the extended part, the diffractograms of the DEP 11 and DEP 21 samples before and after (DEP 11-H and DEP 21-H) the second hydrothermal treatment are reported), (b) DRS-UV spectra and (c) FESEM images (secondary electrons on the left and backscattered electrons on the right) of DEP 31 heterojunction.

From an optical point of view, the three materials exhibit similar behavior, with an appreciable red-shift absorption edge when compared to pristine ZnO, but less pronounced relative to that observed in the heterostructures prepared through the supramolecular precursor (Section 2.3). In this case, the higher visible light absorption tail (around 500 nm) was recorded for DEP 11, indicating better cooperation between the oxide and nitride phases after the second calcination step. In addition, this reveals that for the deposition synthesis route, the identification of an optimal reaction time that strictly depends on the amount of the organic precursor is crucial in order to obtain the desired performance in terms of visible light harvesting exploited by the heterojunction.

2.5. Characterization of C_3N_4-ZnO Heterojunction Synthetized by Ultrasonic/Mechanical Method

Several studies have reported liquid exfoliation as an effective top-down method to obtain thin layers of C3N4 starting from the bulk material [36,43,44]. Such preparation can easily be obtained through the sonication of C_3N_4 dispersed in a medium. Therefore, as a further alternative procedure, the ZnO/C_3N_4 heterojunction was synthetized through the mechanical mixture of C_3N_4 and ZnO, both synthetized independently and treated with ultrasounds, with the purpose of making exfoliated C_3N_4 sheets to deposit on the ZnO surface (US samples, see Table 1). Again, since water was observed to be a good dispersal medium in terms of the homogeneous dispersion of the starting materials, we proceeded with the suspension of two previously prepared phases in an aqueous environment. The bulky starting materials ZnO and C_3N_4 synthetized via hydrothermal and melamine condensation were each used for mechanical mixing.

XRPD characterization in Figure 5a reveals typical patterns for the two bare materials employed in ultrasonic mixing and the materials obtained from them, designated US 21 and US 31: a great difference from the XRPD patterns of the other mixed samples reported in the previous sections immediately emerges, namely, the appearance of the reflection at the 27° 2θ angle for the C_3N_4 matrix. This may indicate that with this synthetic route, the nitride phase can maintain a higher crystallinity degree: this would overcome one of the most challenging issues concerning the employment of C_3N_4 in photocatalysis, that is, rapid charge carrier recombination, mainly due to the low surface area and the high amorphization degree. Indeed, in more crystalline materials, thanks to the long-range order, the photoinduced charge carrier pairs have much more time to travel in the solid, thus preventing their recombination. According to Cao et al., light absorption of semiconductor materials is primarily controlled by the band gap, which in turn can be tuned by changing morphology, crystalline quality and crystal structure. A post-annealing treatment can be used as a method to improve the crystallinity and thus the light absorption of semiconductor materials [45]; moreover, the interlayer region generated by the coupling of the two composites may reduce electron–hole recombination [46]. Conversely, this structural advantage is not exhibited by the DRS UV-vis spectroscopy results in Figure 5b, where an appreciable band gap red-shift is not detected, in contrast to the previous cases, with values around 3.20 eV. This behavior is more evident in the material with lower nitride content. Moreover, no additional shoulders arising from C3N4-induced defects during simultaneous growth with ZnO were detected around 500 nm: this can be considered a further indication of the less intimate contact during the solid-state reaction. In this case, the morphology was also explored, with the FESEM image of sample US 31 reported in Figure 5c. The picture reveals the almost complete cleavage of the hexagonal morphology of the starting ZnO_H material (compare with Figure 3e,f), resulting in a lamellar morphology, thus revealing a drastic effect of ultrasonic irradiation on the particle nanostructures. Surprisingly, the C3N4 phase is not easily visible as in the previous synthesized samples: only the backscattered electron image lets us observe its distribution over ZnO platelets by means of darker spots.

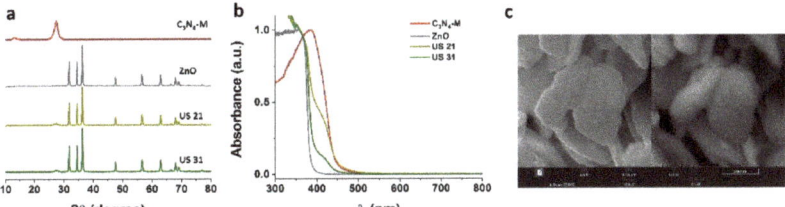

Figure 5. (**a**) XRPD pattern, (**b**) DRS-UV spectra of ZnO, C_3N_4-M, US 31 and US 21 samples and (**c**) FESEM images (secondary electrons on the left and backscattered electrons on the right) of the US 31 heterojunction.

Summarizing, the mechanical mixing achieved via ultrasound irradiation effectively leads to a more crystalline heterojunction characterized by lower visible light harvesting, deriving from the minor cooperation of the two phases at the interface.

3. Materials and Methods

3.1. Sample Preparation

All organic and inorganic reagents employed in the synthesis of the materials were purchased from Sigma-Aldrich and used without any further purification.

The various materials prepared in this work, the different precursors and the synthetic routes employed are summarized in Table 1 (see below).

3.1.1. Bare C_3N_4

Three different bare C_3N_4 samples were synthetized according to the most common procedure, namely, the thermal condensation of the employed precursors, i.e., melamine ($C_3H_6N_6$) and cyanuric acid ($C_3H_3N_3O_3$). Specifically, the material denoted as C_3N_4-M was prepared by the thermal condensation of melamine only (0.977 g) by directly heating the precursor in air at 550 °C for 4 h (5 °C/min ramp). Conversely, two additional pure C_3N_4 materials were produced by developing a supramolecular structure via thermal polycondensation of melamine (0.977 g) and cyanuric acid (1.023 g) dispersed in either H_2O or CH_3CH_2OH: in both cases, the two reagents (in 1:1 molar ratio) were introduced into 80 mL of the respective solvent. The obtained samples were collected and washed (with either water or ethanol, depending on the solvent employed) and dried at 70 °C for two days in the oven; the final materials were obtained by applying a calcination step at 550 °C for 4 h (5 °C/min ramp) and labeled C_3N_4-CMW (water) and C_3N_4-CME (ethanol), respectively.

3.1.2. Bare ZnO

Pristine zinc oxide was synthesized through a hydrothermal procedure, as described in Ref. [47]. Briefly, a 20 mL $Zn(NO_3)_2$ 1M water solution was prepared; following the complete dissolution of the zinc salt, a 4M NaOH solution was added drop by drop to the initial solution until a pH of 11-12 was reached and a dense white precipitate, namely, $Zn(OH)_2$, was visible. Then, the precipitate was inserted into a stainless steel autoclave and treated at 175 °C for 16 h. The obtained ZnO was washed with water and collected by means of centrifugation; finally, the material was dried overnight at 70 °C.

3.1.3. C_3N_4–ZnO Heterojunctions

Inspired by our expertise in the synthesis of transition metal oxide-based semiconductors' photoactive heterojunctions [15,47–49], three different synthetic strategies were employed for the preparation of interfaced C_3N_4-ZnO material, as listed hereafter.

3.1.4. Co-synthesis from Supramolecular Adduct

As described in Section 3.1.1, melamine and cyanuric acid were mixed in either H_2O or CH_3CH_2OH to form the supramolecular adduct to finally obtain C_3N_4 (C_3N_4–CMW and C_3N_4–CME, respectively); in this case, $Zn(NO_3)_2$ (in 1:1:1 molar ratio) was added to the suspension. Once the dissolution of zinc nitrate was completed, a 4M NaOH solution was added dropwise to allow for $Zn(OH)_2$ precipitation. The mixtures were then centrifuged, washed three times with H_2O, dried in the oven at 70 °C, transferred to a crucible and calcined in air at 550 °C for 4 h (5 °C/min ramp). The materials were labeled ZnO-CMW (in water) and ZnO-CME (in ethanol), respectively. An additional sample was realized following the same procedure but starting from melamine and zinc nitrate precursors: this was labeled ZnO-M.

A further modification of this recipe was applied for the material labeled ZnO-CMW_H: it was obtained as described for ZnO-CMW in this section, but before the calcination step in air, the sample was hydrothermally treated in a stainless steel autoclave at 175 °C for 16 h. The sample was then washed with water and collected by means of centrifugation; finally, the material was dried overnight at 70 °C.

3.1.5. Deposition Synthesis

With the aim to allow for the direct polymerization of melamine onto the ZnO-H particle surface, the deposition approach was considered for the preparation of another set of samples with interconnected C_3N_4-ZnO phases. The procedure involved the introduction of melamine to the ZnO-H nanoparticle water suspension; after 30 min of stirring, the samples were dried at 70 °C in the oven for 2 days and eventually calcined at 550 °C for 4 h (5 °C/min ramp). The explored ZnO–melamine molar ratios were 1:1, 2:1 and 3:1, with the obtained samples named DEP 11 (1.56 g of melamine and 1.00 g of ZnO-H), DEP21 (0.76 g of melamine and 1.00 g of ZnO-H) and DEP31 (0.5 g of melamine and 1.00 g of ZnO-H), respectively.

In this case, a further hydrothermal treatment step (175 °C for 16 h) was employed for the DEP11 and DEP21 samples and newly labeled DEP11_H and DEP21_H, respectively.

3.1.6. Ultrasonic/Mechanical Mixture

The last considered synthetic route evaluated the effect of the ultrasonic/mechanical mixture in the constitution of the C_3N_4-ZnO heterojunction. For this procedure, the independently prepared C_3N_4-M and ZnO-H (see Table 1 and Section 3.1.1 and 3.1.2) were dispersed in water with ZnO–C_3N_4 weight ratios of 3:1 and 2:1 and with the corresponding samples referred to as US 31 (0.40 g of melamine) and US 21 (0.55 g of melamine), respectively. The two-phase dispersion was stirred in an ice bath with an immersion sonicator (Bandelin Sonopuls HD 3100 (2022 Merck KGaA, Darmstadt, Germany) equipped with an MS 73 probe, 1 Hz frequency, 30 W power) for 10 min, then centrifuged and dried in the oven at 70 °C for two days.

3.2. Sample Characterization

XRPD (X-ray powder diffraction) was performed to obtain structural information regarding single- and multiple-phase synthetized materials, paying particular attention both to the phase identification and to the possible presence of additional impurity reflections, especially arising from the organic precursors. The acquired patterns were recorded by means of a PANalytical PW3040/60 X'Pert PRO MPD (Malvern Panalytical Ltd., Malvern, UK) with a copper K radiation source (0.15418 nm) in Bragg–Brentano geometry. Samples were scanned continuously in the 2θ range between 10° and 80°. The X'Pert High-Score (Malvern Panalytical Ltd, Malvern, UK) software was used to identify the mineral phases present in the samples.

Diffuse Reflectance UV-vis spectroscopy was employed to characterize the samples from an optical point of view; the optical spectra were recorded using a Varian Cary 5000 spectrophotometer (Agilent, Santa Clara, CA, USA) and the Carywin-UV/scan software (Agilent, Santa Clara, CA, USA). A sample of PTFE with 100% reflectance was used as a reference. The optical band gap energies were calculated by applying the Tauc plot on the obtained spectra, considering that the energy dependence of the absorption coefficient for semiconductors in the region near the absorption edge is proportional to the material energy gap and dependent on the kind of transition (direct or indirect allowed) [50].

FESEM (Field-Emission Scanning Electron Microscope) images were recorded with the instrument FEG-SEM TESCAN S9000G (Source: Schottky emitter; Resolution: 0.7 nm at 15 keV (in-beam SE) (Libušina tř. 21, 623 00 Brno–Kohoutovice, Czech Republic); Accelerating Voltage: 0.2–30 keV; Microanalyzer: OXFORD (Abingdon OX13 5QX, UK); Detector: Ultim Max; Software: AZTEC (Casole Bruzio CS, 87059, Italy)

Surface area measurements were carried out on a Micromeritics Accelerated Surface Area and Porosimetry System (ASAP) 2020/2010 (4356 Communications Drive, Norcross, GA 30093, USA) using the Brunauer–Emmett–Teller (BET) equation on the N_2 adsorption measurement. Prior to the adsorption run, all samples were outgassed at 160 °C for 3 h.

4. Conclusions

In this study, we systematically investigated the structural, optical and morphological features arising from the use of different synthetic routes of the C_3N_4-ZnO heterojunction. Specifically, the different applied preparations for the C_3N_4-ZnO heterostructures included:

- ✓ The production of a supramolecular precursor realized by the dispersion of melamine and cyanuric acid in water or ethanol, subsequently mixed with zinc precursors and thermally calcined. In this case, we demonstrate that the prepared supramolecular complexes show the expected molecular structures. The addition of a further hydrothermal step after the thermal treatment in air was decisive in obtaining a mixed material free from impurities.
- ✓ Direct growth by means of melamine thermal condensation of C_3N_4 on hydrothermally pre-formed ZnO nanoparticles.
- ✓ The ultrasonic/mechanical mixing of both pre-formed ZnO (from hydrothermal synthesis) and C_3N_4 (from melamine-only condensation) nanopowders.
- ✓ A significant trend can be extrapolated from this experimental approach involving the different intimate contacts between the two phases, strictly connected to the particular preparation method adopted: it appears that the heterojunction prepared starting from the supramolecular complex shows a tighter association between the two phases, resulting in much higher visible light harvesting (connected to the appearance of adsorption bands in the visible region); on the other hand, lower crystallinity is observed.
- ✓ An intermediate situation is represented by the material obtained via the deposition method, where improved crystallinity, especially regarding the nitride phase, is accompanied by less optical activity at visible frequencies.
- ✓ Finally, solid-state mixing supported by ultrasound irradiation provided crystalline materials with a completely modified morphology compared to the starting nanostructures.

In conclusion, this study provides a useful tool for obtaining C_3N_4-ZnO heterojunctions by different synthetic procedures, characterized by diverse features concerning structural, optical and morphological peculiarities. We are confident that these materials could be employed in photocatalytic applications thanks to the formation of the designed heterojunctions. The final goal will be to overcome charge carrier pair recombination and extend their lifetime.

Author Contributions: Conceptualization, M.C.P. and E.C.; methodology, A.A.; investigation, F.S. and C.T.; data curation, A.A.; writing—original draft preparation, E.C.; writing—review and editing, M.C.P.; supervision, M.C.P. and E.C.; project administration, M.C.P. All authors have read and agreed to the published version of the manuscript.

Funding: This research received no external funding.

Institutional Review Board Statement: Not applicable.

Informed Consent Statement: Not applicable.

Data Availability Statement: Not applicable.

Acknowledgments: This paper is part of a project that has received funding from the Compagnia di San Paolo, Torino, under the project SusNANOCatch "Sustainable strategies to reduce the presence in the environment of nanoparticles deriving from depollution processes"; from the European Union's Horizon 2020 research and innovation program under Marie Skłodowska-Curie grant agreement no. 765860 (AQUAlity); from the European Union's Horizon 2020 research and innovation program under Marie Skłodowska-Curie grant agreement no. 101007578" (SusWater); and from the Italian MIUR through the PRIN Project 20179337R7, MULTI-e "Multielectron transfer for the conversion of small molecules: an enabling technology for the chemical use of renewable energy". FEG-SEM S9000 by Tescan was purchased with funds from Regione Piemonte (project POR FESR 2014-20 INFRA-P SAX).

Conflicts of Interest: The authors declare no conflict of interest.

References

1. Baly, E.C.C.; Heilbron, I.M.; Barker, W.F. CX.—Photocatalysis. Part I. The Synthesis of Formaldehyde and Carbohydrates from Carbon Dioxide and Water. *J. Chem. Soc. Faraday Trans.* **1921**, *119*, 1025–1035. [CrossRef]
2. Hernandez-Ramirez, A.; Medina-Ramirez, I. *Photocatalytic Semiconductors*; Springer International Publisher: Cham, Switzerland, 2016.
3. Coronado, J.M.; Fresno, F.; Hernàndez-Alonso, M.D.; Portela, R. *Design of Advanced Photocatalytic Materials for Energy and Environmental Applications*; Springer: London, UK, 2013.
4. Ollis, D.F.; Pelizzetti, E.; Serpone, N. Destruction of Water Cntaminants. *Environ. Sci. Technol.* **1991**, *25*, 1522–1529. [CrossRef]
5. Hoffman, M.R.; Martin, S.T.; Choi, W.; Bahnemann, D.W. Environmental Applications of Semiconductor Photocatalysis. *Chem. Rev.* **1995**, *95*, 69–96. [CrossRef]
6. Maeda, K.; Domen, K. Photocatalytic Water Splitting: Recent Progress and Future Challenges. *J. Phys. Chem. Lett.* **2010**, *1*, 2655–2661. [CrossRef]
7. Domen, K.; Kondo, J.N.; Michikazu, H.; Tsuyoshi, T. Photo- and Mechano-Catalytic Overall Water Splitting Reactions to Form Hydrogen and Oxygen on Heterogeneous Catalysts. *Bull. Chem. Soc. Jpn.* **2000**, *73*, 1307–1331. [CrossRef]
8. Chang, Z.; Wang, T.; Gong, J. CO_2 photo-reduction: Insights into CO_2 activation and reaction on surfaces of photocatalysts. *Energy Environ. Sci.* **2016**, *9*, 2177–2196. [CrossRef]
9. Inoue, T.; Fujishima, A.; Konishi, S.; Honda, K. Photoelectrocatalytic Reduction of Carbon Dioxide in Aqueous Suspensions of Semiconductor Powders. *Nature* **1979**, *277*, 637–638. [CrossRef]
10. Jerez, S.; Lopez-Romero, J.M.; Turco, M.; Jimenez-Guerrero, P.; Vautard, R.; Montavez, J.P. Impact of Evolving Greenhouse Gas Forcing on the Warming Signal in Regional Climate Model Experiments. *Nat. Commun.* **2018**, *9*, 1304. [CrossRef]
11. Figures, C.; Schellnhuber, H.J.; Whiteman, G.; Rockström, J.; Hobley, A.; Rahmstorf, S. Three Years to Safeguard our Climate. *Nature* **2017**, *546*, 593–595. [CrossRef]
12. Wang, H.; Zhang, L.; Chen, Z.; Hu, J.; Li, S.; Wang, Z.; Liu, J.; Wang, X. Semiconductor Heterojunction Photocatalysts: Design, Construction, and Photocatalytic Performances. *Chem. Soc. Rev.* **2014**, *43*, 5234–5244. [CrossRef] [PubMed]
13. Yu, J.; Wang, S.; Low, J.; Xiao, W. Enhanced Photocatalytic Performance of Direct Z-Scheme g-C_3N_4-TiO_2 Photocatalysts for the Decomposition of Formaldehyde in Air. *Phys. Chem. Chem. Phys.* **2013**, *15*, 16883–16890. [CrossRef] [PubMed]
14. Kong, L.; Zhang, X.; Wang, C.; Xu, J.; Du, X.; Li, L. Ti^{3+} Defect mediated g-C_3N_4/TiO_2 Z-Scheme System for Enhanced Photocatalytic Redox Performance. *Appl. Surf. Sci.* **2018**, *448*, 288–296. [CrossRef]
15. Cerrato, E.; Paganini, M.C. Mechanism of Visible Photon Absorption: Unveiling of the C_3N_4–ZnO Photoactive Interface by means of EPR Spectroscopy. *Mater. Adv.* **2020**, *1*, 2357. [CrossRef]
16. Zhou, J.; Zhang, M.; Zhu, Y. Preparation of Visible Light-driven g-C_3N_4@ZnO Hybrid Photocatalyst via Mechanochemistry. *Phys. Chem. Chem. Phys.* **2014**, *16*, 17627–17633. [CrossRef]
17. Sun, J.-X.; Yuan, J.-P.; Qia, L.-G.; Jiang, X.; Xie, A.-J.; Shen, Y.-H.; Zhu, J.-F. Fabrication of Composite Photocatalyst g-C_3N_4–ZnO and Enhancement of Photocatalytic Activity under Visible Light. *Dalton Trans.* **2012**, *41*, 6756–6763. [CrossRef] [PubMed]
18. Huang, l.; Xu, H.; Li, Y.; Li, H.; Cheng, X.; Xia, J.; Xu, Y.; Cai, G. Visible-Light-induced WO_3/g-C_3N_4 Composites with Enhanced Photocatalytic Activity. *Dalton Trans.* **2013**, *42*, 8606–8616. [CrossRef] [PubMed]
19. Shalom, M.; Inal, S.; Fettkenhauer, C.; Neher, D.; Antonietti, M. Improving Carbon Nitride Photocatalysis by Supramolecular Preorganization of Monomers. *J. Am. Chem. Soc.* **2013**, *135*, 7118–7121. [CrossRef] [PubMed]
20. Shalom, M.; Guttentag, M.; Fettkenhauer, C.; Inal, S.; Neher, D.; Llobet, A.; Antonietti, M. In Situ Formation of Heterojunctions in Modified Graphitic Carbon Nitride: Synthesis and Noble Metal Free Photocatalysis. *Chem. Mater.* **2014**, *26*, 5812–5818. [CrossRef]
21. Wang, Y.; Zhao, S.; Zhang, Y.; Fang, J.; Chen, W.; Yuan, S.; Zhou, Y. Facile Synthesis of Self-Assembled g-C_3N_4 with Abundant Nitrogen Defects for Photocatalytic Hydrogen Evolution. *ACS Sustain. Chem. Eng.* **2018**, *6*, 10200–10210. [CrossRef]
22. Zhao, S.; Zhang, Y.; Zhou, Y.; Wang, Y.; Qiu, K.; Zhang, C.; Fang, J.; Sheng, X. Facile One-sSep Synthesis of Hollow Mesoporous g-C_3N_4 Spheres with Ultrathin Nanosheets for Photoredox Water Splitting. *Carbon* **2018**, *126*, 247–256. [CrossRef]
23. Jun, Y.-S.; Lee, E.Z.; Wang, X.; Hong, W.H.; Stucky, G.D.; Thomas, A. From Melamine-Cyanuric Acid Supramolecular Aggregates to Carbon Nitride Hollow Spheres. *Adv. Funct. Mater.* **2013**, *23*, 3661–3667. [CrossRef]
24. Goettmann, F.; Fischer, A.; Antonietti, M.; Thomas, A. Chemical Synthesis of Mesoporous Carbon Nitrides Using Hard Templates and Their Use as a Metal-Free Catalyst for Friedel–Crafts Reaction of Benzene. *Angew. Chem. Int. Ed. Engl.* **2006**, *45*, 4467–4471. [CrossRef] [PubMed]
25. Liu, J.; Zhang, T.; Wang, Z.; Dawsona, G.; Chen, W.M. Simple Pyrolysis of Urea into Graphitic Carbon Nitride with Recyclable Adsorption and Photocatalytic Activity. *J. Mater. Chem.* **2011**, *21*, 14398. [CrossRef]
26. Thomas, A.; Fischer, A.; Goettmann, F.; Antonietti, M.; Muller, J.-O.; Schlogl, R.; Carlsson, J.M. Graphitic Carbon Nitride Materials: Variation of Structure and Morphology and their Use as Metal-Free Catalysts. *J. Mater. Chem.* **2008**, *18*, 4893–4908. [CrossRef]
27. Inagaki, M.; Tsumura, T.; Kinumoto, T.; Toyoda, M. Graphitic Carbon Nitrides (g-C_3N_4) with Comparative Discussion to Carbon Materials. *Carbon* **2019**, *141*, 580–607. [CrossRef]
28. Alves, I.; Demazeau, G.; Tanguy, B.; Weill, F. On a New Model of the Graphitic Form of C_3N_4. *Solid State Commun.* **2001**, *109*, 697–701. [CrossRef]

29. Kumar, S.G.; Karthikeyan, S.; Lee, A.F. g-C_3N_4-based Nanomaterials for Visible Light-Driven Photocatalysis. *Catalysts* **2018**, *8*, 74. [CrossRef]
30. Wang, X.; Maeda, K.; Thomas, A.; Takanabe, K.; Xin, G.; Carlsson, J.M.; Domen, K.; Antonietti, M. A Metal-Free Polymeric Photocatalyst for Hydrogen Production from Water under Visible Light. *Nat. Mater.* **2009**, *8*, 76–80. [CrossRef]
31. Xiao, J.; Han, Q.; Cao, H.; Rabeah, J.; Yang, J.; Guo, Z.; Zhou, L.; Xie, Y.; Bruckner, A. Number of Reactive Charge Carriers—A Hidden Linker between Band Structure and Catalytic Performance in Photocatalysts. *ACS Catal.* **2019**, *9*, 8852–8861. [CrossRef]
32. Huda, M.N.; Turner, J.A. Morphology-dependent optical absorption and conduction properties of photoelectrochemical photocatalysts for H2 production: A case study. *J. Appl. Phys.* **2010**, *107*, 123703. [CrossRef]
33. Wróbel, J.; Piechota, J. On the structural stability of ZnO phases. *Solid State Commun.* **2008**, *146*, 324–329. [CrossRef]
34. Fanetti, S.; Citroni, M.; Dziubek, K.; Nobrega, M.M.; Bini, R. The Role of H-bond in the High-Pressure Chemistry of Model Molecules. *J. Phys. Condens. Matter.* **2018**, *30*, 094001. [CrossRef] [PubMed]
35. Yu, W.; Xu, D.; Peng, T. Enhanced Photocatalytic Activity of g-C_3N_4 for Selective CO_2 Reduction to CH_3OH via Facile Coupling of ZnO: A Direct Z-scheme Mechanism. *J. Mater. Chem. A* **2015**, *3*, 19936–19947. [CrossRef]
36. Chen, Q.; Hou, H.; Zhang, D.; Hu, S.; Min, T.; Liu, B.; Yang, C.; Pu, W.; Hu, J.; Yang, J. Enhanced Visible-Light Driven Photocatalytic Activity of Hybrid ZnO/g-C_3N_4 by High Performance Ball Milling. *J. Photochem. Photobiol. A Chem.* **2018**, *350*, 1–9. [CrossRef]
37. Wang, K.; Li, Q.; Liu, B.; Cheng, B.; Ho, W.; Yu, J. Sulfur-doped g-C_3N_4 with Enhanced Photocatalytic CO_2-Reduction Performance. *Appl. Catal. B Environ.* **2015**, *176*, 44–52. [CrossRef]
38. Zhai, J.; Wang, T.; Wang, C.; Liu, D. UV-Light-Assisted Ethanol Sensing Characteristics of g-C_3N_4/ZnO Composites at Room Temperature. *Appl. Surf. Sci.* **2018**, *441*, 317–3323. [CrossRef]
39. Le, S.; Jiang, T.; Li, Y.; Zhao, Q.; Li, Y.; Fang, W.; Gong, M. Highly Efficient Visible-Light-Driven Mesoporous Graphitic Carbon Nitride/ZnO Nanocomposite Photocatalysts. *Appl. Catal. B Environ.* **2017**, *200*, 601–610. [CrossRef]
40. Cao, Q.; Li, Q.; Pi, Z.; Zhang, J.; Sun, L.W.; Xu, J.; Cao, Y.; Cheng, J.; Bian, Y. Metal-Organic-Framework-Derived Ball-Flower-like Porous Co_3O_4/Fe_2O_3 Heterostructure with Enhanced Visible-Light-Driven Photocatalytic Activity. *Nanomaterials* **2022**, *12*, 904. [CrossRef]
41. Cao, Q.; Hao, S.; Wu, Y.; Pei, K.; You, W.; Che, R. Interfacial Charge Redistribution in Interconnected Network of Ni_2P–Co_2P Boosting Electrocatalytic Hydrogen Evolution in both Acidic and Alkaline Conditions. *Chem. Eng. J.* **2021**, *424*, 130444. [CrossRef]
42. Zhu, L.-Y.; Yuan, K.; Yang, J.-G.; Ma, H.-P.; Wang, T.; Ji, X.-M.; Feng, J.-J.; Devi, A.; Lu, H.-L. Fabrication of Heterostructured p-CuO/n-SnO_2 Core-Shell Nanowires for Enhanced Sensitive and Selective Formaldehyde Detection. *Sens. Actuators B Chem.* **2019**, *290*, 233–241. [CrossRef]
43. Zhang, X.; Xie, X.; Wang, H.; Zhang, J.; Pan, B.; Xie, Y. Enhanced Photoresponsive Ultrathin Graphitic-Phase C_3N_4 Nanosheets for Bioimaging. *J. Am. Chem Soc.* **2013**, *135*, 18–21. [CrossRef] [PubMed]
44. Yang, S.; Gong, Y.; Zhang, J.; Zhan, L.; Ma, L.; Fang, Z.; Vaiijtai, R.; Wang, X.; Ajayan, P.M. Exfoliated Graphitic Carbon Nitride Nanosheets as Efficient Catalysts for Hydrogen Evolution Under Visible Light. *Adv. Mater.* **2013**, *25*, 2452–2456. [CrossRef] [PubMed]
45. Cao, Q.; Yu, J.; Cao, Y.; Delaunay, J.-J.; Che, R. Unusual Effects of Vacuum Annealing on Large-Area Ag_3PO_4 Microcrystalline Film Photoanode Boosting Cocatalyst- and Scavenger-free Water Splitting. *J. Mater.* **2021**, *7*, 929–939. [CrossRef]
46. Nocchetti, M.; Pica, M.; Ridolfi, B.; Donnadio, A.; Boccalon, E.; Zampini, G.; Pietrella, D.; Casciola, M. AgCl-ZnAl Layered Double Hydroxides as Catalysts with Enhanced Photodegradation and Antibacterial Activities. *Inorganics* **2019**, *7*, 120. [CrossRef]
47. Cerrato, E.; Gionco, C.; Paganini, M.C.; Giamello, E.; Albanese, E.; Pacchioni, G. Origin of Visible Light Photoactivity of the CeO_2/ZnO Heterojunction. *ACS Appl. Energy Mater.* **2018**, *1*, 4247–4260. [CrossRef]
48. Cerrato, E.; Zickler, G.A.; Paganini, M.C. The Role of Yb Doped ZnO in the Charge Transfer Process and Stabilization. *J. Alloys Compd.* **2019**, *816*, 152555. [CrossRef]
49. Paganini, M.C.; Cerrato, E. Photoactive Systems based on Semiconducting Metal Oxides. In *Material Science in Photocatalysis*; Garcıa-Lopez, E.I., Palmisano, L., Eds.; Elsevier: Amsterdam, The Netherlands, 2021; Volume 1, pp. 221–234.
50. Tauc, J. *The Optical Properties of Solids*; Academic Press: New York, NY, USA, 1966.

Article

Synthesis and Structure of ZnO-Decorated Graphitic Carbon Nitride (g-C$_3$N$_4$) with Improved Photocatalytic Activity under Visible Light

Maria I. Chebanenko [1,*], Sofia M. Tikhanova [1], Vladimir N. Nevedomskiy [2] and Vadim I. Popkov [1]

[1] Hydrogen Energy Laboratory, Ioffe Institute, 194021 Saint Petersburg, Russia
[2] Center of Nanoheterostructure Physics, Ioffe Institute, 194021 Saint Petersburg, Russia
* Correspondence: m_chebanenko@list.ru; Tel.: +7-(931)-213-25-76

Abstract: The volume of dye production in the chemical industry is growing rapidly every year. Given the global importance of clean water resources, new wastewater treatment solutions are required. Utilizing photocatalysis by harvesting solar energy represents a facile and promising solution for removing dangerous pollutants. This study reports the possibility of increasing the photocatalytic activity of g-C$_3$N$_4$ by creating nanocomposites with ZnO. Exfoliated g-C$_3$N$_4$/ZnO nanocomposites were synthesized by heat treatment of urea and subsequent ultrasonic exfoliation of the colloidal solution by introducing zinc acetate. The uniformity of the distribution of ZnO nanoparticles is confirmed by the method of elemental mapping. The obtained X-ray diffractograms of the obtained nanocomposites show typical X-ray reflections for g-C$_3$N$_4$ and ZnO. It was found that the introduction of oxide into g-C$_3$N$_4$ leads to an increase in the specific surface area values due to the developed ZnO surface. The maximum value of the specific surface area was obtained for a sample containing 7.5% ZnO and was 75.2 m^2/g. The g-C$_3$N$_4$/7.5% ZnO sample also demonstrated increased photocatalytic activity during the decomposition of methylene blue under the influence of visible light, which led to a twofold increase in the reaction rate compared to initial g-C$_3$N$_4$.

Keywords: graphitic carbon nitride; photocatalysts; g-C$_3$N$_4$/ZnO; visible light; photocatalytic degradation

Citation: Chebanenko, M.I.; Tikhanova, S.M.; Nevedomskiy, V.N.; Popkov, V.I. Synthesis and Structure of ZnO-Decorated Graphitic Carbon Nitride (g-C$_3$N$_4$) with Improved Photocatalytic Activity under Visible Light. *Inorganics* **2022**, *10*, 249. https://doi.org/10.3390/inorganics10120249

Academic Editor: Roberto Nisticò

Received: 10 November 2022
Accepted: 6 December 2022
Published: 8 December 2022

Publisher's Note: MDPI stays neutral with regard to jurisdictional claims in published maps and institutional affiliations.

Copyright: © 2022 by the authors. Licensee MDPI, Basel, Switzerland. This article is an open access article distributed under the terms and conditions of the Creative Commons Attribution (CC BY) license (https://creativecommons.org/licenses/by/4.0/).

1. Introduction

In the modern world, the rapid development of technology inevitably leads to an increase in the production of an ever-growing number of new chemical compounds. Such compounds are widely used in various industries [1–3]. Organic compounds, which are the basis of industrial dyes, are especially dangerous because their decomposition products can have a detrimental effect on living organisms when released into the environment [4,5]. Spreading through water, they can affect people and wildlife at a considerable distance from their original sources. Organic compounds also decompose at a slow rate and therefore accumulate and spread along the food chain. Consequently, when released into reservoirs with industrial wastewater, they disrupt the overall ecological balance of aquatic ecosystems [6,7]. In this regard, interest in the problem of industrial wastewater treatment has been growing recently.

Currently, physico-chemical, biological, and various other methods are used to solve this problem, which require expensive tools and can also lead to secondary contamination [8–12]. The photocatalytic decomposition process is a safe and effective method that can be used to remove organic pollutants from aquatic environments. Using this method, the photocatalyst converts toxic organic dyes into non-toxic intermediates under the influence of ultraviolet or visible light [13–16]. The development of such catalytic materials is an important task.

In past decade, nanoscale materials gained considerable importance due to their unique properties caused by scale factor absent in their bulk form. The reason for their use in a large number of applications is due to their enhanced catalytic, optical, and electronic properties at the nano level. Nanoscale transition metal oxides are considered the most promising catalysts for wastewater treatment in a simple, reliable, fast, and environmentally friendly way considering their high photocatalytic activity, excellent solubility, and stability [17–22]. Among them, zinc oxide (ZnO) stands out especially. ZnO is a nontoxic, low-cost, and broadly available semiconductor that is widely used in photocatalytic degradation of organic dyes and photocatalytic decomposition of hydrogen generated from water. The nanostructured ZnO as compared to other nanostructures had gained importance because of ease in its fabrication and a number of applications. In order to use it as an efficient photocatalyst for the decomposition of organic pollutants, it is important to extend the work spectrum of ZnO from UV to the visible light region, increase a low specific surface area, and hinder the recombination of the photogenerated electron-hole pairs to improve the photocatalytic efficiency of ZnO [23]. To eliminate this disadvantage, it is proposed that new promising carbon-based materials be used as carriers [24,25]. One such material is graphite-like carbon nitride (g-C_3N_4). Due to its high stability and ability to absorb radiation in the visible region of the spectrum (band gap 2.4–2.9 eV), this compound has been widely used in the field of heterogeneous catalysis [26–28]. The use of g-C_3N_4 as a photocatalyst in the visible region of the spectrum was first reported in [29]. Unlike TiO_2, g-C_3N_4 does not contain noble or rare earth metal atoms in its composition, and it exhibits high photocatalytic activity in processes induced by the action of visible radiation [30–33]. Materials based on g-C_3N_4 can be considered as a promising basis for new functional materials for photocatalysis.

Therefore, the main objective of this paper is the synthesis of a nanocomposite based on g-C_3N_4 and ZnO with improved photocatalytic properties. The novelty of the presented research lies in the use of an improved synthetic route. A special feature is the ultrasonic dispersion of g-C_3N_4, preceding the introduction of zinc acetate. The positive effect of this exfoliation method was previously reported in [34–36]. As a result, the synthesized g-C_3N_4/ZnO powders demonstrate a small band gap and a large surface area, as well as an improved ability to absorb visible radiation and an increase in the photodegradation rate of organic pollutants in composite nanoplanes.

2. Results and Discussion

Figure 1 shows the results of an X-ray diffraction of the obtained powders. The diffractogram of pure g-C_3N_4 contains X-ray peaks typical of graphite-like carbon nitride [37,38]. The most intense peak is located in the region of 27.4° and is associated with the characteristic interlayer stacking of conjugated aromatic rings. A less intense peak at 12.8° corresponds with the layered structure of the packing of tri-s-triazine units, which corresponds with the plane (100) in the JCPDS 87-1526 card. ZnO peaks are traced on diffractograms of the obtained nanocomposites. The intensity of the peaks increases with an increase in the mass fraction of the oxide in the samples. All detectable X-ray reflections of ZnO can be indexed by the wurtzite structure (JCPDS No. 36-1451). It is noted that the presence of oxide does not affect the position of the diffraction peaks of g-C_3N_4, and a slight change in the relative intensity of the peak reflects only the ratio of the components.

The microstructure of the obtained samples is studied using scanning electron microscopy. The pure g-C_3N_4 sample consists of randomly stacked layers consisting of inhomogeneously distributed and agglomerated structures (Figure 2a,b). In the sample containing 7.5% ZnO (Figure 2c,d), the nanolayers are randomly filled with oxide nanoparticles (NPs). Nanocomposite powders look less agglomerated and have a friable structure.

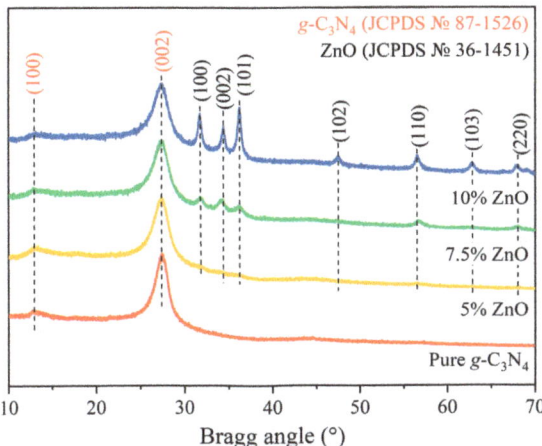

Figure 1. PXRD patterns of the initial g-C$_3$N$_4$, g-C$_3$N$_4$/ZnO (5% ZnO, 7.5% ZnO, and 10% ZnO) nanocomposites.

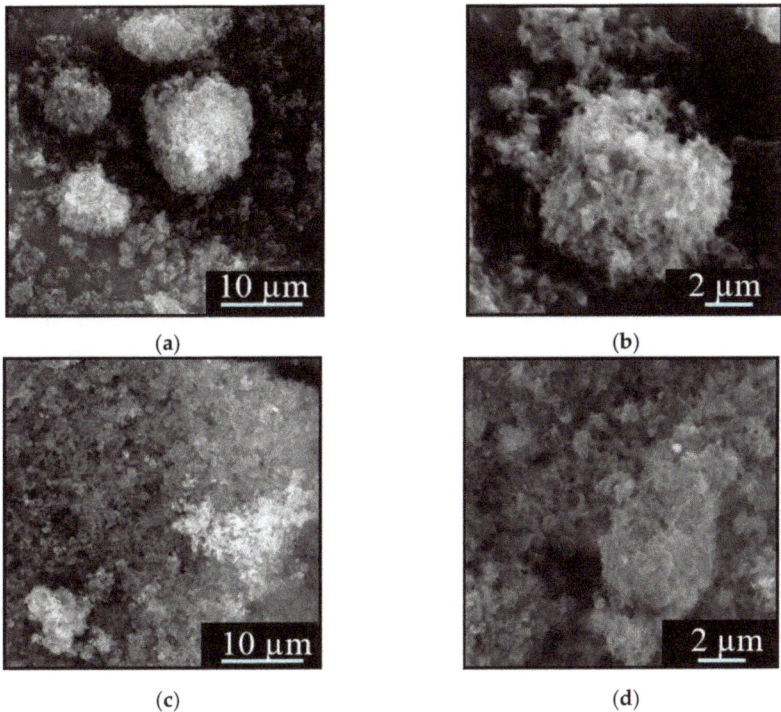

Figure 2. SEM images of (**a**,**b**) pure g-C$_3$N$_4$ and (**c**,**d**) g-C$_3$N$_4$/7.5% ZnO.

X-ray spectroscopy (EDX) elemental mapping results are shown in Figure 3. The elemental mapping results indicate the existence of C, N, Zn, and O elements suggesting that ZnO NPs disperse on the g-C$_3$N$_4$ sheets. A uniform distribution of C, N, Zn, and O atoms could be observed throughout each entire nanoplate.

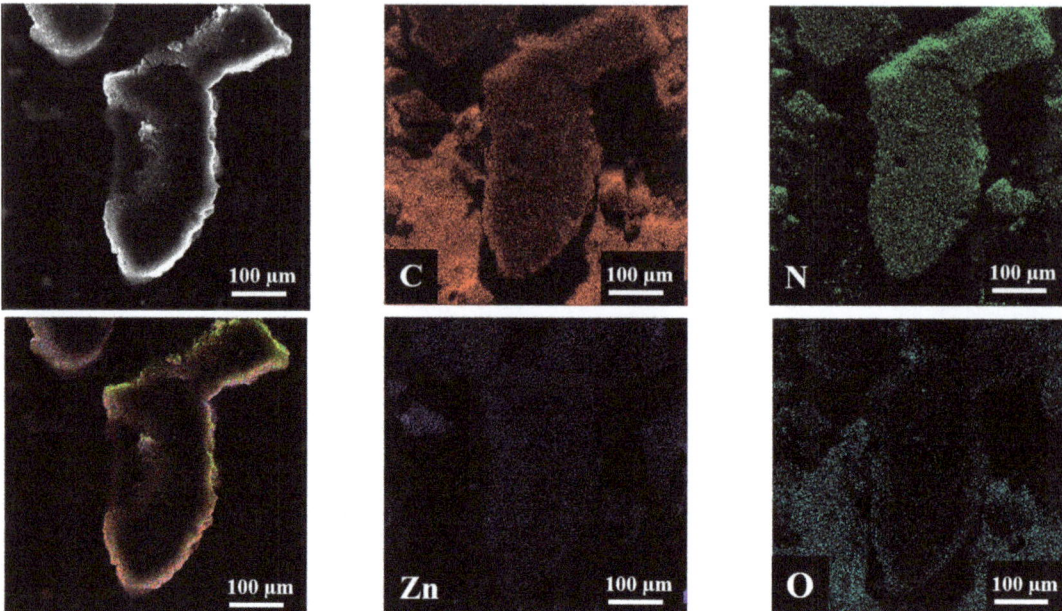

Figure 3. EDX mapping of g-C$_3$N$_4$/7.5% ZnO nanocomposite.

Using TEM and SAED, the surface morphologies and structural characteristics of g-C$_3$N$_4$/7.5% ZnO nanocomposite were analyzed. Previously, a TEM image of pure g-C$_3$N$_4$ was shown in [35], where the morphology consists of agglomerated planes. According to the comparison of the TEM data measured in bright field and dark field modes (Figure 4a and 4b, respectively), ZnO nanocrystals are uniformly located throughout the sample volume, which is consistent with the results of elemental mapping. ZnO can be clearly observed as a cluster of irregularly shaped NPs, and g-C$_3$N$_4$ looks like dimmer sections of two-dimensional layered structures, which indicates the presence of a number of stacked layers. The g-C$_3$N$_4$ layers are compacted with ZnO NPs and have close contact between g-C$_3$N$_4$ and ZnO, which is required to form a heterojunction and achieve better charge separation. The diffraction ring positions in SAED patterns of the nanocomposite (Figure 4c) match the known interplanar distances in ZnO and g-C$_3$N$_4$ and confirm that individual nanocrystals observed in dark-field mode correspond to ZnO. Figure 4d shows an HRTEM image in which the ZnO lattice planes were detected at 7.5 wt% g-C$_3$N$_4$/ZnO composite [39]. The results of X-ray diffraction and TEM confirm that ZnO nanocrystals having (002) crystallographic phases on their surfaces are in direct contact with g-C$_3$N$_4$.

N$_2$ adsorption-desorption isotherms of pure g-C$_3$N$_4$ and g-C$_3$N$_4$/7.5% ZnO were measured to gain information about the specific surface area (Figure 5a). According to the International Union of Pure and Applied Chemistry (IUPAC) classification, obtained isotherms were determined as type III with mesoporous characteristics, which indicates the mesoporous structure of the samples. The pore size distribution of the samples shows that most of the pores fall into the size range from 2 to 50 nm. The specific surface area values are shown in Figure 5b. The surface area for the resulting samples was calculated using the Brunauer–Emmet–Teller (BET) equation in the range of relative pressure (P/P$_0$) 0.05–0.3. The specific surface area of the samples determined by the BET method is 34.4, 54.5, 75.2, and 68.5 m^2/g for g-C$_3$N$_4$, g-C$_3$N$_4$/5% ZnO, g-C$_3$N$_4$/7.5% ZnO, and g-C$_3$N$_4$/10% ZnO, respectively. The optimal amount of oxide in the nanocomposite is 7.5 wt% ZnO, while further increase leads to a decrease in the values of the specific surface area. The surface of the composite samples will absorb more visible light, thereby increasing the photocatalytic efficiency by providing more active centers.

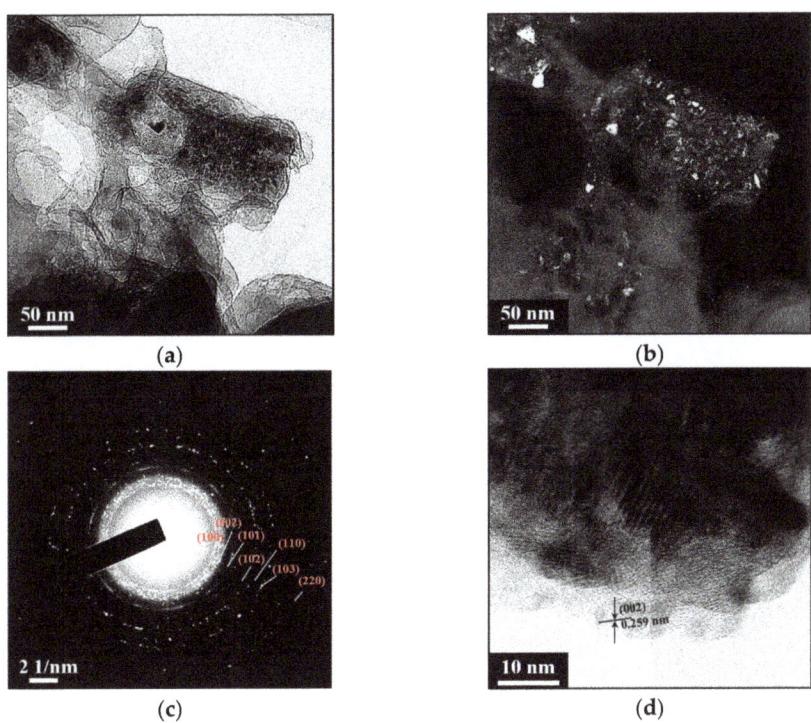

Figure 4. (**a**) The bright-field and (**b**) dark-field transmission electron microscopy (TEM) image of g-C$_3$N$_4$/7.5% ZnO NPs, (**c**) selected area electron diffraction (SAED) pattern, and (**d**) high resolution-TEM (HR-TEM) image.

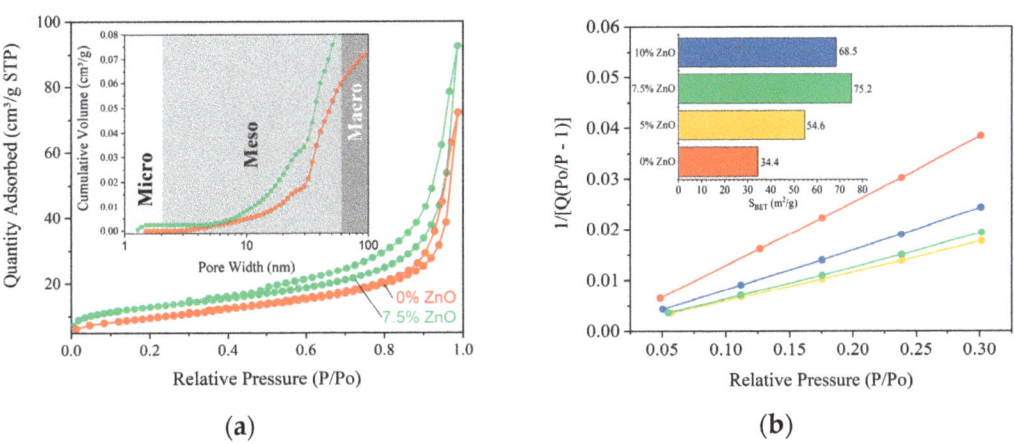

Figure 5. (**a**) N$_2$ adsorption-desorption isotherms of initial g-C$_3$N$_4$ and g-C$_3$N$_4$/7.5% ZnO samples; (**b**) BET surface area plot of initial g-C$_3$N$_4$ and g-C$_3$N$_4$/ZnO nanocomposites.

The UV–vis DRS of all samples is demonstrated in Figure 6a. It can be seen that the absorption of light by g-C$_3$N$_4$ is lower than that of other samples. When g-C$_3$N$_4$ is recombined with ZnO, the light absorption band edge undergoes a slight red shift. Such a modest change in the distance between the bands may indicate a tight contact at the interface with g-C$_3$N$_4$ and ZnO and confirms the synthesis of the nanocomposite. To further study the photocatalytic effect of the composites, a Tauc plot was derived from the UV-vis spectrum.

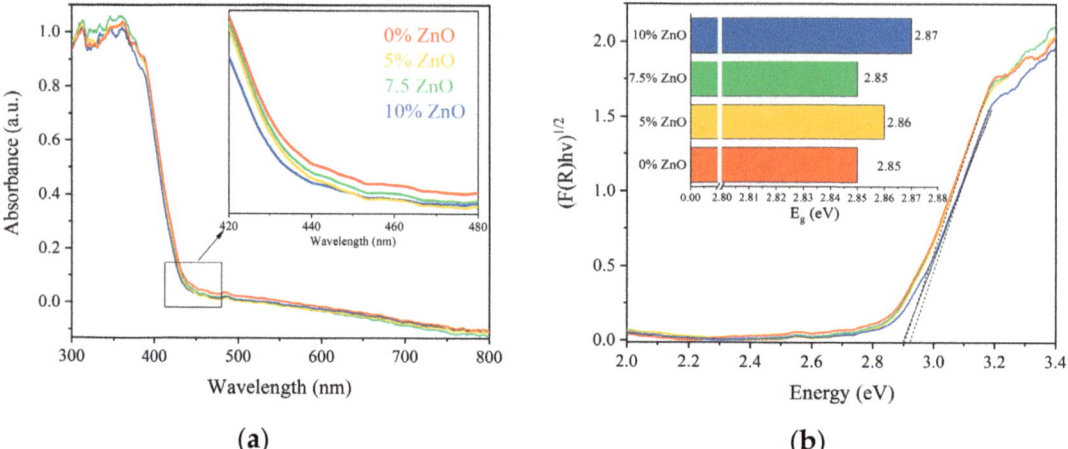

Figure 6. (a) UV-vis DRS and (b) Tauc Plots of the initial g-C$_3$N$_4$ and g-C$_3$N$_4$/ZnO (5% ZnO, 7.5% ZnO and 10% ZnO) nanocomposites.

The addition of ZnO results in a moderate modification of Tauc plots, as illustrated in Figure 6b. As a result of the shift in absorbance edges for the nanocomposite structure, the band gaps for g-C$_3$N$_4$ and g-C$_3$N$_4$/ZnO (5% ZnO, 7.5% ZnO, and 10% ZnO) are 2.85, 2.86, 2.85, and 2.87 eV, respectively. It is reported above that the nanocomposite has a larger contact interface area (Figure 5), which allows faster charging carrier transit and, therefore, suppresses charge recombination, leading to increased photocatalytic degradation efficiency.

All samples were examined for their photocatalytic potential towards photodegradation of MB dye solution. Figure 7a shows the absorption spectra of the dye solution with samples under visible light irradiation at 10-min intervals. It can be seen from Figure 7b that the degradation of methylene blue by g-C$_3$N$_4$ in the dark is negligible. After g-C$_3$N$_4$ is compounded with oxide, its photocatalytic efficiency is further improved. The degradation of MB is analyzed by pseudo-first order kinetics (Figure 7c). The degradation of MB practically did not proceed in the blank experiment, which illustrates MB can not be degraded under visible irradiation. The experiment reveals that MB could be degraded at a rate constant 2.2×10^{-3} min^{-1}, 3.2×10^{-3} min^{-1}, 4.6×10^{-3} min^{-1}, and 3.2×10^{-3} min^{-1} using g-C$_3$N$_4$, g-C$_3$N$_4$/5% ZnO, g-C$_3$N$_4$/7.5% ZnO, and g-C$_3$N$_4$/10% ZnO, respectively. In the present study, a 7.5% ZnO sample shows better photodegradation performance as compared to others. This effect is explained by the optimal ratio of components in the nanocomposite, which positively affects the size of the specific surface area and the availability of active centers. The rate constant of g-C$_3$N$_4$/7.5% ZnO is about two times higher than that of pure g-C$_3$N$_4$. The results reveal that a certain proportion of ZnO has the best photocatalytic activity.

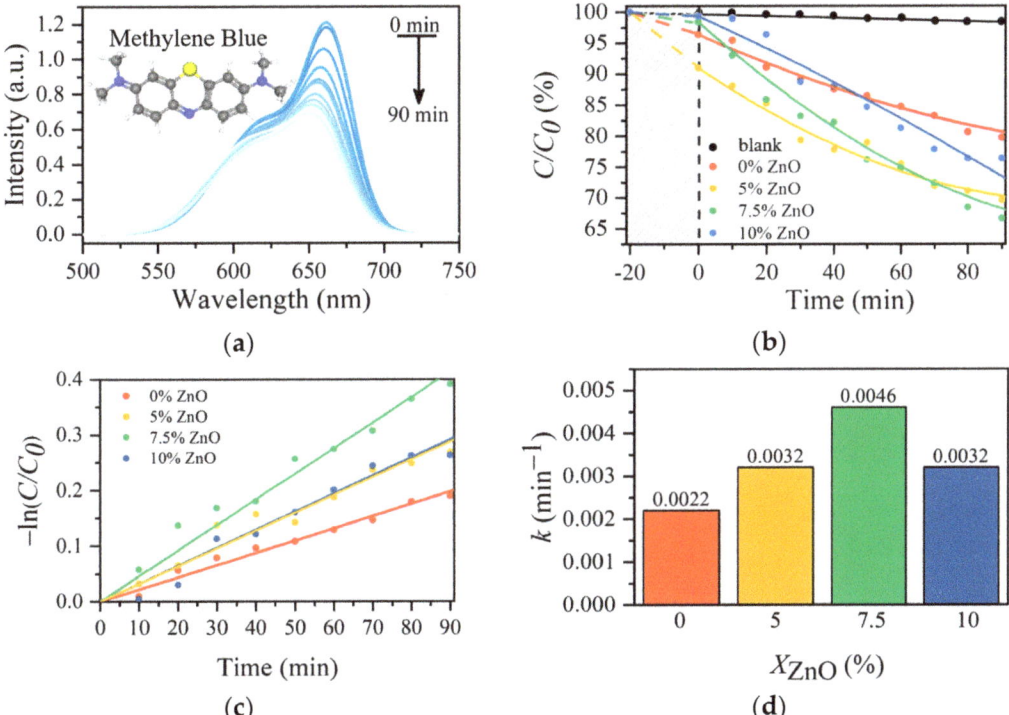

Figure 7. (**a**) Photocatalytic degradation of MB over g-C$_3$N$_4$/7.5% ZnO catalyst under visible light; (**b**) Kinetic curves, (**c**) Logarithmic kinetic curves of the pseudo-first-order process; (**d**) Photodecomposition rate constant versus ZnO content.

Photoluminescence (PL) spectra were obtained to study the processes of charge carrier transfer. Evaluation of the efficiency of carrier separation will additionally confirm the improvement of the photocatalytic characteristics of photocatalysts. Photoluminescence emission spectra of the prepared samples were recorded at an excitation wavelength of 365 nm and are shown in Figure 8a. All samples demonstrated a peak of PL radiation in a similar region between 400 and 500 nm. A decrease in the photoluminescence intensity for g-C$_3$N$_4$/ZnO nanocomposites indicated that the high-efficiency electron transferred from the conduction band of g-C$_3$N$_4$ to the conduction band of ZnO hindered the reorganization of photoinduced charge carriers. This indicates that the recombination rate of electrons and holes under visible light irradiation is lower in this case. This leads to higher charge separation, which can provide more electrons for photocatalytic activity.

The stability of the g-C$_3$N$_4$/7.5% ZnO was examined by recycling the photocatalyst for the photocatalytic degradation of MB dye. After each cycle, the catalyst particles were collected by centrifugation, washed with distilled water, and dried at 60 °C. Then the particles were reused, and the dye solution was replaced with a fresh one. As shown in Figure 8b, no major reduction was observed in the efficiency after four consecutive cycles; as a result, the g-C$_3$N$_4$/7.5% ZnO photocatalysts were photostable.

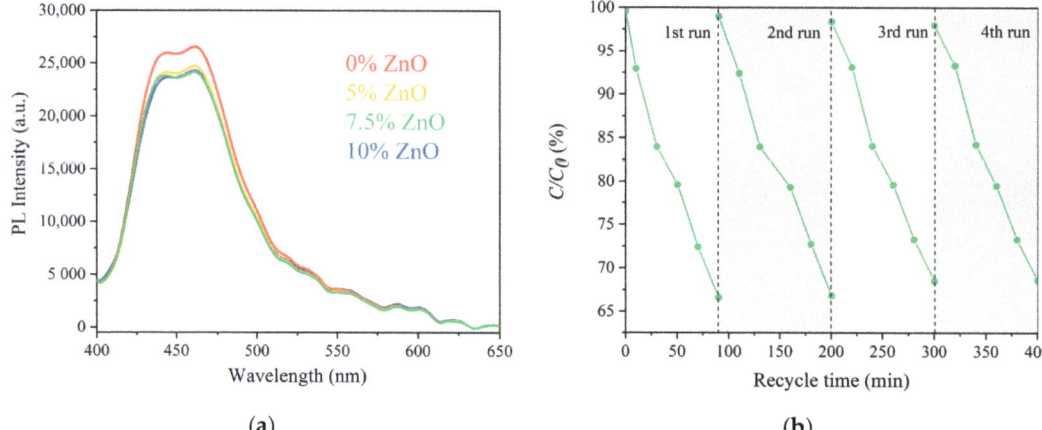

Figure 8. (a) PL spectra of initial g-C$_3$N$_4$ and g-C$_3$N$_4$/ZnO (5% ZnO, 7.5% ZnO and 10% ZnO) nanocomposites; (b) Recycling photocatalytic test of g-C$_3$N$_4$/7.5% ZnO.

Figure 9 shows the mechanism of photocatalytic degradation of MB by g-C$_3$N$_4$/ZnO nanocomposites. ZnO has a band gap of about 3.37 eV [40,41] and cannot be excited by visible light. The spectrum of the g-C$_3$N$_4$/ZnO composite can be broadened to the visible region in contrast to pure ZnO. g-C$_3$N$_4$ is activated by sunlight, causing the electrons to shift from the valence band (VB) of each sample to the conduction band (CB) and create h$^+$ on the VB of g-C$_3$N$_4$. These excited electrons are then transferred to the CB of ZnO, which further promoted it to the surface, leading to the generation of O$_2^-$ radicals. The holes in the VB of ZnO are transferred to the VB of g-C$_3$N$_4$, which ensure effective charge separation in the system. These VB holes in g-C$_3$N$_4$ are further transferred to the surface to react with surrounding molecules and generate ·OH radicals [42,43]. In this way, these generated radicals are actively involved in the degradation of dye molecules.

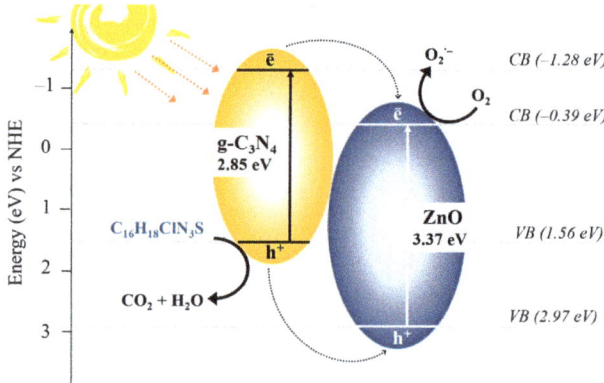

Figure 9. Schematic mechanism of photocatalytic degradation of MB of g-C$_3$N$_4$/ZnO under visible light.

3. Materials and Methods

3.1. Materials

Zinc acetate dihydrate (Zn(CH$_3$COO)$_2$·H$_2$O, NevaReaktiv, Saint-Petersburg, Russia) was used as a precursor for ZnO. Graphitic carbon nitride was prepared from urea (CH$_4$N$_2$O, NevaReaktiv, Saint-Petersburg, Russia). Methylene blue (C$_{16}$H$_{18}$ClN$_3$S, LenReaktiv, Saint-Petersburg, Russia) was used as model dye. All the reagents used in this research work are analytical grade.

3.2. Synthesis of Initial g-C₃N₄ Nanopowder

The initial g-C_3N_4 nanopowder was synthesized via heat treatment of 10 g urea for 1 h in a muffle furnace in air atmosphere. A dry powder of urea was placed in a quartz tube heated to 550 °C at a fixed heating rate of 5°/min. As a result of the heat treatment, a powder of pale-yellow color was obtained.

3.3. Synthesis of Exfoliated g-C₃N₄/ZnO Nanocomposites

To obtain nanocomposites based on g-C_3N_4 and ZnO, the following procedure was performed. The initial sample was treated by ultrasonic dispersion in distilled water for 2 h until a milk-colored homogenized suspension was obtained. The resulting suspension was divided into four equal portions, to each of which the necessary amount of $Zn(CH_3COO)_2 \cdot 2H_2O$ was added and thoroughly mixed using a magnetic stirrer. Further, the obtained suspensions were dried at 100 °C until the water was completely removed, thoroughly ground, and treated thermally at 400 °C. As a result, a series of four samples was obtained: g-C_3N_4, g-C_3N_4-ZnO 5%, g-C_3N_4-ZnO 7.5%, and g-C_3N_4-ZnO 10%.

3.4. Physico-Chemical Characterization

The crystal phases of the catalysts were analyzed by powder X-ray diffraction patterns (XRD) using a Rigaku SmartLab 3 diffractometer (Rigaku Corporation, Tokyo, Japan) (CuKα radiation, λ = 0.154051 nm). The measurement was conducted by varying the Bragg angle in the range of 10–80°.

The elemental composition and morphology of nanopowders was studied with scanning electron microscopy (SEM) using a Tescan Vega 3 SBH (Tescan, Czech Republic) scanning electron microscope with an Oxford INCA x-act X-ray microanalysis device (Oxford Instruments, UK). An Oxford Instruments INCA system was used for chemical analysis, including elemental mapping utilizing a 10 nm probe size.

The microstructure and morphology evaluation were investigated by Transmission Electron Microscopy (TEM) using a JEOL JEM2100F microscope (JEOL Ltd., Tokyo, Japan) in the image and selected area electron diffraction (SAED) mode.

Measurement of N_2 adsorption-desorption was carried out for specific surface area and pore structure analysis (ASAP 2020 Surface Area and Porosity Analyzer Micromeritics, Micromeritics Instrument Corporation, Norcross, GA, USA) in standard mode at 78 K, the temperature of liquid nitrogen.

Diffuse reflectance spectra of samples were obtained using an Avaspec-ULS2048CL-EVO spectrometer (Avantes, Apeldoorn, The Netherlands) equipped with an AvaSphere-30-REFL refractometric integration sphere in the 350–700 nm region.

3.5. Photocatalytic Measurements

Photocatalytic activity of synthesized nanocomposites was studied by measuring the change in the absorption spectra of methylene blue ($C_{16}H_{18}ClN_3S$, MB) under visible light irradiation. For the test, 0.01 g of the nanocomposite was placed in a solution of MB (10 mg/l). The prepared suspensions were continuously mixed for 20 min in the dark to establish sorption equilibrium. Then the samples were constantly mixed at a temperature of 25 °C throughout the analysis in a closed box under the illumination of an Xe lamp (power consumption 70 W, λ_{max} = 410 nm). Aliquots of samples were taken at intervals of 10 min during the 90 min of the experiment. All samples were centrifuged to remove solid particles, and the supernatant was analyzed using an Avaspec-ULS2048CL-EVO spectrometer equipped with an Avalight-XE Pulsed Xenon Lamp (Avantes, Apeldoorn, The Netherlands).

4. Conclusions

In this work, nanocrystalline g-C_3N_4/ZnO powders were obtained by heat treatment of a mixture of g-C_3N_4 and zinc acetate. The resulting series of samples (pure g-C_3N_4, g-C_3N_4/5% ZnO, g-C_3N_4/7.5% ZnO, and g-C_3N_4/10% ZnO) was studied using a complex of physico-chemical analysis methods. The ZnO NPs were evenly distributed on the surface and represented a crystalline modification with a wurtzite structure. The specific surface area of the powder reached its maximum value at 7.5% of ZnO content. A further increase in the amount of ZnO in the composition led to a decrease in the values of the specific surface area. This structure of the Z-scheme led to an increase in the photocatalytic activity of the studied nanocomposite. Compared with pure material, g-C_3N_4/7.5% ZnO showed a much greater photocatalytic activity. Thus, g-C_3N_4/ZnO is practically suitable for the photocatalytic decomposition of organic materials under visible light.

Author Contributions: Conceptualization, M.I.C. and V.I.P.; methodology, M.I.C.; software, M.I.C. and S.M.T.; validation, M.I.C.; formal analysis, M.I.C. and S.M.T.; investigation, M.I.C. and S.M.T.; resources, V.I.P. and V.N.N.; data curation, M.I.C.; writing—original draft preparation, M.I.C. and S.M.T.; writing—review and editing, M.I.C. and V.I.P.; visualization, V.N.N.; supervision, V.I.P.; project administration, V.I.P. All authors have read and agreed to the published version of the manuscript.

Funding: This research received no external funding.

Acknowledgments: The author would like to thank M.I. Tenevich for help with scanning electron microscopy studies.

Conflicts of Interest: The authors declare no conflict of interest.

References

1. Shannon, M.A.; Bohn, P.W.; Elimelech, M.; Georgiadis, J.G.; Marin, B.J.; Mayes, A.M. Science and technology for water purification in the coming decades. *Nature* **2008**, *452*, 301–310. [CrossRef] [PubMed]
2. Nasrollahzadeh, M.; Sajjadi, M.; Iravani, S.; Varma, R.S. Green-synthesized nanocatalysts and nanomaterials for water treatment: Current challenges and future perspectives. *J. Hazard. Mater.* **2020**, *401*, 123401. [CrossRef] [PubMed]
3. Nayak, S.; Mohapatra, L.; Parida, K. Visible light-driven novel g-C_3N_4/NiFe-LDH composite photocatalyst with enhanced photocatalytic activity towards water oxidation and reduction reaction. *J. Mater. Chem. A* **2015**, *3*, 18622–18635. [CrossRef]
4. Mahmood, A.; Muhmood, T.; Ahmad, F. Carbon nanotubes heterojunction with graphene like carbon nitride for the enhancement of electrochemical and photocatalytic activity. *Mater. Chem. Phys.* **2022**, *278*, 125640. [CrossRef]
5. Muhmood, T.; Xia, M.; Lei, W.; Wang, F. Under vacuum synthesis of type-I heterojunction between red phosphorus and graphene like carbon nitride with enhanced catalytic, electrochemical and charge separation ability for photodegradation of an acute toxicity category-III compound. *Appl. Catal. B Environ.* **2018**, *238*, 568–575. [CrossRef]
6. Ren, Q.; Nie, M.; Yang, L.; Wei, F.; Ding, B.; Chen, H.; Liu, Z.; Liang, Z. Synthesis of MOFs for RhB Adsorption from Wastewater. *Inorganics* **2022**, *10*, 27. [CrossRef]
7. Lee, Y.-G.; Chon, K. Green Technologies for Sustainable Water and Wastewater Treatment: Removal of Organic and Inorganic Contaminants. *Separations* **2022**, *9*, 335. [CrossRef]
8. Klaczyński, E.; Ratajczak, P. Oczyszczalnie ścieków—Układy technologiczne (Waste water treatment plants—Process systems). *Wodociągi Kanaliz.* **2013**, *4*, 36–39.
9. Narayanan, C.M.; Narayan, V. Biological wastewater treatment and bioreactor design: A review. *Sustain. Environ. Res.* **2019**, *29*, 33. [CrossRef]
10. Vukšić, M.; Kocijan, M.; Ćurković, L.; Radošević, T.; Vengust, D.; Podlogar, M. Photocatalytic Properties of Immobilised Graphitic Carbon Nitride on the Alumina Substrate. *Appl. Sci.* **2022**, *12*, 9704. [CrossRef]
11. Kocijan, M.; Ćurković, L.; Radošević, T.; Podlogar, M. Enhanced Photocatalytic Activity of Hybrid rGO@TiO_2/CN Nanocomposite for Organic Pollutant Degradation under Solar Light Irradiation. *Catalysts.* **2021**, *11*, 1023. [CrossRef]
12. Kocijan, M.; Ćurković, M.; Gonçalves, G.; Podlogar, M. The Potential of rGO@TiO_2 Photocatalyst for the Degradation of Organic Pollutants in Water. *Sustainability* **2022**, *14*, 12703. [CrossRef]
13. Srivastava, R.R.; Vishwakarma, P.K.; Yadav, U.; Rai, S.; Umrao, S.; Giri, R.; Saxena, P.S.; Srivastava, A. 2D SnS_2 Nanostructure-Derived Photocatalytic Degradation of Organic Pollutants Under Visible Light. *Front. Nanotechnol.* **2021**, *3*, 711368. [CrossRef]
14. Lebedev, L.A.; Chebanenko, M.I.; Dzhevaga, E.V.; Martinson, K.D.; Popkov, V.I. Solvothermal modification of graphitic C_3N_4 with Ni and Co phthalocyanines. *Mendeleev Commun.* **2022**, *32*, 317–319. [CrossRef]
15. Kocijan, M.; Ćurković, L.; Ljubas, D.; Mužina, K.; Bačić, I.; Radošević, T.; Podlogar, M.; Bdikin, I.; Otero-Irurueta, G.; Hortigüela, M.J.; et al. Graphene-Based TiO_2 Nanocomposite for Photocatalytic Degradation of Dyes in Aqueous Solution under Solar-Like Radiation. *Appl. Sci.* **2021**, *11*, 3966. [CrossRef]

16. Kocijan, M.; Ćurković, L.; Bdikin, I.; Otero-Irurueta, G.; Hortigüela, M.J.; Gonçalves, G.; Radošević, T.; Vengust, T.; Podlogar, M. Immobilised rGO/TiO$_2$ Nanocomposite for Multi-Cycle Removal of Methylene Blue Dye from an Aqueous Medium. *Appl. Sci.* **2022**, *12*, 385. [CrossRef]
17. Nazim, M.; Parwaz Khan, A.A.; Asiri, A.M.; Kim, J.H. Exploring Rapid Photocatalytic Degradation of Organic Pollutants with Porous CuO Nanosheets: Synthesis, Dye Removal, and Kinetic Studies at Room Temperature. *ACS Omega* **2021**, *6*, 2601–2612. [CrossRef]
18. Lasio, B.; Malfatti, L.; Innocenzi, P. Photodegradation of Rhodamine 6G dimers in silica sol-gel films. *J. Photochem. Photobiol.* **2013**, *271*, 93–98. [CrossRef]
19. Buthiyappan, A.; Abdul Aziz, A.R.; Wan Daud, W.M.A.; Daud, W. Recent advances and prospects of catalytic advanced oxidation process in treating textile effluents. *Rev. Chem. Eng.* **2016**, *32*, 1–47. [CrossRef]
20. Shi, S.; Xu, J.; Li, L. Preparation and Photocatalytic Activity of ZnO Nanorods and ZnO/Cu$_2$O Nanocomposites. *Main Group Chem.* **2017**, *16*, 47–55. [CrossRef]
21. Chebanenko, M.I.; Martinson, K.D.; Matsukevich, I.V.; Popkov, V.I. The effect of MgO additive on the g-C$_3$N$_4$ performance in electrochemical reforming of water-ethanol solution. *Nanosyst. Phys. Chem. Math.* **2020**, *11*, 474–479. [CrossRef]
22. Pant, B.; Park, M.; Hee Lee, J.; Kim, H.-Y.; Park, S.-J. Novel magnetically separable silver-iron oxide nanoparticles decorated graphitic carbon nitride nano-sheets: A multifunctional photocatalyst via one-step hydrothermal process. *J. Colloid Interface Sci.* **2017**, *496*, 343–352. [CrossRef]
23. Pant, B.; Park, M.; Kim, H.-Y.; Park, S.-J. Ag-ZnO photocatalyst anchored on carbon nanofibers: Synthesis, characterization, and photocatalytic activities. *Synth. Met.* **2016**, *220*, 533–537. [CrossRef]
24. Elshafie, M.; Younis, S.A.; Serp, P.; Gad, E.A.M. Preparation characterization and non-isothermal decomposition kinetics of different carbon nitride sheets. *Egypt. J. Pet.* **2019**, *29*, 21–29. [CrossRef]
25. Huang, R.; Wu, J.; Zhang, M.; Liu, B.; Zheng, Z.; Luo, D. Strategies to enhance photocatalytic activity of graphite carbon nitride-based photocatalysts. *Mater. Des.* **2019**, *210*, 110040. [CrossRef]
26. Muhmood, T.; Asim Khan, M.; Xia, M.; Lei, W.; Wang, F.; Ouyang, Y. Enhanced photo-electrochemical, photo-degradation and charge separation ability of graphitic carbon nitride (g-C$_3$N$_4$) by self-type metal free heterojunction formation for antibiotic degradation. *J. Photochem. Photobiol. A Chem.* **2017**, *348*, 118–124. [CrossRef]
27. Zhao, L.; Zhang, L.; Lin, H.; Nong, Q.; Cui, M.; Wu, Y.; He, Y. Fabrication and characterization of hollow CdMoO$_4$ coupled g-C$_3$N$_4$ heterojunction with enhanced photocatalytic activity. *J. Hazard. Mater.* **2015**, *299*, 333–342. [CrossRef]
28. Sun, S.; Ding, H.; Mei, L.; Chena, Y.; Hao, Q.; Chen, W.; Xu, Z.; Chen, D. Construction of SiO$_2$-TiO$_2$/g-C$_3$N$_4$ composite photocatalyst for hydrogen production and pollutant degradation: Insight into the effect of SiO$_2$. *Chin. Chem. Lett.* **2020**, *31*, 2287–2294. [CrossRef]
29. Mo, Z.; She, X.; Li, Y.; Liu, L.; Huang, L.; Chen, Z.; Zhang, Q.; Xu, H.; Li, H. Synthesis of g-C$_3$N$_4$ at different temperatures for superior visible/UV photocatalytic performance and photoelectrochemical sensing of MB solution. *RSC Adv.* **2015**, *5*, 101552–101562. [CrossRef]
30. Maeda, K.; Wang, X.; Nishihara, Y.; Lu, D.; Antonietti, M.; Domen, K. Photocatalytic Activities of Graphitic Carbon Nitride Powder for Water Reduction and Oxidation under Visible Light. *J. Phys. Chem. C* **2009**, *113*, 4940–4947. [CrossRef]
31. Dong, F.; Wu, L.; Sun, Y.; Fu, M.; Wu, Z.; Lee, S.C. Efficient synthesis of polymeric g-C$_3$N$_4$ layered materials as novel efficient visible light driven photocatalysts. *J. Mater. Chem.* **2011**, *21*, 15171–15174. [CrossRef]
32. Kharlamov, A.; Bondarenko, M.; Kharlamova, G.; Gubareni, N. Features of the synthesis of carbon nitride oxide (g-C$_3$N$_4$)O at urea pyrolysis. *Diam. Relat. Mater.* **2016**, *66*, 16–22. [CrossRef]
33. Chidhambaram, N.; Ravichandran, K. Single step transformation of urea into metal-free g-C$_3$N$_4$ nanoflakes for visible light photocatalytic applications. *Mater. Lett.* **2017**, *207*, 44–48. [CrossRef]
34. Chebanenko, M.I.; Zakharova, N.V.; Lobinsky, A.A.; Popkov, V.I. Ultrasonic-Assisted Exfoliation of Graphitic Carbon Nitride and its Electrocatalytic Performance in Process of Ethanol Reforming. *Semiconductors* **2019**, *53*, 28–33. [CrossRef]
35. Chebanenko, M.I.; Lobinsky, A.A.; Nevedomskiy, V.N.; Popkov, V.I. NiO-decorated Graphitic Carbon Nitride toward Electrocatalytic Hydrogen Production from Ethanol. *Dalton Trans.* **2020**, *49*, 12088–12097. [CrossRef] [PubMed]
36. Chebanenko, M.I.; Lebedev, L.A.; Ugolkov, V.L.; Prasolov, N.D.; Nevedomskiy, V.N.; Popkov, V.I. Chemical and structural changes of g-C$_3$N$_4$ through oxidative physical vapor deposition. *Appl. Surf. Sci.* **2022**, *600*, 154079. [CrossRef]
37. Muhmood, T.; Xia, M.; Lei, W.; Wang, F.; Mahmood, A. Fe-ZrO$_2$ imbedded graphene like carbon nitride for acarbose (ACB) photo-degradation intermediate study. *Adv. Powder Technol.* **2018**, *29*, 3233–3240. [CrossRef]
38. Muhmood, T.; Xia, M.; Lei, W.; Wang, F.; Khan, M.A. Design of Graphene Nanoplatelet/Graphitic Carbon Nitride Heterojunctions by Vacuum Tube with Enhanced Photocatalytic and Electrochemical Response. *ur. J. Inorg. Chem.* **2018**, *2018*, 1726–1732. [CrossRef]
39. Xiuling Guo, X.; Duan, J.; Li, C.; Zhang, Z.; Wang, W. Highly efficient Z-scheme g-C$_3$N$_4$/ZnO photocatalysts constructed by co-melting-recrystallizing mixed precursors for wastewater treatment. *J. Mater. Sci.* **2020**, *55*, 2018–2031. [CrossRef]
40. Wang, M.; Jiang, L.; Kim, E.J.; Hahn, S.H. Electronic structure and optical properties of Zn(OH)$_2$: LDA+U calculations and intense yellow luminescence. *RSC Adv.* **2015**, *5*, 87496. [CrossRef]
41. Guan, R.; Li, J.; Zhang, J.; Zhao, Z.; Wang, D.; Zhai, H.; Sun, D. Photocatalytic Performance and Mechanistic Research of ZnO/g-C$_3$N$_4$ on Degradation of Methyl Orange. *ACS Omega* **2019**, *4*, 20742–20747. [CrossRef]

42. Gayathri, M.; Sakar, M.; Satheeshkumar, E.; Sundaravadivel, E. Insights into the mechanism of ZnO/g-C_3N_4 nanocomposites toward photocatalytic degradation of multiple organic dyes. *J. Mater. Sci. Mater. Electron.* **2022**, *33*, 9347–9357. [CrossRef]
43. Alharthi, F.A.; Alghamdi, A.A.; Alanazi, H.S.; Alsyahi, A.A.; Ahmad, N. Photocatalytic Degradation of the Light Sensitive Organic Dyes: Methylene Blue and Rose Bengal by Using Urea Derived g-C_3N_4/ZnO Nanocomposites. *Catalysts.* **2020**, *10*, 1457. [CrossRef]

Article

Uptake of BF Dye from the Aqueous Phase by CaO-g-C_3N_4 Nanosorbent: Construction, Descriptions, and Recyclability

Ridha Ben Said [1,2,*], Seyfeddine Rahali [1], Mohamed Ali Ben Aissa [1], Abuzar Albadri [3] and Abueliz Modwi [1]

[1] Department of Chemistry, College of Science and Arts, Qassim University, Ar Rass, Saudi Arabia
[2] Laboratoire de Caractérisations, Applications et Modélisations des Matériaux, Faculté des Sciences de Tunis, Université Tunis El Manar, Tunis 2092, Tunisia
[3] Department of Chemistry, College of Science, Qassim University, Buraydah 52571, Saudi Arabia
* Correspondence: 141255@qu.edu.sa

Abstract: Removing organic dyes from contaminated wastewater resulting from industrial effluents with a cost-effective approach addresses a major global challenge. The adsorption technique onto carbon-based materials and metal oxide is one of the most effective dye removal procedures. The current work aimed to evaluate the application of calcium oxide-doped carbon nitride nanostructures (CaO-g-C_3N_4) to eliminate basic fuchsine dyes (BF) from wastewater. CaO-g-C_3N_4 nanosorbent were obtained via ultrasonication and characterized by scanning electron microscopy, X-ray diffraction, TEM, and BET. The TEM analysis reveals 2D nanosheet-like nanoparticle architectures with a high specific surface area (37.31 m^2/g) for the as-fabricated CaO-g-C_3N_4 nanosorbent. The adsorption results demonstrated that the variation of the dye concentration impacted the elimination of BF by CaO-C_3N_4 while no effect of pH on the removal of BF was observed. Freundlich isotherm and Pseudo-First-order adsorption kinetics models best fitted BF adsorption onto CaO-g-C_3N_4. The highest adsorption capacity of CaO-g-C_3N_4 for BF was determined to be 813 mg·g^{-1}. The adsorption mechanism of BF is related to the π-π stacking bridging and hydrogen bond, as demonstrated by the FTIR study. CaO-g-C_3N_4 nanostructures may be easily recovered from solution and were effectively employed for BF elimination in at least four continuous cycles. The fabricated CaO-g-C_3N_4 adsorbent display excellent BF adsorption capacity and can be used as a potential sorbent in wastewater purification.

Keywords: calcium oxide-doped carbon nitride nanostructures; basic fuchsine; elimination mechanism; π-π stacking

1. Introduction

Water pollution is one of the most important environmental hazards in the modern world, caused by wastewater discharge, insufficient treatment methods, and leakage into the natural water cycle [1,2]. Depending on the source, such as industrial plants, wastewater streams can contain excessively polluting components. Organics [3] (phenolic compounds, dyes, halogenated compounds, oils, etc.) and heavy metals (Hg, Cd, Pb, Cr, Ag, etc.) [4] are potential contaminants in wastewater, as they are biodegradable, volatile, and recycled organic compounds, suspended particles, pathogens, and parasites. Most chemical dyes are probable carcinogens [5]. Thus, before discharging wastewater, it is important to lessen or remove the presence of these potentially fatal substances.

Among these dyes, basic fuchsin BF is a triarylmethane dye that is inflammable and has antibacterial and fungicidal characteristics [6,7]. It is commonly employed as a colorant in textile and leather goods as well as in the staining of collagen and tubercle bacillus [8]. Because of its low biodegradability and its toxicity, carcinogenicity, and unsightliness [9–11], Basic Fuchsin removal from wastewater systems is a major concern that should be studied and executed as soon as possible.

In addition to the traditional biological, electrochemical, and photocatalytic oxidation and decomposition routes, physical processes (such as adsorption) are common methods that have also been developed and are used to remove organic pollutants from wastewater streams [12–14]. Even though these technologies can turn organic pollutants into non-hazardous molecules and can be used in various ways, their inability to be scaled up is a significant problem from an engineering point of view.

More specifically, the adsorption method was widely regarded as the most effective way to treat dye wastewater because of its significant adsorption capacity, low cost, good selectivity, and ease of operation [15–18]. Therefore, many researchers invest a lot of time and effort into creating new adsorbents, as well as adsorption mechanisms and treatment technology, in the hopes that they would be more useful in the treatment of dye wastewater [19–21].

Besides, graphitic carbon nitride (g-C_3N_4) nanosheet has been identified as an indispensable material for two-dimensional structures due to its graphitic-like structure and high stability under ambient circumstances [22]. It is composed of carbon and nitrogen and is most commonly employed for energy conversion and storage. Its π conjugated polymeric metal-free semiconducting 2D structure is composed of graphitic planes composed of sp2-hybridized carbon and nitrogen [23]. Because g-C_3N_4 contains a sufficient number of edge amino and amino groups (NH/NH_2), it can supply several binding sites. Therefore, g-C_3N_4 is regarded as a suitable adsorbent for removing pollutants from wastewater. Nevertheless, g-C_3N_4 nanosheets capability to adsorb is limited by its small surface area and few functional groups [24].

Therefore, the development of g-C_3N_4-containing compounds with higher photonic efficiency, such as TiO_2 and ZnO, piqued the curiosity of a vast number of researchers [25,26]. This was accomplished by combining g-C_3N_4 with another semiconductor and decorating g-C_3N_4 with noble metals [27–31]. Construction of heterojunctions comprised of g-C_3N_4 mixed with another type of compound, such as CaO nanomaterials, and preparation of a Ca-O doped with g-C_3N_4 with an improved surface texture by selecting the optimal preparation method are the most beneficial means of enhancing the adsorption properties of g-C_3N_4.

In the current study, a mesoporous CaO@g-C_3N_4 nanocomposite was successfully produced using a simple sonochemical process and evaluated as a promising adsorbent material for adsorbing the basic fuchsin dye from a contaminated aqueous phase. The physicochemical relationship between characterizations and measurements of equilibrium and kinetics was studied. Adsorption isotherm data were also modeled, and the adsorption performance of CaO@g-C_3N_4 nanocomposite for basic fuchsin was investigated.

2. Experimental

2.1. Chemicals

Sodium hydroxide (NaOH, ≥99%), sodium chloride (NaCl, ≥99%), basic fuchsin (BF, ≥85%), urea (CH_4N_2O, ≥98%) and calcium carbonate ($CaCO_3$, ≥99%) purchased from Merck Company were used without further purification. The required dyes concentrations (25, 50, 100, 150, 200, and 300 ppm) were obtained by diluting BF stock solution (500 ppm).

2.2. CaO-g-C_3N_4 Nanosorbent Fabrication

The nanosheets of g-C_3N_4 were produced through the thermal breakdown of urea. In a typical technique, 0.075 moles of a carbamide compound were placed in a covered pot and tempered with a heating rate of 10 °C/min at 723 K for 120 min. The produced yellow raw g-C_3N_4 was then cooled, pulverized, and stored in a dark container. Thermally decomposing carbonate salts created calcium oxide (CaO) nanoparticles. Two grams of calcium carbonate salts were weighed, placed in a crucible, and annealed at 1073 K for one hour. CaO-g-C_3N_4 nanoparticles were produced using a step-by-step ultrasonication technique aided by an organic solvent (ethanol). In 125 mL of ethanol, 2.76 mg of g-C_3N_4 was sonicated for 15 min. CaO nanoparticles were added to the g-C_3N_4 ethanolic solution along with an additional 45 min of sonication. The yellowish solution generated was evaporated at 368 K for 1440 min. CaO-g-C_3N_4 nanosorbent was ultimately tempered at 453 K for 60 min.

2.3. CaO-g-C_3N_4 Nanosorbent Characterizations

The nanosorbent CaO-g-C_3N_4 was studied using a variety of analytical and spectroscopic techniques. Energy dispersive X-ray (EDX) spectroscopy was used to calculate the elemental composition of the CaO-g-C_3N_4 nanosorbent. The transmission electron microscope (Tecnai G20-USA) was used to make morphological observations, and the stimulating voltage was set at 200 kV. X-ray diffraction (XRD) was used to analyze the phase structure using a Bruker D8 Advance diffractometer Cu-K (λ = 1.540) radiation source. An ASAP 2020 device was used to evaluate the accurate analysis of the surface area. Before and after the BF dye elimination, Fourier transformed infrared (FTIR) spectra were recorded using a Nicolet 5700 spectrometer equipped with a KBr pellet.

2.4. BF Dye Removal Experiments

By mixing 25 mL of BF dye solution with 10 mg of CaO-g-C_3N_4 nanosorbent at varying starting concentrations (5–300 mg/L), batch removal tests of BF dye were conducted. In order to attain equilibrium, the set mixture was stirred for 24 h at 400 rpm. After centrifugation, a clear solution was produced. The dye volumes and beginning concentrations were 100 mL and 250 ppm, respectively, for the kinetic experiment, and the CaO-g-C_3N_4 nanosorbent mass was 40 mg. The test was conducted in the dark with magnetic stirring. Later, 10 mL of the suspension was withdrawn and centrifuged for 10 min to measure the remaining concentration of BF dye.

Using a spectrophotometer, the concentration of dye was determined, and equilibrium dye capacity (q_e) was calculated using the following equation:

$$q_e = \frac{C_0 - C_e}{m} \cdot v \tag{1}$$

where q_t (mg·g^{-1}) is the quantity of dye removed by a unit mass of nanosorbent m (g) at a specified interval of time (min), V is the volume of the solution (L), C_0 is the initial dye concentration, and C_t is the concentration at time t (mg L^{-1}). A similar calculation was used to compute the amount adsorbed at equilibrium, q_t:

$$q_t = \frac{C_0 - C_t}{m} \cdot v \tag{2}$$

The influence of the dye's elimination on the pH of the aqueous media was investigated by setting the initial pH value of dye solution from 3 to 11 pH range by using either NaOH (0.1 mole·L^{-1}) or HCl (0.1 mole·L^{-1}). The pH of zero-point charge (pHzpc) for CaO-g-C_3N_4 nanosorbent was evaluated by the salt addition approach. A fixed amount of CaO-g-C_3N_4 nanosorbent (20 mg) was added in each flask containing 20 mL NaCl solution (0.01 M) with pH initial (pHi) values raised from 2 to 12 (by adjustments using 0.1 M NaOH or HCl solutions). The mixture was stirred for one hour, and the final pH (pH$_f$) was calculated after eliminating CaO-g-C_3N_4 by filtration. For the reusability test, the CaO-g-C_3N_4 nanosorbent used after the adsorption experiments was recovered by filtration and then calcined at 773 K. for one hour. After that, the recovered CaO-g-C_3N_4 nanosorbent was reused for further adsorption tests.

3. Results and Discussion

3.1. CaO-g-C_3N_4 Nanosorbent Characterizations

The scanning elemental mapping analysis for Ca, N, O, and C in the CaO-g-C_3N_4 nanosorbent aggregates (Figure 1a–e) indicates an overall homogeneous dispersion, as shown in Figure 1b–e. On the elemental maps, a brighter zone implies a higher elemental ratio. The CaO-g-C_3N_4 nanocomposite has created a homogenous distribution, according to this observation. An image taken using EDX identifies the individual components that are present in CaO-g-C_3N_4 nanosorbent material. As a result, it is clear from the findings of the EDX performed on CaO-g-C_3N_4 nanosorbent that the surface is composed of carbon (C), nitrogen (N), calcium (Ca), and oxygen (O). These findings are because the results depict bands corresponding to each component (Figure 1a–f).

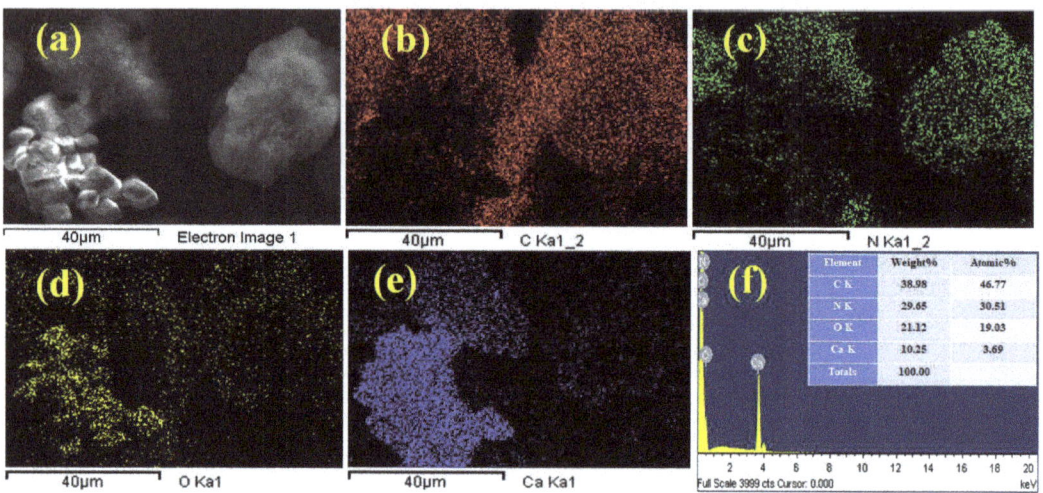

Figure 1. (a–e) Elemental mapping distribution and (f) EDX graph of CaO-g-C_3N_4 nanosorbent.

The TEM micrograph was utilized to investigate the textural qualities of the fabricated CaO-g-C_3N_4 nanosorbent. The TEM photographs of the CaO, g-C_3N_4, and CaO-g-C_3N_4 nanostructures are depicted in Figure 2. The TEM image of CaO (Figure 2a) presents like-sheets shape nanoparticles. On the other hand, the TEM image of g-C_3N_4 (Figure 2b) displays layers with soft surface sheets as a typical graphitic nanostructure [32]. Furthermore, the images with different magnifications obtained from the TEM of the CaO-g-C_3N_4 demonstrated that the morphology had an apparently random appearance, as given in (Figure 2a–d). The CaO-g-C_3N_4 nanosorbent exhibits characteristic 2D nanosheet-like nanoparticle architectures, as seen in (Figure 2a–d). The average particle size of the CaO nanoparticles integrated into the CaO-g-C_3N_4 nanosorbent is around 20–60 nm. CaO are observed to be well disseminated on the g-C_3N_4 surface, forming an abundance of self-active sites on the nanosorbent surface.

Figure 3 depicts the typical peaks at 12.92° and 27.64° for g-C_3N_4 in the XRD pattern. The first peak corresponds to the in-plane packing of tris-triazine units with a d-spacing of 0.685 nm, which agrees with the distance between holes in the nitride pores. However, the great peak at 27.64° is associated with C–N aromatic stacking units separated by 0.322 nm, which corresponds to the 002 interlayer layering plane of the connected aromatic system [33]. Alternatively, the peaks 32.14, 37.25, 53.77, 64.00, and 67.30° correspond to the (110), (200), (202), (311), and (222) planes of the cubic phase of CaO (XRD file JCPDS 77-2376) [34]. Besides the CaO peaks, Ca (OH)$_2$ and $CaCO_3$ peaks are seen at 18,00°, 29,32°, 47.39°, and 48.40°, respectively. The presence of calcite ($CaCO_3$) and hydroxide peaks indicates incomplete pyrolysis of the precursor and fast carbonation and hydrolysis of CaO by ambient CO_2 and water vapor. Literature indicates that CaO nanoparticles have a strong potential for capturing greenhouse gas CO_2 [35]. Finally, the XRD pattern obtained from the fabricated indicates the presence of the g-C_3N_4 and CaO peaks (Figure 3), implying the construction of the target nanosorbent.

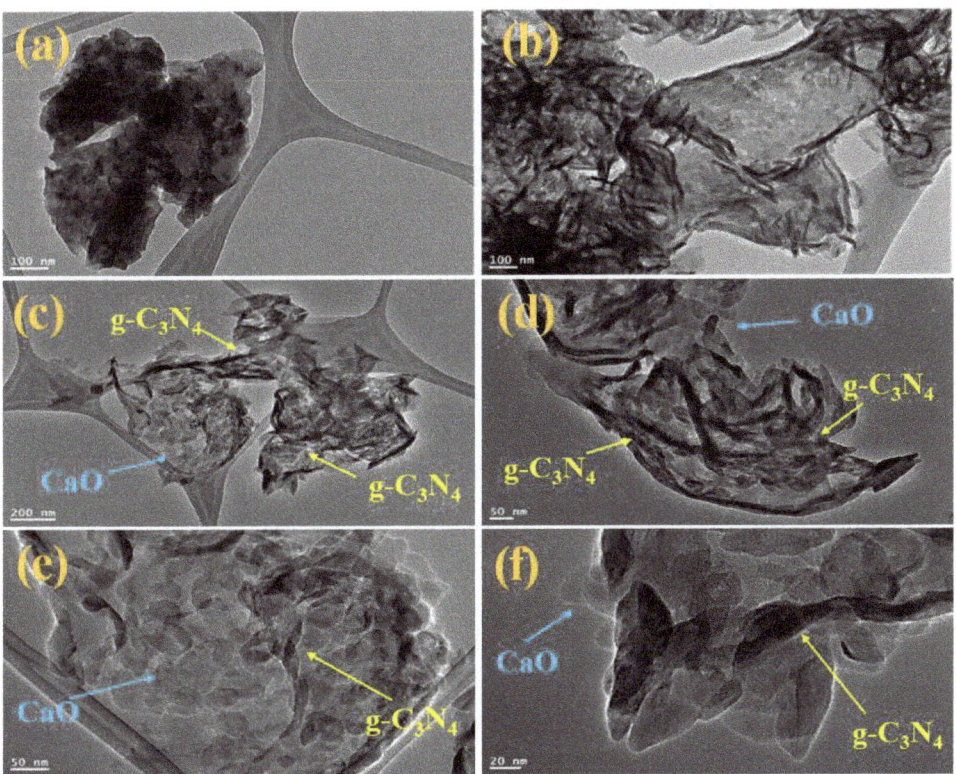

Figure 2. TEM images of (**a**) CaO, (**b**) g-C$_3$N$_4$, and (**c**–**f**) CaO-g-C$_3$N$_4$ nanosorbent with different magnifications.

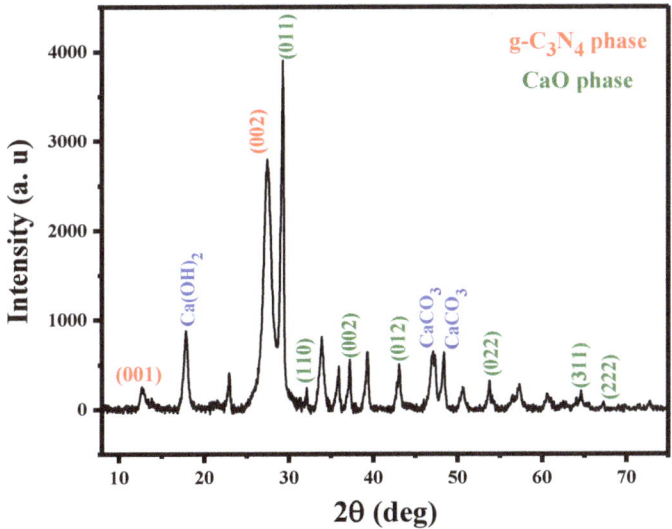

Figure 3. XRD patterns of CaO-g-C$_3$N$_4$ nanosorbent.

Nanomaterials utilized as adsorbents are profoundly influenced by their particular surface area and porous structure, which can provide additional adsorption and reactive sites. The N2 absorption-desorption isotherms of CaO-g-C_3N_4 nanosorbent, which may be categorized as type IV according to the IUPAC system, were determined [36]. Figure 4a,b displays the BET surface isotherms and pore size distribution of the CaO-g-C_3N_4 nanosorbent as manufactured. According to the results, the CaO-g-C_3N_4 nanosorbent absorption-desorption graphs fit isotherm type IV and the hysteresis loop (H_2) for relative pressures between 0.0 and 1.0. This result confirmed the mesoporous nature of the CaO-g-C_3N_4 nanosorbent [37,38]. Due to the presence of several active sites on the surface, the CaO-g-C_3N_4 nanosorbent increased surface area, demonstrated by a higher specific surface area (37.31 m^2/g) with a pore volume of 0.136 cc/g, will improve the sorption capacity [39].

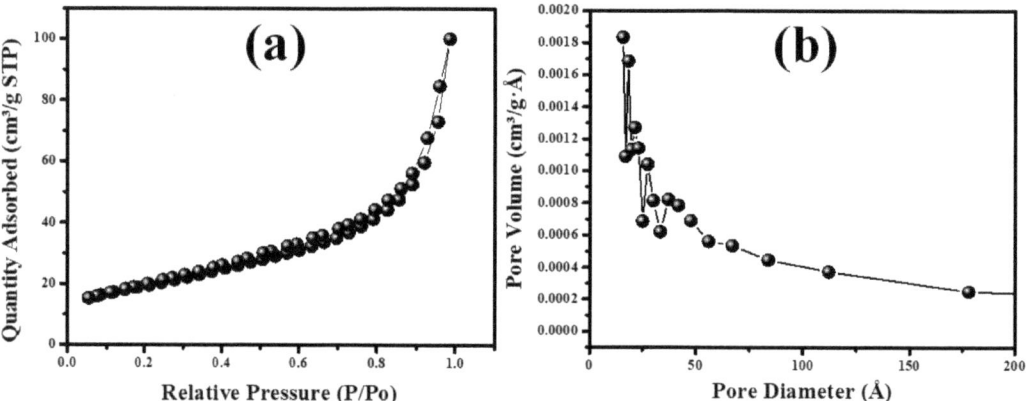

Figure 4. (a) BET surface isotherms and (b) pore sizes distribution of the CaO-g-C_3N_4 nanosorbent.

The chemical condition of the elements on the surface of the CaO@g-C_3N_4 nanostructure was determined by XPS analysis; see Figure 5a–d. CaO exists because the Ca 2p peaks at 349 and 352.6 eV correspond to the divalent oxidation states of calcium oxygen molecules (Ca-2p3/2 and Ca-2p1/2) [40,41]. Ca-O and hydroxyl groups in water molecules correspond to the three O 1s peaks detected at 532.6, 533.7, and 534.4 eV which correspond to the lattice oxygen of the layer-structured Ca-O, and adsorbed H_2O or surface hydroxyl oxygen, respectively (Figure 5b) [42]. As shown in Figure 4c, carbon in the C-C and N-C=N states is attributed with two distinct contributions at 285.8 and 288.2 eV, respectively. According to the XPS analysis, only Ca, O, C, and N are present. The peaks at 398.8 and 400.3 eV (Figure 5d) for N 1s which indicate, respectively, the sp2 hybridized carbon–nitrogen bonding in (C–N) and N-O of the CN and the binding energy of the N atoms in C-N-C [43]. The absence of other impurity peaks supports the results of the XRD and EDS studies.

3.2. BF Removal onto CaO-g-C_3N_4 Nanosorbent

3.2.1. Impact of Variation pH on BF Removal by CaO-g-C_3N_4 Nanosorbent

The solution's pH controls the adsorbent's sorption affinity by adjusting the surface charge and the ionizing strength of the adsorbent [44]. Adsorption experiments of CaO-g-C_3N_4 nanosorbent were conducted under various initial pH values in order to demonstrate the impact of pH value on the adsorption of BF dyes (from 3 to 11). Figure 6a illustrates the influence of pH on BF uptake. It was discovered that BF dyes may be stably adsorbed without observable alterations. The zero-point charge experiment was performed to explain the acquired results. The pHzpc of CaO-g-C_3N_4 nanosorbent is determined to be 10.6, as shown in Figure 6b. At lower pH, the surface of the CaO-g-C_3N_4 nanosorbent is positively charged and becomes negative at pH greater than pHpzc (=10.6). The pH studies showed that the adsorption is pH-independent, indicating that the electrostatic interactions do not control the adsorption mechanism. The adsorption was likely due to the formation of H bonding between -OH and -NH$_2$ onto the CaO-g-C$_3$N$_4$ surface with BF dye molecules [45].

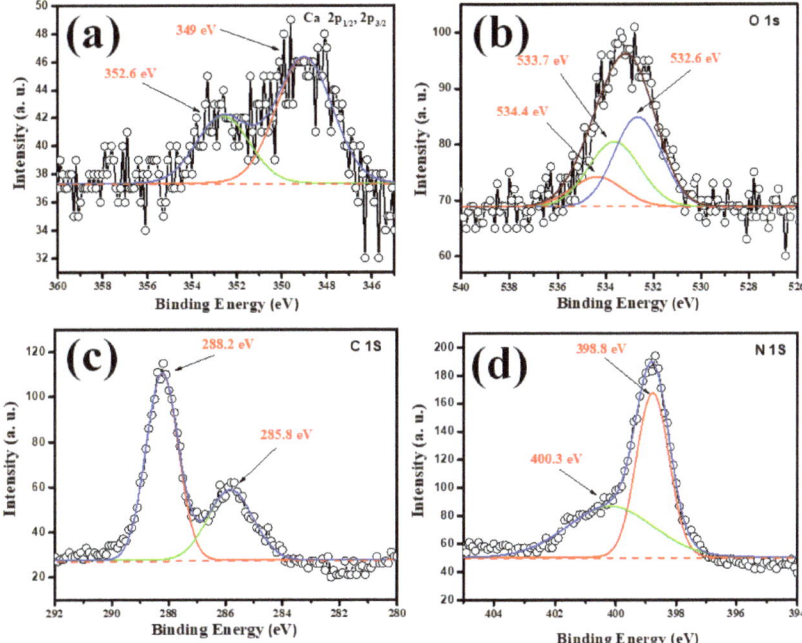

Figure 5. XPS spectra of (**a**) Ca-2p; (**b**) O 1s; (**c**) C 1s and (**d**) N 1s for CaO@g-C$_3$N$_4$ nanosorbent.

3.2.2. Influence of the Initial BF Dye Concentration and Doping

The influence of the initial BF dye concentration in the range of 5–300 mg/L on the adsorption efficiency of CaO-g-C$_3$N$_4$ nanosorbent was scrutinized under the following operating conditions: contact time 1440 min, room temperature, pH 7, 400 rpm stirring speed, and a CaO-g-C$_3$N$_4$ sorbent dose of 10 mg. As shown in Figure 6c, increasing the initial BF concentration from 5 to 300 mg/L improved the adsorption capacity significantly from 60.61 to 738.08 mg/g. These results indicate that BF molecules in the reaction medium interact more strongly with the top layer of the CaO-C$_3$N$_4$ sorbent particles at lower concentrations due to a large amount of vacant active sites. Conversely, the ratio of accessible sites for BF molecules declines with a further rise in the concentration attributed to saturation. To compare the BF dye adsorption capacities of g-C$_3$N$_4$ and CaO-g-C$_3$N$_4$, a series of adsorption experiments were conducted at different initial BF concentrations and a pH value of 7. The obtained results are also shown in Figure 6c. It is interesting to remark that CaO-g-C$_3$N$_4$ exhibits higher BF adsorption capacity than the respective capacities of pure g-C$_3$N$_4$ for the different initial BF concentrations. This result demonstrates that g-C$_3$N$_4$ nanosheets have the capability to adsorb is limited by its small surface area and few functional groups. The doping of g-C$_3$N$_4$ by CaO enhances the adsorption properties of g-C$_3$N$_4$ by improving surface texture.

3.2.3. Adsorption Equilibrium of BF Dyes onto CaO-C$_3$N$_4$ Nanosorbent

The greatest amount of BF absorbed by CaO-g-C$_3$N$_4$ nanosorbent is a crucial characteristic for assessing the high adsorption capacity exhibited. To estimate the absorption capacity of CaO-g-C$_3$N$_4$ nanosorbent, two adsorption isotherm models (Freundlich and Langmuir) were utilized to assess the adsorption data, as depicted in Figure 7a,b. Table 1 contains the formulas corresponding to each isotherm model and the derived parameters.

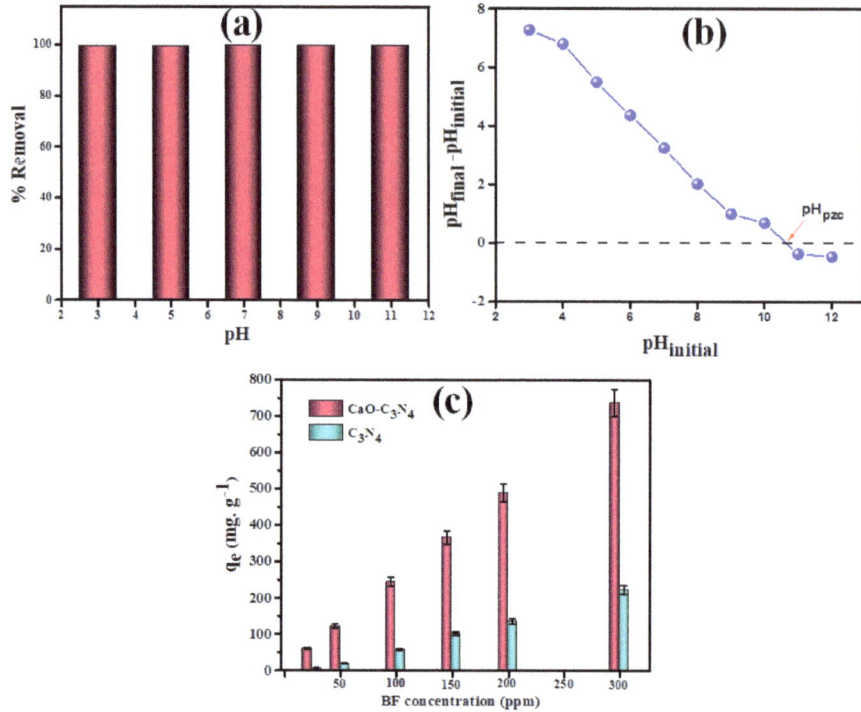

Figure 6. (**a**) Effect of pH on % removal of BF, (**b**) plot for the determination of pH$_{pzc}$ for CaO-g-C$_3$N$_4$ nanosorbent, and (**c**) influence of dye concentration onto the adsorption capacity of CaO-g-C$_3$N$_4$ and g-C$_3$N$_4$ nanosorbent.

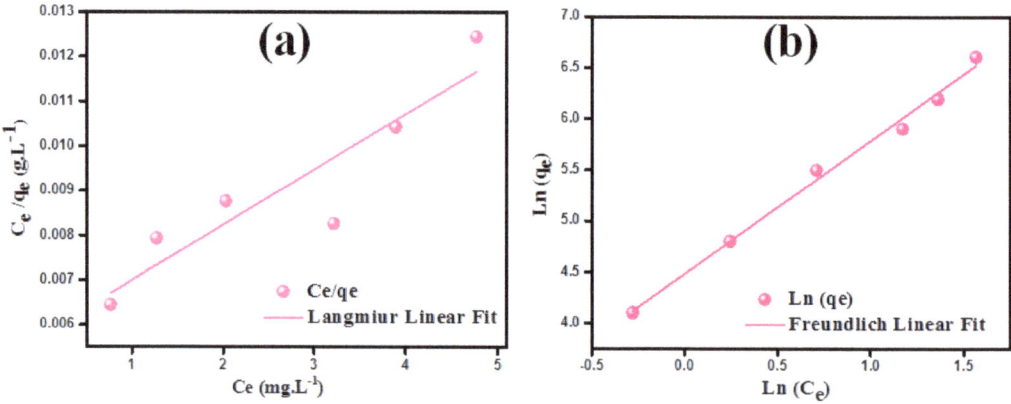

Figure 7. Experimental and fitted adsorption data using the Freundlich (**a**) and Langmuir (**b**) models.

As can be noted from the isotherm graphs and the experimental data for BF adsorption over CaO-C$_3$N$_4$ nanosorbent, the Freundlich adsorption isotherm has the highest $R^2 = 0.996$. These findings indicate that the Freundlich adsorption isotherm curve is more accurate to fit the experimental data.

Table 1. Used equilibrium isotherm models for the adsorption of BF onto CaO-g-C$_3$N$_4$ nanosorbent.

Equilibrium Models	Linear Form	Parameter	Value
Langmuir [46]	$\frac{C_e}{q_e} = \frac{1}{q_m K_L} + \frac{C_e}{q_m}$	q_m (mg/g) K_L (mg/g) R_L (L/mg) R^2	813.0 0.212 0.0015 0.842
Freundlich [47]	$ln q_e = ln K_F + \frac{1}{n} ln C_e$	n K_F (L/mg) R^2	0.97 88.89 0.996

The greatest sorption capacity of CaO-g-C$_3$N$_4$ nanosorbent for BF dyes is found to be 813 mg·g^{-1}, as given in Table 1.

3.2.4. BF Contact Time and Adsorption Kinetic Studies

Figure 8a shows the relationship between contact time and BF adsorption on the surface of the CaO-C$_3$N$_4$ nanosorbent. Adsorption capacity is seen to increase with longer contact times, reaching equilibrium after 25.9 min. Beyond this equilibrium threshold, the adsorption capacity and the amount of BF adsorbed are in dynamic equilibrium. BF molecules were rapidly adsorbed by CaO-C$_3$N$_4$ nanosorbent for the first 25.9 min, after which the adsorption rate declined until it reached its maximum value at around 1440 min. The initial adsorption rates are relatively high due to the abundance of active sites on the surface of the CaO-g-C$_3$N$_4$. After attaining equilibrium, the active site concentration falls, and dye adsorption does not occur.

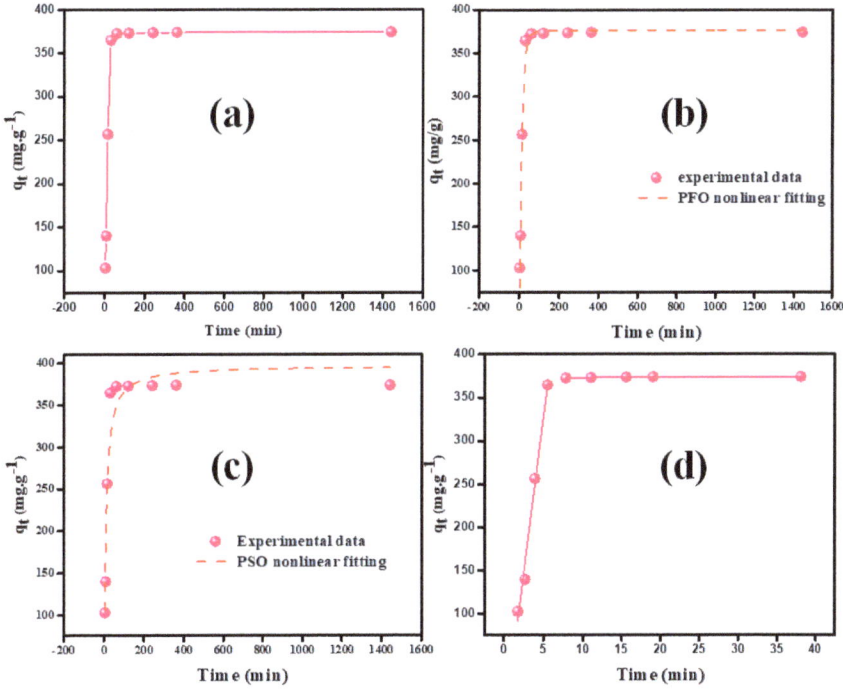

Figure 8. (a) Equilibrium time study (b) PFO, (c) PSO and (d) Intra-particle diffusion graphs for the uptake of BF onto CaO-g-C$_3$N$_4$.

The adsorption kinetics measures the rate of solute adsorption at the solid–liquid interface and gives essential information on the equilibrium period for the design and management of an adsorption process [48]. As shown in Figure 8b,c, the BF adsorption kinetics by the CaO-g-C$_3$N$_4$ nanosorbent was investigated using pseudo-first-order (PFO) and pseudo-second-order (PSO) kinetic models (b and c). The form of the relevant nonlinear equations is shown in Table 2.

Table 2. Kinetics models for BF adsorption onto CaO-g-C$_3$N$_4$ nanosorbent.

Kinetics Models	Kinetic Equations	Parameter	Value
PFO [49]	$q_t = q_e\left(1 - e^{-1k_1 t}\right)$	Q_m (exp) (mg/g) Q_e (mg/g) K_1 (min^{-1}) R^2	375.69 376.49 0.080 0.984
PSO [49]	$q_t = \dfrac{t\, k_2 q_e^2}{k_2 q_e t + 1}$	Q_e (cal) (mg/g) K_2 (g/mg·min) R^2	397.13 0.0003 0.930
Intra-particle Diffusion [50]	$q_t = k_{dif}\sqrt{t} + C$	K_{dif1} (mg. min$^{1/2}$/g) C_1 R^2 K_{dif2} (mg·min$^{1/2}$/g) C_2 R^2	72.78 33.53 0.987 0.04 372.83 0.643

The computed model parameters under the experimental conditions tested are summarized in Table 2. The PFO model might adequately describe the experimental adsorption kinetics data. It claims that the ratio of the square of the number of accessible sites to the rate of adsorption site occupancy. The form formula for the PFO nonlinear linear model is shown in Table 2. Using the computed model parameters in Table 2, the extraordinarily high R^2 value of 0.984 is determined. Compared to the PSO, the PFO equation provides a perfect fit, as shown by the findings. For the tested BF concentrations, there is only a small difference between the experimental Q_{max} values and the model-estimated Q_{max} values. As a result, the PFO model's best fit implies that the kinetic adsorption may be mathematically described using the concentration of BF in solution [44,51].

Through the intra-particle diffusion/transport mechanism, the BF elimination may be transferred from the bulk of the solution to the solid phase of the CaO-g-C$_3$N$_4$ nanosorbent. In some circumstances, the step of the adsorption process known as intra-particular diffusion is restrictive. The diffusion pattern developed by Weber and Morris supports the notion of intra-particulate diffusion [52,53]. As q_t and $t^{1/2}$ are compared linearly, the removal of BF onto the CaO-C$_3$N$_4$ surface demonstrates the efficacy of the intra-particle diffusion kinetic pattern. In addition, the intra-particle mode of diffusion is characterized by the regression coefficient (R^2 = 0.987). The diameter of the boundary layer is represented by parameter C's value. The higher percentages of the constants in Table 2 demonstrate the solution boundary layer's strong influence on the removal of BF dyes [52,53]. It can be seen that the first stage of sorption has a larger rate than the second phase, which is shown by k_{dif1} > k_{dif2} (Table 2). The high value of the rate produced by the first step may be explained by the movement of the dye mover through the solution and onto the surface of the outer CaO-g-C$_3$N$_4$ that is generated by the boundary layer. Comparatively, the subsequent phase describes the last equilibrium step, when intra-particle diffusion begins to diminish due to the solute's modest concentration gradient and the restricted number of holes and pores available for diffusion [54].

3.3. Regeneration and Reusability Study

By removing BF from the surface of the nanomaterial, the reusability and regeneration of CaO-g-C$_3$N$_4$ sorbent were investigated. Following the adsorption experiment, the BF was removed from the CaO-g-C$_3$N$_4$ by heating it in an oven at 773 K for a one hour. The recovered CaO-g-C$_3$N$_4$ was then reapplied to the BF elimination process. CaO-g-C$_3$N$_4$ has been used efficiently for the removal of BF for at least four continuous cycles, as shown by the reusability results (Figure 9a). As shown, there was no obvious decrease in the elimination effectiveness during four adsorption–desorption cycles, and only 4%, 7%, and 9% of the adsorption capacity for BF declined at the second, third, and fourth cycles, respectively.

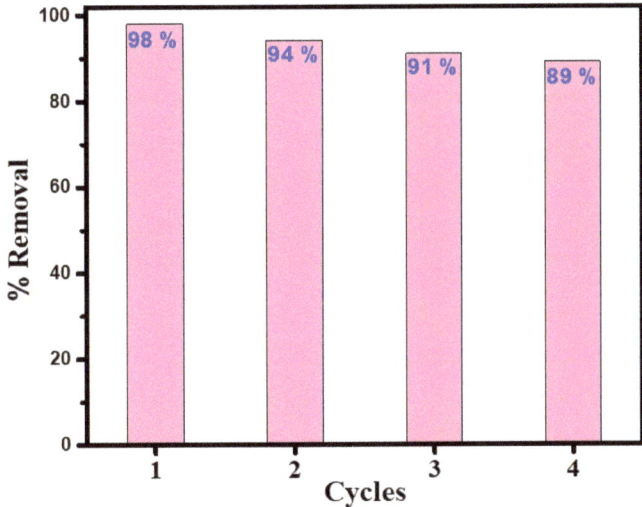

Figure 9. Reusability effectiveness of CaO-g-C$_3$N$_4$.

3.4. Comparison Study

As shown in Table 3, the calculated adsorption capacity of CaO-g-C$_3$N$_4$ for BF using the Langmuir isotherm model is 813.00 mg·g^{-1}. It is to one's advantage to evaluate the CaO-g-C$_3$N$_4$ adsorption capacity in relation to the diverse sorbents that can be utilized for BF elimination. Table 3 shows the various sorbents with high adsorption capacities for BF removal. Compared to previously reported sorbents like MgO and modified activated carbons, CaO-g-C$_3$N$_4$ has a higher capacity for adsorption. This finding confirmed that CaO-g-C$_3$N$_4$ is an efficient BF dye adsorbent.

Table 3. Observation of adsorption capacities of the CaO-g-C$_3$N$_4$ using various nanomaterial adsorbents.

Adsorbents	q_e (mg g^{-1})	Best pH	BET Surface Area (m^2/g)	References
Fe-MgO/kaolinite	10.36	9.0	-	[55]
YZnO nanoparticles	75.53	11	20.26	[56]
Al/MCM-41	54.44	3–9	997	[57]
Euryale ferox Salisbury seed shell	19.48	6.0	-	[58]
ESM	47.85	6.0	11.56	[59]
Fe/ZSM-5	251.87	5.0	399	[60]
Modified activated carbons	238.10	8.5	613	[61]
MgO	493.90	11	12.22	[62]
MgOg-C$_3$N$_4$	1250	7.0	84.2	[63]
CaO-g-C$_3$N$_4$	813.00	Independent	37.31	Current study

3.5. Adsorption Mechanism

The adsorption mechanism of BF dyes by nanosorbent has been elucidated using FTIR analysis. Figure 10a depicts the FTIR spectra of nanosorbent prior to and following BF adsorption. CaO-g-C$_3$N$_4$ spectrum reveals a number of distinguishable bands: the bandwidth between 3000 and 3600 cm^{-1} corresponds to the stretching vibration modes of O–H and NH. The absorption bands at 1242, 1326, and 1412 cm^{-1} are associated with aromatic C–N stretching, while the absorption bands at 1578 and 1640 cm^{-1} are associated with C≡N stretching. The band at 884 cm^{-1} corresponds to the triazine ring mode, a frequent carbon nitride mode. The characteristic band located at 805 cm^{-1} is assigned to Ca-O vibration mode [64]. After adsorption, as can be observed in Figure 10a, many typical BF bands form and move around with respect to the free molecules, suggesting that CaO-g-C$_3$N$_4$ and BF molecules may interact. Also, following the adsorption of BF dyes, several vibration bonds of CaO–C$_3$N$_4$, such as aromatic C–N stretching and triazine ring modes, have shifted position. This study demonstrated that delocalized electron systems of C$_3$N$_3$ and functional groups of CaO-g-C$_3$N$_4$ were responsible for the adsorption of BF molecules. In addition, the O–H and NH stretching vibration modes were shifted, demonstrating the interaction of BF molecules with CaO-g-C$_3$N$_4$ nanosorbent via hydrogen bonds. Lie et al. demonstrate that sorbents containing pyrazine and imine groups are beneficial to the formation of π-π stacking and hydrogen bonds interactions with organics dyes [65]. Also, the examination of the pH's effect indicates that the electrostatic attraction could not dominate (control) the adsorption mechanism of BF onto the CaO-g-C$_3$N$_4$ nanosorbent. The suggested BF adsorption mechanism (Figure 10b) onto the CaO-g-C$_3$N$_4$ involves hydrogen bonds and the π-π stacking bridging [66].

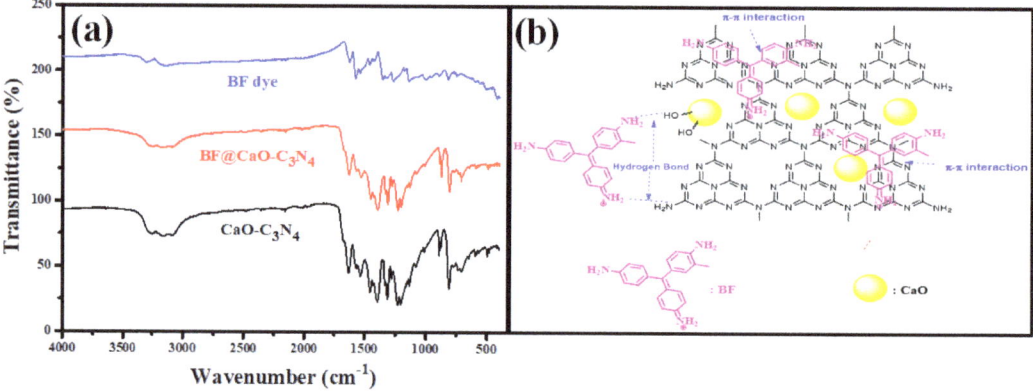

Figure 10. (**a**) FTIR spectra of BF, CaO-g-C$_3$N$_4$ and CaO-g-C$_3$N$_4$ @BF and (**b**) Possible adsorption mechanism of BF dyes onto CaO-g-C$_3$N$_4$.

4. Conclusions

Mesoporous CaO-g-C$_3$N$_4$ nanosorbent was created using the ultrasonication technique, and it was subsequently employed as an adsorbent to remove BF dyes from wastewater. CaO-g-C$_3$N$_4$ nanosorbent removal efficiency was studied by adjusting pH, contact time, and BF concentration. The greatest adsorption capacity observed was 813 mg·g^{-1}, indicating that the reported data demonstrated outstanding elimination effectiveness toward BF dye. The BF elimination by CaO-g-C$_3$N$_4$ nanosorbent was evaluated employed different adsorption and kinetic models, and the best-fitting was committed by the Freundlich adsorption isotherm and PFO kinetics models. The suggested BF adsorption mechanism onto the CaO-g-C$_3$N$_4$ involves hydrogen bonds and the π-π stacking bridging. CaO-g-C$_3$N$_4$ nanostructures may be easily recovered from solution and were effectively employed for BF elimination in at least four continuous cycles.

Author Contributions: R.B.S.: Conceptualization, Methodology, Formal analysis, Investigation, Visualization. S.R.: Methodology, Formal analysis, Investigation. M.A.B.A.: Methodology, Formal analysis, Investigation. A.A.: Conceptualization, Methodology, Formal analysis, Investigation. A.M.: Conceptualization, Methodology, Formal analysis, Investigation, Visualization. All authors have read and agreed to the published version of the manuscript.

Funding: The authors extend their appreciation to the Deputyship for Research & Innovation, Ministry of Education, Saudi Arabia for funding this research work through the project number (QU-IF-05-01-28461).

Acknowledgments: The authors extend their appreciation to the Deputyship for Research & Innovation, Ministry of Education, Saudi Arabia for funding this research work through the project number (QU-IF-05-01-28461). The authors give thanks to Qassim University for technical support.

Conflicts of Interest: The authors declare no conflict of interest.

References

1. Zamora-Ledezma, C.; Negrete-Bolagay, D.; Figueroa, F.; Zamora-Ledezma, E.; Ni, M.; Alexis, F.; Guerrero, V.H. Heavy metal water pollution: A fresh look about hazards, novel and conventional remediation methods. *Environ. Technol. Innov.* **2021**, *22*, 101504. [CrossRef]
2. Brown, D. Effects of colorants in the aquatic environment. *Ecotoxicol. Environ. Saf.* **1987**, *13*, 139–147. [CrossRef] [PubMed]
3. Hasan, Z.; Jhung, S.H. Removal of hazardous organics from water using metal-organic frameworks (MOFs): Plausible mechanisms for selective adsorptions. *J. Hazard. Mater.* **2015**, *283*, 329–339. [CrossRef]
4. Zeitoun, M.M.; Mehana, E. Impact of water pollution with heavy metals on fish health: Overview and updates. *Glob. Vet.* **2014**, *12*, 219–231.
5. La Farre, M.; Pérez, S.; Kantiani, L.; Barceló, D. Fate and toxicity of emerging pollutants, their metabolites and transformation products in the aquatic environment. *TrAC Trends Anal. Chem.* **2008**, *27*, 991–1007. [CrossRef]
6. Thakur, R.H.; Fung, D. Effect of Dyes on the Growth of Food MOLDS 1. *J. Rapid Methods Autom. Microbiol.* **1995**, *4*, 1–35. [CrossRef]
7. Mirzaei, F.; Nilash, M.M.; Fakhari, A.R. Development of a new electromembrane extraction combined with ion mobility spectrometry for the quantification of malachite green in water samples. *Int. J. Ion Mobil. Spectrom.* **2020**, *23*, 153–160. [CrossRef]
8. Rajumon, R.; Anand, J.C.; Ealias, A.M.; Desai, D.S.; George, G.; Saravanakumar, M. Adsorption of textile dyes with ultrasonic assistance using green reduced graphene oxide: An in-depth investigation on sonochemical factors. *J. Environ. Chem. Eng.* **2019**, *7*, 103479. [CrossRef]
9. Ritter, C. Studies of the toxicity of basic fuchsin for certain bacteria. *Am. J. Public Health Nations Health* **1940**, *30*, 59–65. [CrossRef]
10. Fernandes, E.P.; Silva, T.S.; Carvalho, C.M.; Selvasembian, R.; Chaukura, N.; Oliveira, L.M.; Meneghetti, S.M.P.; Meili, L. Efficient adsorption of dyes by γ-alumina synthesized from aluminum wastes: Kinetics, isotherms, thermodynamics and toxicity assessment. *J. Environ. Chem. Eng.* **2021**, *9*, 106198. [CrossRef]
11. Casas, N.; Parella, T.; Vicent, T.; Caminal, G.; Sarrà, M. Metabolites from the biodegradation of triphenylmethane dyes by Trametes versicolor or laccase. *Chemosphere* **2009**, *75*, 1344–1349. [CrossRef] [PubMed]
12. Moumen, A.; Belhocine, Y.; Sbei, N.; Rahali, S.; Ali, F.A.M.; Mechati, F.; Hamdaoui, F.; Seydou, M. Removal of Malachite Green Dye from Aqueous Solution by Catalytic Wet Oxidation Technique Using Ni/Kaolin as Catalyst. *Molecules* **2022**, *27*, 7528. [CrossRef] [PubMed]
13. Adam, F.A.; Ghoniem, M.; Diawara, M.; Rahali, S.; Abdulkhair, B.Y.; Elamin, M.; Aissa, M.A.B.; Seydou, M. Enhanced adsorptive removal of indigo carmine dye by bismuth oxide doped MgO based adsorbents from aqueous solution: Equilibrium, kinetic and computational studies. *RSC Adv.* **2022**, *12*, 24786–24803. [CrossRef]
14. Attia, G.; Rahali, S.; Teka, S.; Fourati, N.; Zerrouki, C.; Seydou, M.; Chehimi, S.; Hayouni, S.; Mbakidi, J.-P.; Bouquillon, S. Anthracene based surface acoustic wave sensors for picomolar detection of lead ions. Correlation between experimental results and DFT calculations. *Sens. Actuators B Chem.* **2018**, *276*, 349–355. [CrossRef]
15. Pandey, S.; Son, N.; Kim, S.; Balakrishnan, D.; Kang, M. Locust Bean gum-based hydrogels embedded magnetic iron oxide nanoparticles nanocomposite: Advanced materials for environmental and energy applications. *Environ. Res.* **2022**, *214*, 114000. [CrossRef]
16. Doondani, P.; Jugade, R.; Gomase, V.; Shekhawat, A.; Bambal, A.; Pandey, S. Chitosan/Graphite/Polyvinyl Alcohol Magnetic Hydrogel Microspheres for Decontamination of Reactive Orange 16 Dye. *Water* **2022**, *14*, 3411. [CrossRef]
17. Gomase, V.; Jugade, R.; Doondani, P.; Deshmukh, S.; Saravanan, D.; Pandey, S. Dual modifications of chitosan with PLK for amputation of cyanide ions: Equilibrium studies and optimization using RSM. *Int. J. Biol. Macromol.* **2022**, *223*, 636–651. [CrossRef]
18. Rahali, S.; Ben Aissa, M.A.; Khezami, L.; Elamin, N.; Seydou, M.; Modwi, A. Adsorption behavior of Congo red onto barium-doped ZnO nanoparticles: Correlation between experimental results and DFT calculations. *Langmuir* **2021**, *37*, 7285–7294. [CrossRef]

19. Wu, Y.; Zuo, F.; Zheng, Z.; Ding, X.; Peng, Y. A novel approach to molecular recognition surface of magnetic nanoparticles based on host–guest effect. *Nanoscale Res. Lett.* **2009**, *4*, 738–747. [CrossRef]
20. Khezami, L.; Aissa, M.A.B.; Modwi, A.; Guesmi, A.; Algethami, F.K.; Bououdina, M. Efficient removal of organic dyes by Cr-doped ZnO nanoparticles. *Biomass Convers. Biorefin.* **2022**, 1–14. [CrossRef]
21. Khezami, L.; Aissa, M.A.B.; Modwi, A.; Ismail, M.; Guesmi, A.; Algethami, F.K.; Ticha, M.B.; Assadi, A.A.; Nguyen-Tri, P. Harmonizing the photocatalytic activity of g-C3N4 nanosheets by ZrO_2 stuffing: From fabrication to experimental study for the wastewater treatment. *Biochem. Eng. J.* **2022**, *182*, 108411. [CrossRef]
22. Aldaghri, O.; Modwi, A.; Idriss, H.; Ali, M.; Ibnaouf, K. Cleanup of Cd II from water media using Y_2O_3@ gC3N4 (YGCN) nanocomposite. *Diam. Relat. Mater.* **2022**, *129*, 109315. [CrossRef]
23. Modwi, A.; Elamin, M.R.; Idriss, H.; Elamin, N.Y.; Adam, F.A.; Albadri, A.E.; Abdulkhair, B.Y. Excellent Adsorption of Dyes via MgTiO3@ g-C3N4 Nanohybrid: Construction, Description and Adsorption Mechanism. *Inorganics* **2022**, *10*, 210. [CrossRef]
24. Modwi, A.; Basith, N.; Ghoniem, M.; Ismail, M.; Aissa, M.B.; Khezami, L.; Bououdina, M. Efficient Pb (II) adsorption in aqueous solution by hierarchical 3D/2D TiO_2/CNNS nanocomposite. *Mater. Sci. Eng. B* **2023**, *289*, 116191. [CrossRef]
25. Toghan, A.; Abd El-Lateef, H.M.; Taha, K.K.; Modwi, A. Mesoporous TiO_2@ g-C3N4 composite: Construction, characterization, and boosting indigo carmine dye destruction. *Diam. Relat. Mater.* **2021**, *118*, 108491. [CrossRef]
26. Qamar, M.A.; Shahid, S.; Javed, M.; Iqbal, S.; Sher, M.; Akbar, M.B. Highly efficient g-C3N4/Cr-ZnO nanocomposites with superior photocatalytic and antibacterial activity. *J. Photochem. Photobiol. A Chem.* **2020**, *401*, 112776. [CrossRef]
27. Gogoi, D.; Makkar, P.; Ghosh, N.N. Solar light-irradiated photocatalytic degradation of model dyes and industrial dyes by a magnetic CoFe2O4–gC3N4 S-scheme heterojunction photocatalyst. *ACS Omega* **2021**, *6*, 4831–4841. [CrossRef]
28. Varapragasam, S.J.; Andriolo, J.M.; Skinner, J.L.; Grumstrup, E.M. Photocatalytic Reduction of Aqueous Nitrate with Hybrid Ag/g-C3N4 under Ultraviolet and Visible Light. *ACS Omega* **2021**, *6*, 34850–34856. [CrossRef]
29. Cai, W.; Tang, J.; Shi, Y.; Wang, H.; Jiang, X. Improved in situ synthesis of heterostructured 2D/2D BiOCl/g-C3N4 with enhanced dye photodegradation under visible-light illumination. *ACS Omega* **2019**, *4*, 22187–22196. [CrossRef]
30. Alhaddad, M.; Mohamed, R.M.; Mahmoud, M.H. Promoting Visible Light Generation of Hydrogen Using a Sol–Gel-Prepared MnCo2O4@ g-C3N4 p–n Heterojunction Photocatalyst. *ACS Omega* **2021**, *6*, 8717–8725. [CrossRef]
31. Paul, D.R.; Gautam, S.; Panchal, P.; Nehra, S.P.; Choudhary, P.; Sharma, A. ZnO-modified g-C3N4: A potential photocatalyst for environmental application. *ACS Omega* **2020**, *5*, 3828–3838. [CrossRef] [PubMed]
32. Li, Y.; Wang, J.; Yang, Y.; Zhang, Y.; He, D.; An, Q.; Cao, G. Seed-induced growing various TiO_2 nanostructures on g-C3N4 nanosheets with much enhanced photocatalytic activity under visible light. *J. Hazard. Mater.* **2015**, *292*, 79–89. [CrossRef] [PubMed]
33. Ngullie, R.C.; Alaswad, S.O.; Bhuvaneswari, K.; Shanmugam, P.; Pazhanivel, T.; Arunachalam, P. Synthesis and Characterization of Efficient ZnO/g-C3N4 Nanocomposites Photocatalyst for Photocatalytic Degradation of Methylene Blue. *Coatings* **2020**, *10*, 500. [CrossRef]
34. Madhu, B.; Bhagyalakshmi, H.; Shruthi, B.; Veerabhadraswamy, M. Structural, AC conductivity, dielectric and catalytic behavior of calcium oxide nanoparticles derived from waste eggshells. *SN Appl. Sci.* **2021**, *3*, 637. [CrossRef]
35. Mirghiasi, Z.; Bakhtiari, F.; Darezereshki, E.; Esmaeilzadeh, E. Preparation and characterization of CaO nanoparticles from Ca(OH)2 by direct thermal decomposition method. *J. Ind. Eng. Chem.* **2014**, *20*, 113–117. [CrossRef]
36. Tonelli, M.; Gelli, R.; Giorgi, R.; Pierigè, M.I.; Ridi, F.; Baglioni, P. Cementitious materials containing nano-carriers and silica for the restoration of damaged concrete-based monuments. *J. Cult. Herit.* **2021**, *49*, 59–69. [CrossRef]
37. Yue, Y.; Zhang, P.; Wang, W.; Cai, Y.; Tan, F.; Wang, X.; Qiao, X.; Wong, P.K. Enhanced dark adsorption and visible-light-driven photocatalytic properties of narrower-band-gap Cu2S decorated Cu2O nanocomposites for efficient removal of organic pollutants. *J. Hazard. Mater.* **2020**, *384*, 121302. [CrossRef]
38. Modwi, A.; Abbo, M.; Hassan, E.; Houas, A. Effect of annealing on physicochemical and photocatalytic activity of Cu 5% loading on ZnO synthesized by sol–gel method. *J. Mater. Sci. Mater. Electron.* **2016**, *27*, 12974–12984. [CrossRef]
39. Li, Y.; Lv, K.; Ho, W.; Dong, F.; Wu, X.; Xia, Y. Hybridization of rutile TiO_2 (rTiO2) with g-C3N4 quantum dots (CN QDs): An efficient visible-light-driven Z-scheme hybridized photocatalyst. *Appl. Catal. B Environ.* **2017**, *202*, 611–619. [CrossRef]
40. Liu, J.; Liu, M.; Chen, S.; Wang, B.; Chen, J.; Yang, D.-P.; Zhang, S.; Du, W. Conversion of Au (III)-polluted waste eggshell into functional CaO/Au nanocatalyst for biodiesel production. *Green Energy Environ.* **2020**, *7*, 352–359. [CrossRef]
41. Karuppusamy, I.; Samuel, M.S.; Selvarajan, E.; Shanmugam, S.; Kumar, P.S.M.; Brindhadevi, K.; Pugazhendhi, A. Ultrasound-assisted synthesis of mixed calcium magnesium oxide (CaMgO2) nanoflakes for photocatalytic degradation of methylene blue. *J. Colloid Interface Sci.* **2021**, *584*, 770–778. [CrossRef] [PubMed]
42. Rojas, J.; Toro-Gonzalez, M.; Molina-Higgins, M.; Castano, C. Facile radiolytic synthesis of ruthenium nanoparticles on graphene oxide and carbon nanotubes. *Mater. Sci. Eng. B* **2016**, *205*, 28–35. [CrossRef]
43. Thomas, A.; Fischer, A.; Goettmann, F.; Antonietti, M.; Müller, J.-O.; Schlögl, R.; Carlsson, J.M. Graphitic carbon nitride materials: Variation of structure and morphology and their use as metal-free catalysts. *J. Mater. Chem.* **2008**, *18*, 4893–4908. [CrossRef]
44. Mustafa, B.; Modwi, A.; Ismail, M.; Makawi, S.; Hussein, T.; Abaker, Z.; Khezami, L. Adsorption performance and Kinetics study of Pb (II) by RuO_2–ZnO nanocomposite: Construction and Recyclability. *Int. J. Environ. Sci. Technol.* **2022**, *19*, 327–340. [CrossRef]
45. Li, N.; Dang, H.; Chang, Z.; Zhao, X.; Zhang, M.; Li, W.; Zhou, H.; Sun, C. Synthesis of uniformly distributed magnesium oxide micro-/nanostructured materials with deep eutectic solvent for dye adsorption. *J. Alloys Compd.* **2019**, *808*, 151571. [CrossRef]

46. Al-Ghouti, M.A.; Razavi, M.M. Water reuse: Brackish water desalination using Prosopis juliflora. *Environ. Technol. Innov.* **2020**, *17*, 100614. [CrossRef]
47. Ayawei, N.; Ebelegi, A.N.; Wankasi, D. Modelling and interpretation of adsorption isotherms. *J. Chem.* **2017**, *2017*, 3039817. [CrossRef]
48. Oladipo, B.; Govender-Opitz, E.; Ojumu, T.V. Kinetics, Thermodynamics, and Mechanism of Cu (II) Ion Sorption by Biogenic Iron Precipitate: Using the Lens of Wastewater Treatment to Diagnose a Typical Biohydrometallurgical Problem. *ACS Omega* **2021**, *6*, 27984–27993. [CrossRef]
49. Lagergren, S.K. About the theory of so-called adsorption of soluble substances. *Sven. Vetenskapsakad. Handingarl* **1898**, *24*, 1–39.
50. Chien, S.; Clayton, W. Application of Elovich equation to the kinetics of phosphate release and sorption in soils. *Soil Sci. Soc. Am. J.* **1980**, *44*, 265–268. [CrossRef]
51. Khezami, L.; Elamin, N.; Modwi, A.; Taha, K.K.; Amer, M.; Bououdina, M. Mesoporous Sn@ TiO_2 nanostructures as excellent adsorbent for Ba ions in aqueous solution. *Ceram. Int.* **2022**, *48*, 5805–5813. [CrossRef]
52. Hameed, B.; Salman, J.; Ahmad, A. Adsorption isotherm and kinetic modeling of 2, 4-D pesticide on activated carbon derived from date stones. *J. Hazard. Mater.* **2009**, *163*, 121–126. [CrossRef] [PubMed]
53. El-Sikaily, A.; El Nemr, A.; Khaled, A.; Abdelwehab, O. Removal of toxic chromium from wastewater using green alga Ulva lactuca and its activated carbon. *J. Hazard. Mater.* **2007**, *148*, 216–228. [CrossRef]
54. Ali, I.; Peng, C.; Ye, T.; Naz, I. Sorption of cationic malachite green dye on phytogenic magnetic nanoparticles functionalized by 3-marcaptopropanic acid. *RSC Adv.* **2018**, *8*, 8878–8897. [CrossRef] [PubMed]
55. Khan, T.A.; Khan, E.A. Removal of basic dyes from aqueous solution by adsorption onto binary iron-manganese oxide coated kaolinite: Non-linear isotherm and kinetics modeling. *Appl. Clay Sci.* **2015**, *107*, 70–77. [CrossRef]
56. Aissa, B.; Khezami, L.; Taha, K.; Elamin, N.; Mustafa, B.; Al-Ayed, A.; Modwi, A. Yttrium oxide-doped ZnO for effective adsorption of basic fuchsin dye: Equilibrium, kinetics, and mechanism studies. *Int. J. Environ. Sci. Technol.* **2021**, *19*, 9901–9914. [CrossRef]
57. Guan, Y.; Wang, S.; Wang, X.; Sun, C.; Wang, Y.; Hu, L. Preparation of mesoporous Al-MCM-41 from natural palygorskite and its adsorption performance for hazardous aniline dye-basic fuchsin. *Microporous Mesoporous Mater.* **2018**, *265*, 266–274. [CrossRef]
58. Kalita, S.; Pathak, M.; Devi, G.; Sarma, H.; Bhattacharyya, K.; Sarma, A.; Devi, A. Utilization of Euryale ferox Salisbury seed shell for removal of basic fuchsin dye from water: Equilibrium and kinetics investigation. *RSC Adv.* **2017**, *7*, 27248–27259. [CrossRef]
59. Bessashia, W.; Berredjem, Y.; Hattab, Z.; Bououdina, M. Removal of Basic Fuchsin from water by using mussel powdered eggshell membrane as novel bioadsorbent: Equilibrium, kinetics, and thermodynamic studies. *Environ. Res.* **2020**, *186*, 109484. [CrossRef]
60. Mohammed, B.B.; Hsini, A.; Abdellaoui, Y.; Abou Oualid, H.; Laabd, M.; El Ouardi, M.; Addi, A.A.; Yamni, K.; Tijani, N. Fe-ZSM-5 zeolite for efficient removal of basic Fuchsin dye from aqueous solutions: Synthesis, characterization and adsorption process optimization using BBD-RSM modeling. *J. Environ. Chem. Eng.* **2020**, *8*, 104419. [CrossRef]
61. Huang, L.; Kong, J.; Wang, W.; Zhang, C.; Niu, S.; Gao, B. Study on Fe (III) and Mn (II) modified activated carbons derived from Zizania latifolia to removal basic fuchsin. *Desalination* **2012**, *286*, 268–276. [CrossRef]
62. Ghoniem, M.G.; Ali, F.A.M.; Abdulkhair, B.Y.; Elamin, M.R.A.; Alqahtani, A.M.; Rahali, S.; Ben Aissa, M.A. Highly selective removal of cationic dyes from wastewater by MgO nanorods. *Nanomaterials* **2022**, *12*, 1023. [CrossRef] [PubMed]
63. AbuMousa, R.A.; Khezami, L.; Ismail, M.; Ben Aissa, M.A.; Modwi, A.; Bououdina, M. Efficient Mesoporous MgO/g-C_3N_4 for Heavy Metal Uptake: Modeling Process and Adsorption Mechanism. *Nanomaterials* **2022**, *12*, 3945. [CrossRef] [PubMed]
64. Margaretha, Y.Y.; Prastyo, H.S.; Ayucitra, A.; Ismadji, S. Calcium oxide from Pomacea sp. shell as a catalyst for biodiesel production. *Int. J. Energy Environ. Eng.* **2012**, *3*, 33. [CrossRef]
65. Li, J.; Wang, B.; Chang, B.; Liu, J.; Zhu, X.; Ma, P.; Sun, L.; Li, M. One new hexatungstate-based binuclear nickel (II) complex with high selectivity adsorption for organic dyes. *J. Mol. Struct.* **2021**, *1231*, 129674. [CrossRef]
66. Feng, Y.; Chen, G.; Zhang, Y.; Li, D.; Ling, C.; Wang, Q.; Liu, G. Superhigh co-adsorption of tetracycline and copper by the ultrathin g-C_3N_4 modified graphene oxide hydrogels. *J. Hazard. Mater.* **2022**, *424*, 127362. [CrossRef]

Disclaimer/Publisher's Note: The statements, opinions and data contained in all publications are solely those of the individual author(s) and contributor(s) and not of MDPI and/or the editor(s). MDPI and/or the editor(s) disclaim responsibility for any injury to people or property resulting from any ideas, methods, instructions or products referred to in the content.

Article

Preparation, Microstructural Characterization and Photocatalysis Tests of V^{5+}-Doped TiO_2/WO_3 Nanocomposites Supported on Electrospun Membranes

Michel F. G. Pereira [1], Mayane M. Nascimento [1], Pedro Henrique N. Cardoso [1], Carlos Yure B. Oliveira [2], Ginetton F. Tavares [3] and Evando S. Araújo [1,*]

[1] Research Group on Electrospinning and Nanotechnology Applications (GPEA-Nano), Department of Materials Science, Federal University of San Francisco Valley, Juazeiro 48902-300, Brazil
[2] Department of Fishing and Aquaculture, Federal Rural University of Pernambuco, Recife 52171-900, Brazil
[3] Research and Extension Center, Laboratory of Fuels and Materials (NPE/LACOM), Department of Chemistry, Federal University of Paraíba, Campus I, João Pessoa 58051-900, Brazil
* Correspondence: evando.araujo@univasf.edu.br; Tel.: +55-74-2102-7645

Abstract: Metal oxide nanocomposites (MON) have gained significant attention in the literature for the possibility of improving the optical and electronic properties of the hybrid material, compared to its pristine constituent oxides. These superior properties have been observed for TiO_2 — based MON, which exhibit improved structural stability and photoactivity in environmental decontamination processes. In addition, the use of polymer membrane-supported MON is preferable to prevent further aggregation of particles, increase the surface area of the semiconductor in contact with the contaminant, and enable material reuse without considerable efficiency loss. In this work, V^{5+}-doped TiO_2/WO_3 MON nanostructures were prepared by the sintering process at 500 °C and supported in electrospun fiber membranes for application as photocatalyst devices. Microstructural characterization of the samples was performed by XRD, SEM, EDS, Raman, and DSC techniques. The reflectance spectra showed that the bandgap of the MON was progressively decreased (3.20 to 2.11 eV) with the V^{5+} ions doping level increase. The fiber-supported MON showed photoactivity for rhodamine B dye degradation using visible light. In addition, the highest photodegradation efficiency was noted for the systems with 5 wt% vanadium oxide dispersed in the fibers (92% dye degradation in 120 min of exposure to the light source), with recyclability of the composite material for use in new photocatalysis cycles. The best results are directly related to the microstructure, lower bandgap and aggregation of metal oxide nanocomposite in the electrospun membrane, compared to the support-free MON.

Keywords: metal oxide nanocomposites; microstructural analysis; electrospinning; polymer substrate; photodegradation

1. Introduction

The increase in industrial activity associated with the accelerated growth of the world population in recent decades is one of the main factors that contribute to the pollution of the physical and biological components on earth. In particular, the inappropriate disposal of by-products from the production of medicines, agrochemicals, heavy metals, and dyes can contaminate aquatic environments, due to its toxic and carcinogenic components resulting from the decomposition of its molecules. The scarcity of effective and low-cost procedures that ensure large-scale discarding and treatment of industrial waste contributes to intensifying this current problem [1,2].

Although the advanced water treatment technologies, such as sedimentation, coagulation, aerobic-activated sludge-based treatment, and nitrification–denitrification have been widely used after conventional primary and secondary treatments, they have some drawbacks, such as high energy consumption, carbon emission, excess sludge discharge, ineffectiveness in removing contaminants, such as organic dyes, and considerable additional cost [3,4]. In this sense, the search for new technologies and functional materials for water treatment that are able to meet current needs, with lower cost and environmental impact, is an open field of research today [2,5–7].

Metal oxides (semiconductors that respond to external stimulus—light, electric potential, pressure, among others) have received increased attention from academic researchers and industrial developers as potential materials for the production of efficient and environmentally friendly devices for the remediation of organic-contaminated water [8–10]. This ability is directly related to their intrinsic properties, such as high chemical and structural stability, high surface area/volume ratio, excellent electrical/electronic response, surface interactions with contaminant molecules, and the possibility of reusing the material in new decontamination cycles [8].

In particular, the use of metal oxide nanocomposites (MON) (resulting from the mixture between two or more selective nanoscale metal oxides [11,12] has been preferable for these applications, due to the possibility of having better decontamination performance when compared to the use of their constituent oxides, in isolation. Currently, among the environmentally friendly techniques for removing contaminants using MON, heterogeneous photocatalysis (HP) stands out [13,14].

Heterogeneous photocatalysis is a photochemical process for the formation of free radicals, such as the hydroxyl HO·, which is a highly oxidizing agent, and degrades organic substances present in water. When the semiconductor in solution is exposed to a light source with sufficient energy to activate it, the absorption of photons with higher energy than the semiconductor bandgap occurs. At the electronic level, this process causes an electron (e^-) to be transferred from the valence band to the conduction band of the material, leaving a hole (h^+) in the valence band. These holes generate HO· free radicals from the water adsorbed on the metal oxide surface, which are used in the organic contaminant oxidation reactions [14].

Titanium dioxide in its allotropic anatase form (anatase TiO_2, ~3.2 eV bandgap) is one of the most widely used oxides in HP using MON, due to its non-toxicity, high chemical stability, excellent photoactivity and photostability, and lower cost than other oxides [15,16]. A limitation for photocatalytic processes using anatase is the demand of ultraviolet radiation as a light stimulus. The light source must have a higher energy than the material bandgap energy, so that electrons from its valence band can be ejected into its conduction band, which would not occur using visible light. Advances in this field of research involve the incorporation of selective metal oxides into TiO_2, such as tungsten, zinc, indium, copper, niobium, and vanadium oxides, in order to increase the process efficiency [1,15,17,18].

Tungsten trioxide (WO_3, 2.6–2.8 eV bandgap) has excellent electrical charge transfer properties, broadens the excitation wavelength range of the final material, in addition to promoting effective photocharge separation of the composite catalyst and inhibiting electron-hole recombination, which significantly improves the MON photocatalytic activity [19,20]. The interaction of WO_3 with TiO_2 increases the conductivity of the TiO_2/WO_3 system in comparison to pure TiO_2 [21,22], while TiO_2 doping with niobium pentoxide (Nb_2O_5) [23] or vanadium pentoxide (V_2O_5) [24] delays the phase transition from anatase to rutile (phase with less active surface) and improves the electronic properties of the resulting material [17]. In addition, the combined action between TiO_2, WO_3, and V_2O_5 results in nanocomposites with superior performance (sensitivity, electric and electronic properties, catalysis and phase stability), which demonstrate the emerging possibilities of using this MON configuration for environmental decontamination applications [21,25,26].

Its particles must be available in a small enough size to provide a large surface area of action, in order to enhance the MON photocatalytic activity. In practice, as the concentration of nanoparticles in solution increases, the tendency of these particles to agglomerate can also increase [27], which is the opposite of what is expected. In these studies, the impossibility of filtration or complete removal of oxide particles from the aqueous environment after the reaction is observed. This represents a limitation for the reuse of the functional material in new photocatalysis cycles (or even to prevent water contamination) [28]. In recent years, these limitations have motivated the development of supported nanoparticles [29,30]. In this configuration, different types of substrates are used, such as porous glasses, lamellar compounds, zeolites, and polymer membranes, in order to disperse and stabilize the metal oxides [29,30]. The advantages of using an inert polymeric matrix as support for photocatalysts are the possibility of having flexible devices, as well as the ease of oxides removal from the aqueous environment, without significant loss of photocatalyst for reuse [29,31].

In this work, V^{5+}-doped TiO_2/WO_3 (1/1 in mol) nanostructures were prepared with different dopant concentrations, dispersed in a fibrous membrane (with high surface area to volume ratio) from the electrospinning process and tested as photocatalyst device. The materials microstructure was characterized by scanning electron microscopy (SEM), energy dispersive spectroscopy (EDS), X-ray diffraction (XRD), Raman spectroscopy, differential scanning calorimetry (DSC), and UV–vis with diffuse reflectance spectroscopy (DRS) methods, and their properties were related to the results obtained in photocatalysis tests, from the degradation of rhodamine B (RhB) model dye in aqueous solution. The results indicated that the interaction between the constituent oxides and their dispersion in the electrospun membrane were factors that significantly improved the efficiency of the degradation process using visible light, in addition to enabling the reuse of these functional materials in new photocatalysis cycles.

2. Results and Discussion

2.1. Microstructural Analysis

The XRD data of the TW, TWV3, and TWV5 MON are shown in Figure 1. The X-Pert HighScore software was used to identify the crystal phases of the studied materials. The crystallographic parameters were obtained from the International Center for Diffraction Data (ICDD). The tetragonal crystal phase of the anatase TiO_2 (a = b = 3.7821 Å, c = 9.5022 Å, and $\alpha = \beta = \gamma = 90°$ lattice constants [32], 01-071-1166 ICDD card) and monoclinic WO_3 (a = 7.306 Å, b = 7.540 Å, c = 7.692 Å, $\alpha = \gamma = 90°$ and $\beta \neq 90°$ [33], 01-083-0951 ICDD card) were detected by XRD measurements. This demonstrates that the oxide mixture sintering (with the selected parameters) has not resulted in a phase change or crystalline solid solution formation in the resulting MON.

In addition, no diffraction peaks of V_2O_5 were detected in the XRD spectra of the TWV3 and TWV5 composites, in comparison to the TW sample [34]. The V_2O_5 nanopowders were used with low weight percentages and are uniformly dispersed in the TiO_2/WO_3 base mixture. This corroborates that the crystallographic planes of V_2O_5 (concentrated in the region of $2\theta = 20–35°$ [35]) were superimposed by the peaks of TiO_2 and WO_3 (with higher intensities).

All sintered oxide mixtures show similar values of WO_3 parameters, with a = 7.30 Å, b = 7.53 Å, c = 7.68 Å, $\alpha = \gamma = 90°$, and $\beta = 90.59°$, with no changes in comparison to lattice constants of the starting material.

The TiO_2 crystal constants in the TW sample (Table 1) also show similar results to the pristine material (a = b = 3.783 Å, c = 9.509 Å, and $\alpha = \beta = \gamma = 90°$ [22]), since the 1:1 in mol $TiO_2:WO_3$ configuration (WO_3 wt% close to that of TiO_2) provides a greater surface interaction between the oxides [34]. On the other hand, the a, b, and c values increased in vanadium-doped samples, in comparison to the TW sample (Table 1). This also demonstrates an increase in the TiO_2 unit cell volume (with no significant difference in this variable from TWV3 to the TWV5 sample). This suggests an atomic interaction degree, directly proportional to the dopant concentration. In fact, the ionic radius of the V^{5+} ions (0.68 Å [36]) is approximate to the T^{4+} ions, which step up the occurrence of vanadium ions doping in the atomic structure of anatase, because of the sintering process.

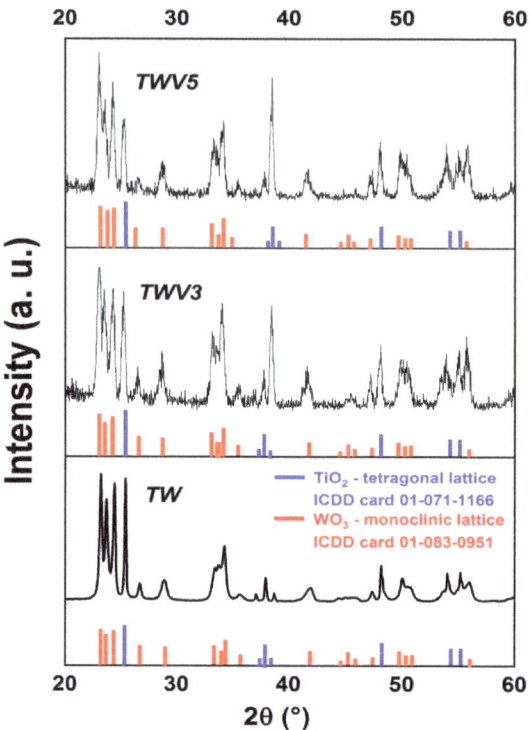

Figure 1. XRD spectra of the TW, TWV3 and TWV5 nanostructures. The colored bars represent the crystallographic planes described in the TiO$_2$ and WO$_3$ ICDD cards.

Table 1. Lattice constants and crystallite size of titanium dioxide in TW, TWV3, and TWV5 MON.

MON	Lattice Parameters (Å)		Average Crystallite Size, D (nm)
	a = b	c	
TW	3.783	9.509	20.23 (±2.09)
TWV3	3.803	9.704	25.31 (±2.64)
TWV5	3.804	9.708	25.43 (±2.67)

The crystallite average sizes (D) of the WO$_3$ and TiO$_2$ in the sintered MON were calculated by the modified Scherrer equation (MSE). By the equation, $D = \lambda/(3\beta \cos(\theta))$ [37], where λ is the wavelength of X-ray radiation, β is the width at half height of the peaks detected in the XRD spectrum, θ is the half of the diffraction angle in these crystallographic planes, and 1/3 is the correction factor for the limit of application of the Sherrer law [38].

The WO$_3$ presented D values of 15.67 (±1.22) nm, 15.84 (±1.04) nm, and 15.91 (±1.62) nm for the TW, TWV3, and TWV5 samples, respectively. In addition, in these systems, the crystallite size values for TiO$_2$ (Table 1) were 20.23 (±2.09) nm, 25.31 (±2.64) nm, and 25.43 (±2.67) nm, in increasing order of V$_2$O$_5$ concentration.

The results for the hypothesis test of means equality (n = 3) demonstrated that the average size of the WO$_3$ crystallite does not indicate significant changes in the three MON configurations. On the other hand, in MON doped with 3 wt% and 5 wt% of vanadium oxide, a significant increase in TiO$_2$ crystallite size is confirmed when compared to the TW sample. There is no statistical difference between the D values calculated for these last two systems.

The Raman spectra of MON are presented in Figure 2. The spectrum of pristine anatase TiO$_2$ was also analyzed in order to compare with composite samples. In Figure 2a is shown that the molecular vibration bands of anatase (141, 196, 397, 514, and 638 cm^{-1} wavenumbers [39]) occur in all spectra of the metal oxide nanocomposites, although shifted to lower wavenumbers (redshift), by 7–11 cm^{-1}. In addition, the monoclinic WO$_3$ bands were determined, typically at 268, 319, 710, and 802 cm^{-1}, with no wavenumber variation in comparison to the pure oxide [34,40].

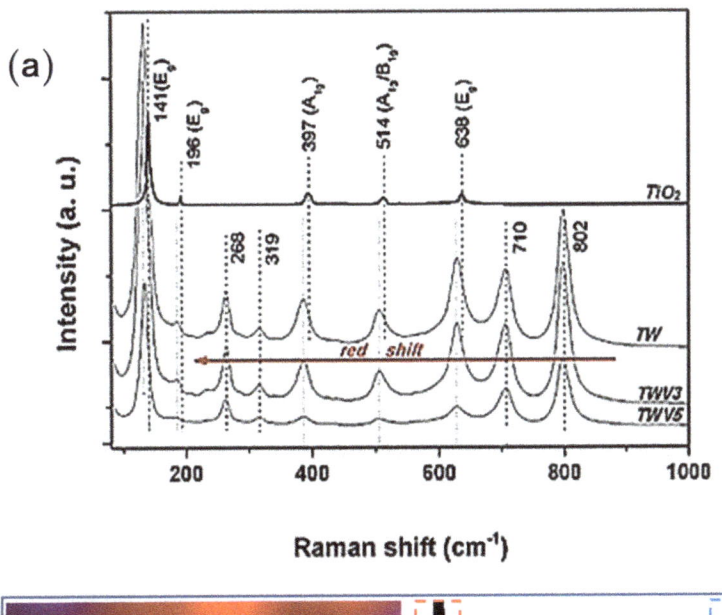

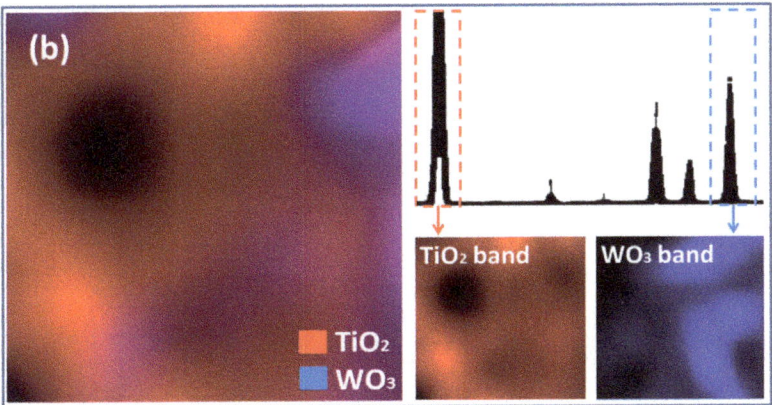

Figure 2. (a) Raman spectra of the metal oxide nanocomposites compared to pristine anatase TiO$_2$; (b) representative Raman mapping of MON samples, with indication of TiO$_2$ and WO$_3$ vibrational bands.

The redshift effect occurs due to the process of surface relaxation, distortions, and additional vibration energy in the crystal lattice (phonon confinement) of the material [41]. In particular, the occurrence of this phenomenon in the MON Raman spectra indicates that W^{6+} and V^{5+} ions occupied vacancies of available Ti atoms in the crystal lattice of anatase (as previously expected). The occupation of these sites by W and V species allows a greater degree of local vibrations and facilitates the molecules collective excitation, which favors photocatalysis processes [41,42]. The representative Raman mapping (Figure 2b) of V^{5+}-doped samples is an important tool that confirms the regular distribution of TiO_2 and WO_3 bands in the materials surface, with no additional active vibration modes in these MON.

The combined results of SEM images, XRD (with anatase unit cell volume invariance in V^{5+}-doped MON), and Raman analysis (not detected V_2O_5 bands, with redshift of TiO_2 bands) indicate that at 5 wt% dopant concentration, a greater amount of the vanadium species is arranged in the sites initially unoccupied by titanium atoms, showing the effectiveness of the performed doping process.

The study of variations in crystallite sizes in the nanostructured systems was important to determine the doping level in the atomic structure of TiO_2 and their effects for application of these materials as photocatalysts. In fact, when the MON crystallites are in nanometer scale, the movement of electrons and holes in an excited semiconductor also depends on the quantum confinement effect. In other words, the phonons resulting from the doping process also influence material electronic properties [43].

Surface micrographs of TWV3 and TWV5 MON (resulting from sintering at 500 °C) were obtained by SEM (Figure 3a,b). Initial analysis returns mostly spherical particles (100–230 nm range), with significant overall contact surface, with no significant differences when compared to the dopant-free TW matrix [22]. The EDS mapping of V^{5+}-doped samples (Figure 3c) confirms the Ti and W metals matrix and vanadium dopant evenly distributed over the entire material structure, which favors the application of these nanostructures as photocatalysts.

Figure 3. SEM micrographs of (**a**) TWV3 and (**b**) sintered TWV5 MON; (**c**) representative EDS mapping of V^{5+}-doped samples, with indication of Ti, W, and V elements.

The calculated mean sample diameters were 145.7 ± 47.0 nm (n = 135) [20], 160.1 ± 74.1 nm (n = 123), and 169.4 ± 63.1 nm (n = 141), for TW, TWV3, and TWV5 MON, respectively. The equality of means tests results (n = 3) demonstrate that there is no significant difference between the average particle size calculated for these produced MON (p-value > α for all comparison of means).

The bandgap energy of the MON semiconductors was calculated from the DRS data (Figure 4a), using the Tauc plot (Figure 4b). By the Tauc relation, the curve of the product of the absorption coefficient (α) and photon energy ($E = h\nu$) to the power of $1/r$, $(\alpha h\nu)^{1/r}$, is plotted as a function of the photon energy (E). The intersection of the auxiliary line in the linearity region of the graph with the x axis represents the measurement of the material optical energy gap (E_g) [44]. Thus, the type of electronic transition (direct or indirect electron mobility from the valence to the conduction band) in a semiconductor is defined by the r value (equal to 1/2 or 2, for direct or indirect bandgap materials, respectively).

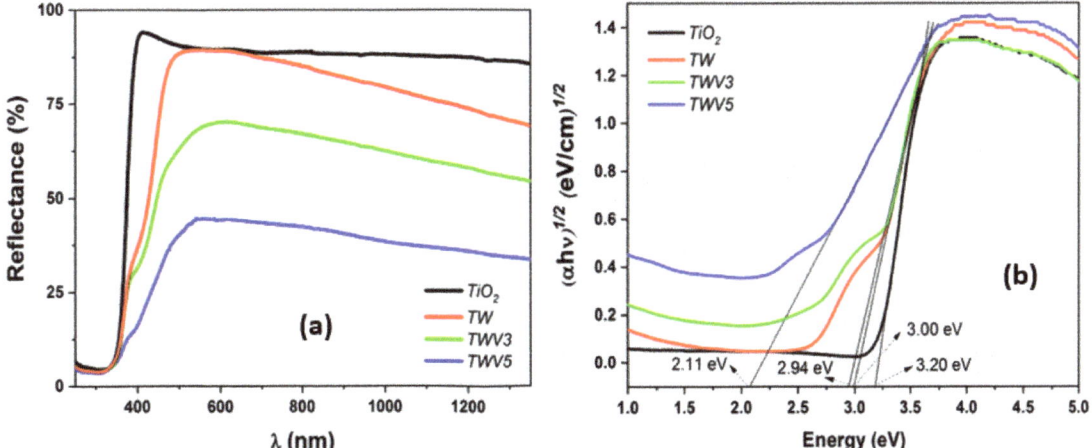

Figure 4. (a) DRS data and (b) Tauc plot for bandgap measurements of the nanostructured metal oxides.

The diffuse reflectance data of all tested nanostructures were modeled by the Tauc relation with r = 2, which indicates that the semiconductor excitation occurs by indirect transition. In other words, for our nanostructured systems, the electrons can be excited from the highest energy state in the valence band to the lowest level in the conduction band, with different momentum values. In this configuration, the electron mobility is facilitated by phonon assistance, which was detected and discussed for the MON samples in the Raman characterization.

The E_g experimental values of the hybrid semiconductors were progressively decreased with the V^{5+} doping level increase (gap energies estimated at 3.00, 2.94 and 2.11 eV for TW, TWV3, and TWV5 samples, respectively), compared to the pristine TiO$_2$ (a typical indirect bandgap semiconductor [44], with E_g= 3.2 eV).

As verified, the progressive introduction of vanadium oxide to the TiO$_2$/WO$_3$ mixture (1/1 in mol) significantly decreased the TiO$_2$ bandgap (3.20 eV to 2.11 eV). It is known that the presence of dopant ions in the system generates an extrinsic-type semiconductor (in which the bandgap can be controlled by purposefully adding impurities to the materials atomic structure).

In the case of our nanostructures, the TiO$_2$ (with tetravalent metal) is doped with V^{5+} ions (pentavalent elements), which will result in an n-type extrinsic semiconductor. In this case, free electrons are the main charge carriers, since each V^{5+} ion provides an additional electron to the system and thus the total number of electrons becomes greater than the number of holes. In this configuration, the added electrons tend to occupy additional energy levels in the bandgap region, very close to the bottom of conduction band (CB), called donor energy levels. Thus, it can be proposed from our results that the progressive increase in the concentration of vanadium in the nanostructures promoted a significant increase in the doping degree, with additional donor levels, which explains the decrease in the TiO$_2$ bandgap to 2.11 eV in the TWV5 nanocomposite.

In this sense, for the TWV5 system (with lower indirect bandgap), the electronic transition occurs with a lower photon energy linear, which makes it difficult for electron/hole recombination to occur, and consequently favors photocatalysis processes with visible light.

The SEM images and EDS analysis of the MON-loaded fibers are shown in Figure 5. It is noted that there is a predominance of regular diameter fibers, without apparent surface defects, which proves that the parameters of the electrospinning process were properly chosen for the production of TWV3- and TWV5-loaded electrospun membranes (Figure 5a,b).

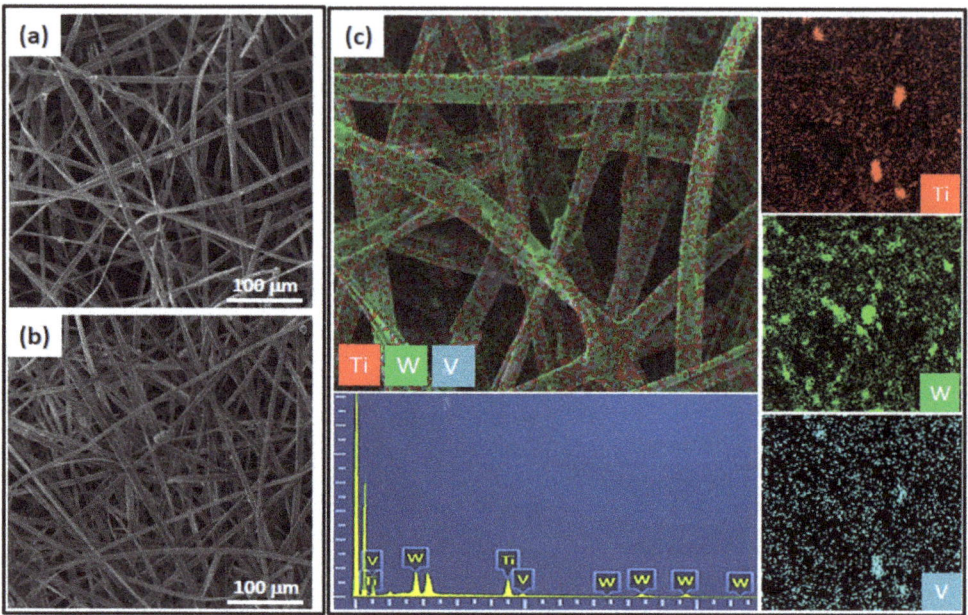

Figure 5. SEM micrographs of (**a**) TWV3 and (**b**) TWV5 MON-loaded electrospun fibers; (**c**) EDS mapping and elemental composition of V^{5+}-doped samples, with indication of Ti, W, and V elements.

The mean diameters of composite fibers were estimated as 5.4 ± 1.7 μm (n = 59) [22], 5.8 ± 2.3 μm (n = 73), and 6.0 ± 1.9 μm (n = 69), with a progressive increase in V_2O_5 concentration. The population mean difference tests revealed that the MMO-loaded fibers have statistically similar mean diameters (n = 3 samples, p-value $> \alpha$).

The EDS mapping and elemental composition of these composite fibers (Figure 5c) confirmed that the Ti, W, and V elements are distributed throughout the fibrous polymer membrane. In other words, the electrospun fiber matrix promotes the dispersion of oxides, reducing the particle aggregation.

Thus, photocatalytic devices can be produced with better performance, due to the greater surface contact area of the MON, with the contaminated aqueous solution. In addition, the immobilization of MON in fibrous template allows the composite material reuse (without considerable material loss) in other photocatalysis cycles. Thus, electrospinning proves to be an effective and low-cost technique for producing potential composites for application in the decontamination of aqueous environments.

In summary, the similarity in morphology (particle size and fiber diameter) with increasing dopant concentration indicates that the better performance of these structures as a photocatalyst should predominantly refer to the interaction degree between the constituent oxides.

The DSC curves of the pristine fibrous membrane and the MON-loaded polymer fibers are given in Figure 6. In MON samples, two broad endotherm peaks are observed in the analyzed temperature range. The first one (centered at 75–90 °C) is assigned to the loss of water and the solvent remaining evaporation from the polymer electrospinning process [45,46].

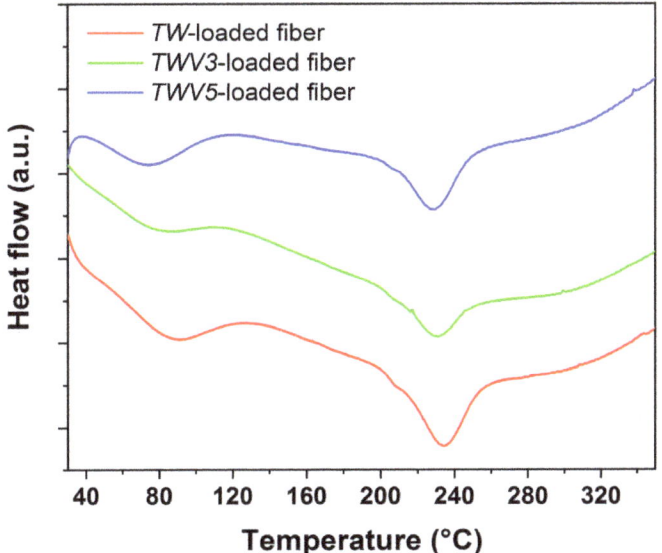

Figure 6. DSC curves of MON-supported fibers.

The second peak, centered at 231–234 °C, is related to the melting point of the polymer crystalline phase [46]. This peak is also described in the literature as the relaxation temperature of the polymer chains that follows the glass transition [47].

In addition, the transition temperatures for the constituent oxides are detected at higher temperatures than those observed in the experiment. In resume, the introduction of semiconductors preserves the polymer melting point, without significant changes in the homogeneity and crystallinity degree of the composites.

2.2. Photocatalysis Tests

During the experiment, aliquots of the RhB solution were removed from the reactor at defined time intervals so that variations in the intensity of the dye absorption peak (at $\lambda = 555$ nm) (Figure 7a–c) could be investigated. The photodegradation kinetic curves of Rhodamine B (RhB) dye using MON-loaded fibers, under visible light excitation, are demonstrated in Figure 7d. In the graph, the y-axis represents the change in the relative concentration (C/C_0) of the dye, where C_0 is the initial concentration and C is the remaining concentration of the contaminant in the reactor (as a function of time). All samples were preserved in a dark environment in contact with the dye in solution for 1 h before exciting the oxides with continuous radiation.

The use of the fibrous membrane (without additives) in solution returned negligible variation in the initial concentration of RhB (1 µM) throughout the analyzed time interval. This proves that the polymer matrix acts as an inert support to prevent the particle aggregation, and provides a greater oxides surface area to interact with the contaminant.

The pristine TiO_2 nanopowder was used in the experimentation and has not demonstrated considerable photocatalytic action in visible light, since the energy of the incident light photons was not sufficient to overcome the anatase bandgap energy (3.2 eV) and promote electron/hole mobility, which is necessary for the dye degradation process to occur.

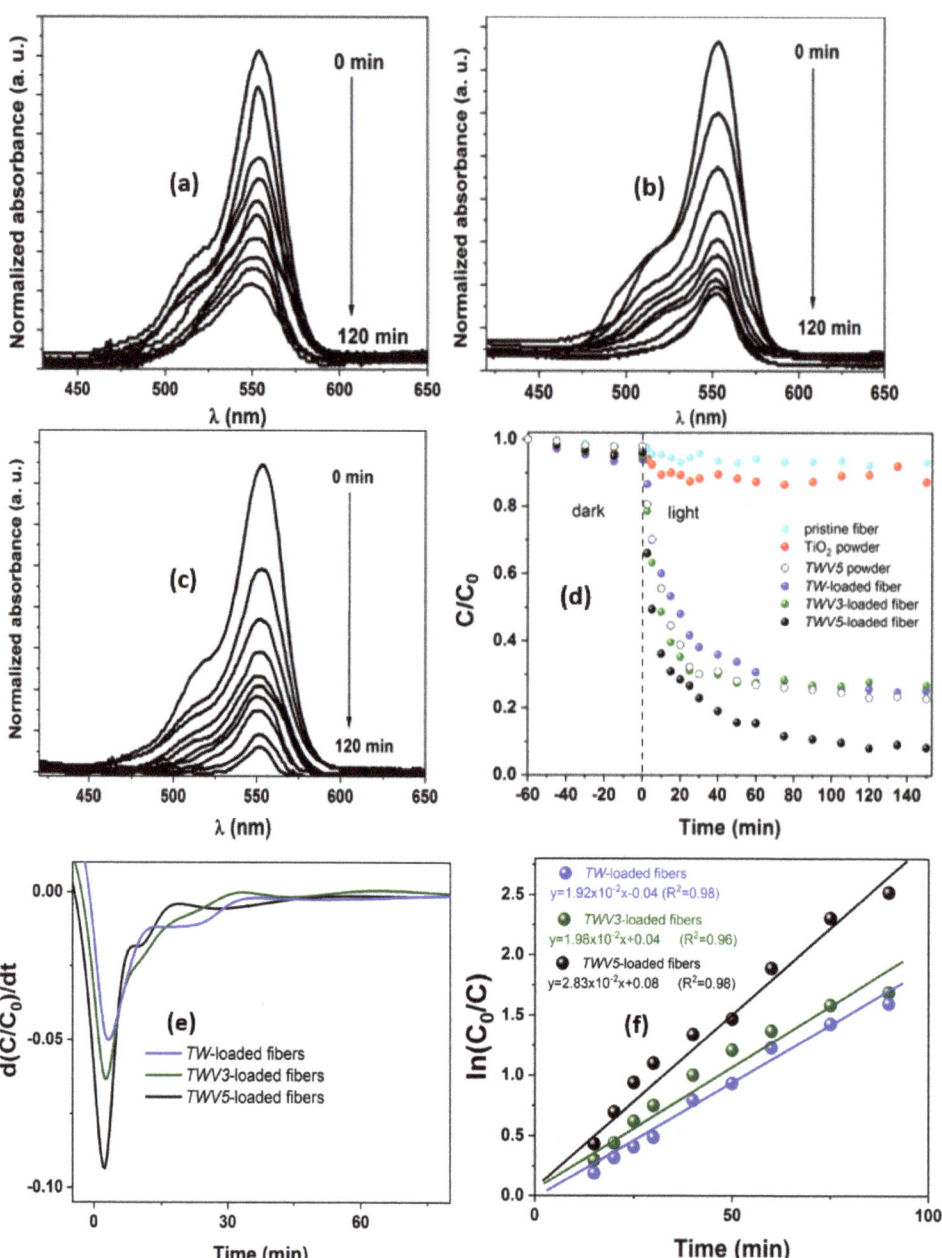

Figure 7. Change in absorbance spectra of RhB dye, using (**a**) TW-, (**b**) TWV3- and (**c**) TWV5-loaded fibers nanocomposites. (**d**) Dye photodegradation dynamics using MON-loaded fibers, under visible light excitation. The TiO_2 and TWV5 powders (with higher and lower bandgap, respectively) were also tested for comparison purposes. (**e**) Derivative of the relative concentration of RhB with time. (**f**) Application of the pseudo-first-order kinetic model to the experimental data.

Differently, the MON supported in the fibers showed photocatalytic activity under the action of visible light (using a lower energy hv). The dye degradation percentage achieved after 120 min of visible light irradiation were 73%, 75%, and 92%, using TW- to TWV5-containing fibers, respectively.

The derivative of the relative concentration of RhB with time $\left(\frac{d(C/C_0)}{dt}\right)$ (Figure 7e) showed that the greatest excursion in the rate of dye degradation, in the time interval of greatest decay of C, occurs with the use of fibers loaded with TWV5 nanostructures (followed by TWV3- and TW-loaded fibers samples, respectively), in a relationship directly proportional to the pollutant photodegradation kinetics.

For the studied metal oxide nanostructures, the RhB photodegradation kinetics was given in terms of the reaction rate coefficient (k), obtained from the application of the pseudo first order (PFO) model [48] to the experimental data. According to the model, k is equivalent to the value of the slope of the linear fit in the graph ln (C_0/C) versus t (Figure 7f). In fact, the dye degradation process is well-fitted by PFO kinetics, since the correlation constants for the three MON were close to 1 (best fit results).

The values obtained for the reaction rate coefficient were 1.92×10^{-2}, 1.98×10^{-2}, and 2.83×10^{-2} min^{-1}, using the samples TW-, TWV3-, and TWV5-loaded fibers. In other words, it was proven that the highest reaction rate was obtained with the sample with the highest concentration of vanadium.

Especially, the increase in the doping level provided a decrease in the TiO$_2$ bandgap (as discussed previously). This allowed the occurrence of free electrons in the conduction band and holes in the valence band, essential to the dye degradation reactions in solution.

The TWV5 powder was also tested and returned a percentage of degradation similar to the noted for TW and TWV3 nanostructures dispersed in the fibers. This demonstrates that the performance difference between the TWV5 powder and fiber-supported TWV5 forms is associated to the greater surface area of these MON induced by the high level of nanoparticles dispersion in the electrospun fibers. More specifically, at the atomic level, vanadium ion substituents induce stress on the TiO$_2$ (host metal oxide) crystal structure. This configuration results in the formation of a large amount of defects in its crystal lattice, providing superior chemical reactivity for photocatalysis processes.

Based on the values of optical bandgap energies and photodegradation results, we present a possible photocatalytic mechanism for the RhB degradation in water, using the TWV5-loaded fibers nanocomposites (illustrated in Figure 8). The authors suggest that the larger surface area of the V-doped TiO$_2$/WO$_3$ (1/1 in mol) nanostructures (that is obtained with their dispersion in the fibers) induces more reactive sites for the photocatalytic reactions that occur on the surface of these semiconductors, promotes a lower probability of e−/h+ recombination, and enhances the charge transport [49]. In the illustrative scheme (Figure 8), the (V vs. NHE)-axis means the band potential versus normal hydrogen electrode scale [49]. The positions of the potentials of the TiO$_2$ and WO$_3$ band edges at pH 7.0 were given according to the literature [50]. Note that there is a difference in the edge positions of both oxides. This creates a potential gradient at the nanocomposite interface, which would facilitate the charge separation and inhibit the e−/h+ recombination [12,49,51].

The RhB photodegradation using TiO$_2$/WO$_3$-based nanocomposites typically involves the action of the hydroxyl (•OH) and superoxide (•O$_2^-$) radicals [11,12,49,52]. The continuous incidence of visible light in the reactor environment generates available electrons and holes (hv → e$^-$ + h$^+$) in both WO$_3$ and V-doped TiO$_2$ oxides for the photocatalysis reactions.

Part of the photogenerated electrons also tend to be transferred from WO$_3$ to V-doped TiO$_2$, due to the potential difference [53,54] (which is influenced by V donor levels below the TiO$_2$ CB, as previously discussed). Similarly, as the potential of the V-doped TiO$_2$ VB edge is more positive than that of WO$_3$, it is suggested that the holes generated in the BV of the first oxide can also move into the BV of this last one. In summary, a hole can migrate to the semiconductor nanocomposite interface either directly or after their transfer from one oxide to another. These charge transfers in opposite directions can decrease recombination rates, in addition to promoting sufficient electron/hole separation for the photocatalysis process to occur more efficiently [49,53]. Thus, the electrons and holes that reach the semiconductor-solution interface can react with the redox species O_2 and H_2O and form the superoxide ($\bullet O_2^-$, $O_2 + e^- \rightarrow \bullet O_2^-$) and hydroxyl ($\bullet OH$, $H_2O + h^+ \rightarrow \bullet OH$) radicals. The analysis of these processes indicates that these radicals and the photogenerated holes are the main agents of oxidation of the organic pollutant RhB into carbon dioxide and water. Thus, the best catalytic activity, obtained with TWV5-loaded fibers, stems from the combination of improved photoresponse and charge separation efficiency, and the presence of a higher active sites density of the hybrid nanostructures [55].

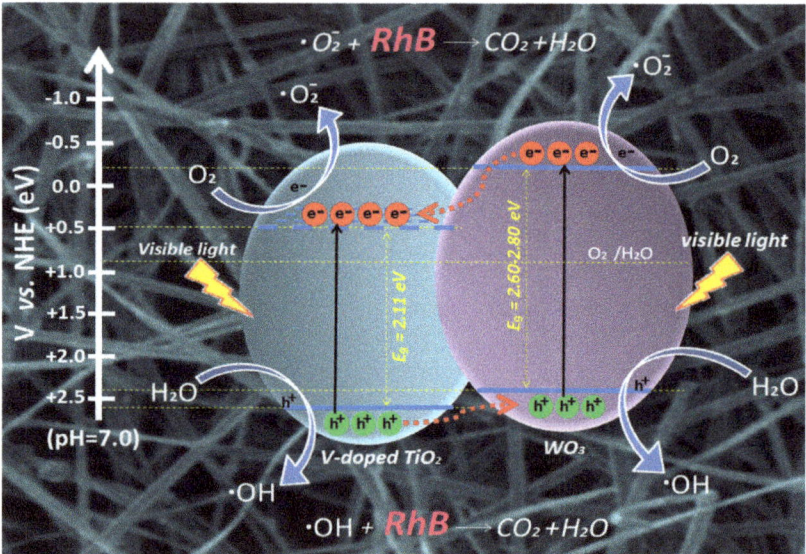

Figure 8. Possible photocatalytic mechanism for the RhB degradation in water, using the TWV5-loaded fibers nanocomposites.

Based on these aspects, the TWV5-loaded fibers were reused in two new photodegradation cycles (Figure 9). The second cycle (Figure 9b) returned a percentage of RB dye degradation of 90% after 120 min of radiation incidence. In the third cycle (Figure 9c), the percentage of 85% was obtained, and with a slight decrease in the reaction rate, with degradation saturation after 140 min of exposure of the material to white light.

The XRD spectrum of TWV5-loaded fibers (Figure 9d), after the third cycle of photodegradation, showed that all the characteristic peaks of TiO_2 and WO_3 (previously described in Figure 1) were preserved after the electrospinning processes and the photocatalysis cycles, with no phase change in the constituent oxides. It is important to highlight that the recyclability of the material is feasible due to the hydrophobic nature of the polymeric membrane and the maintenance of the MON photoactivity, without significant loss of the photocatalyst for reuse.

The photocatalysis process was discontinued with three cycles, since the polymer membrane showed imminence of mechanical wear (behavior expected by the acrylic nature of the polymer) caused by the handling and rearrangement of the material in the reactor during the preparation for its reuse.

In summary, the increasing level of vanadium doping in the crystalline structure of TiO_2, the smaller bandgap value given to the TWV5 nanostructures, associated with their greater surface area when dispersed in electrospun fibers, prove the best results of dye degradation obtained with visible light.

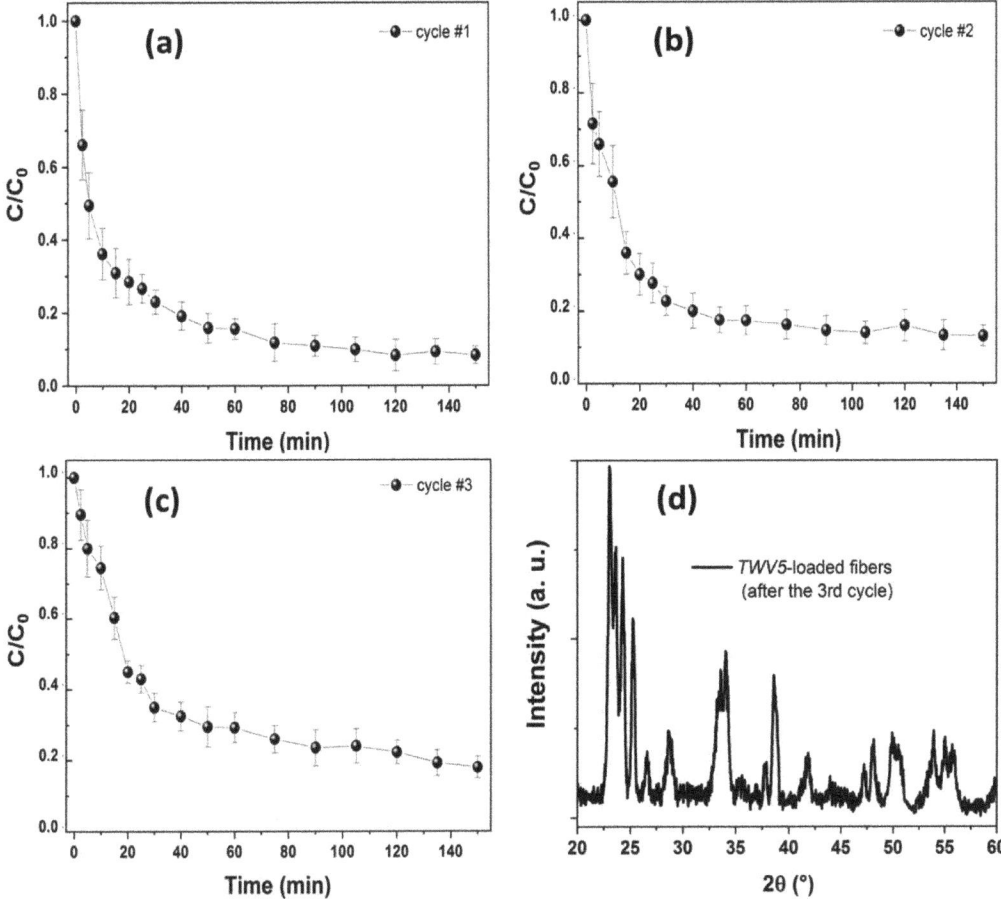

Figure 9. (**a–c**) Dye photodegradation cycles using TWV5-loaded fibers, under visible light excitation; (**d**) XRD spectrum of TWV5-loaded fibers, after the third cycle of photodegradation.

3. Materials and Methods

3.1. Materials

The chemicals titanium dioxide (anatase TiO_2, 99.7% purity, <25 nm particle size, Sigma Aldrich, Burlington, VT, USA), tungsten trioxide (WO_3, 99.9% purity, <100 nm particle size, Sigma Aldrich), vanadium pentoxide (V_2O_5, 99.9% purity, 80 nm particle size, Nanoshell, Layton, UT, USA), acrylic copolymer Eudragit® L100 (Evonik Industries, Essen, NW, Germany), ethanol PA (99.8%, Neon Comercial, Suzano, Brazil), and rhodamine B (Vetec, São Paulo, SP, Brazil) were used as received. All these reagents are analytically pure.

3.2. Sample Preparation

The material samples for the photocatalysis tests were prepared in two steps: (i) sintering of metal oxide nanocomposites (MON) and (ii) dispersion of MON in electrospun polymeric membrane. The tested MON configurations were defined as: TiO_2/WO_3 (1/1 in mol, TW base sample); and TWV3 and TWV5 samples (TW mixture with 3 and 5 wt% of V_2O_5, respectively).

For the sintering process, each metal oxide mixture was sintered in a muffle oven at 500 °C for 2 h [22]. For the electrospinning process, 0.25 g of each sintered MON was mechanically dispersed in 4 mL of alcoholic solution (with a polymer concentration of 0.27 g/mL) and inserted into the reservoir of a conventional syringe (volume of 10 mL and 0.7 mm of diameter of the metal needle). Thus, simultaneously, a pressure of 100 μLmin^{-1} on the syringe plunger and an electric potential difference (ddp) of 15 kV (applied to the tip of the metal capillary) promote the composite material to be electrospun towards a grounded plane metal collector (area of 0.1 × 0.1 m^2, separated by 10 cm from the needle tip), with respective evaporation of the solvent. An illustration of the electrospinning process is shown in Figure 10a. The MON-contained fibrous membrane is then collected for the photocatalysis tests.

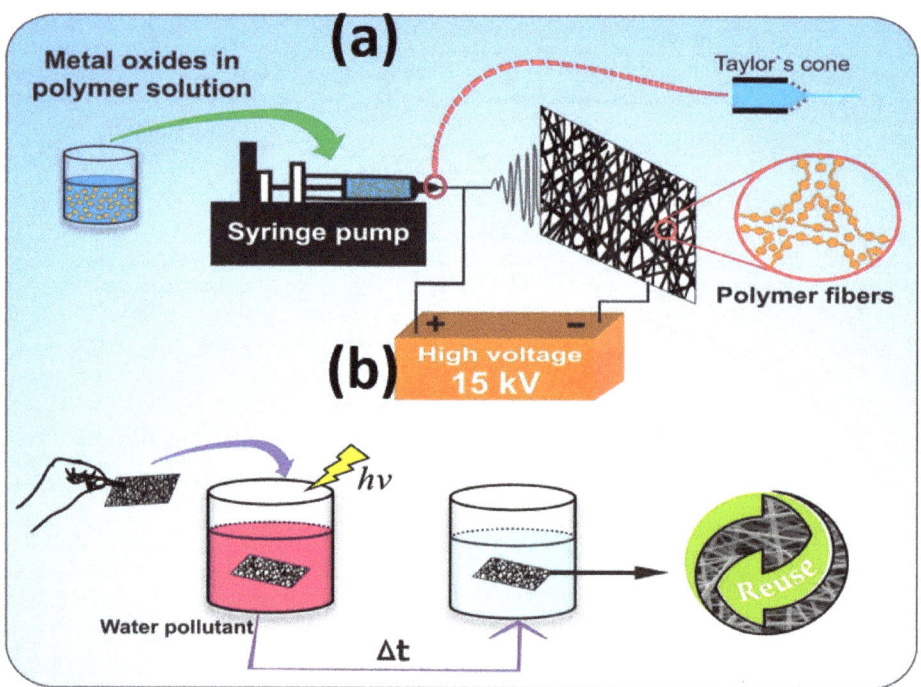

Figure 10. Illustrative scheme of the (**a**) MON dispersed in electrospun fibers and their use in (**b**) photodegradation of the dye in aqueous solution.

3.3. Microstructural Characterization

The sample microstructure was investigated by: scanning electron microscopy (SEM) and energy-dispersive X-ray spectroscopy (EDS) with 10 to 20 kV accelerating voltage (Vega 3XM Tescan equipment, Tescan, Brno, Czech Republic); X-ray diffraction (XRD) in the 20–60° range, scan rate of 0.02 °s^{-1} and Cu Kα radiation (Miniflex Rigaku equipment, Rigaku Corporation, Tokyo, Japan); Raman spectroscopy, excitation with $\lambda = 532$ nm laser, 20–25 mW (HORIBA Scientific, Osaka, Japan); differential scanning calorimetry (DSC), 30–300 °C range, nitrogen flow of 50 mL min^{-1}, heating temperature rate of 10 °C min^{-1} (Schimadzu DSC-60 equipment, Shimadzu Corporation, Kyoto, Japan); and UV–vis with diffuse reflectance spectroscopy (DRS), 200–1400 nm wavelength range (Schimadzu UV-2600i spectrophotometer, Shimadzu Corporation, Kyoto, Japan).

3.4. Photocatalysis Tests

The temporal dynamics of the photocatalysis process was investigated from the variations in the UV-vis absorption of Rhodamine B dye (peak centered at 555 nm) solution (neutral pH) (on a Hach DR 5000 UV-vis spectrophotometer, Hach Company, Ames, IA, USA), under the action of the MON-loaded fibers (19 mg) (scheme in Figure 10b). The photocatalysis tests were performed in a closed cubic box (volume of 40 cm^3), equipped with lateral heat dissipating fans, reactor (capacity of 150 mL, placed on a magnetic stirrer and connected to a sample collection system), and light source (Figure 11a). The metal oxide nanostructures were activated by continuous white light (60 W power, 650 lm/m^2 intensity, spectrum in Figure 11b). The solution in the reactor (surface top) was maintained at 16.5 cm from the light source end. The reuse of composites in new photocatalysis cycles was possible by washing the material in aqueous solution (ultrapure water, neutral pH) under constant agitation for 3 h followed by drying in an oven at 40 °C for 12 h. All measurements were made in triplicate.

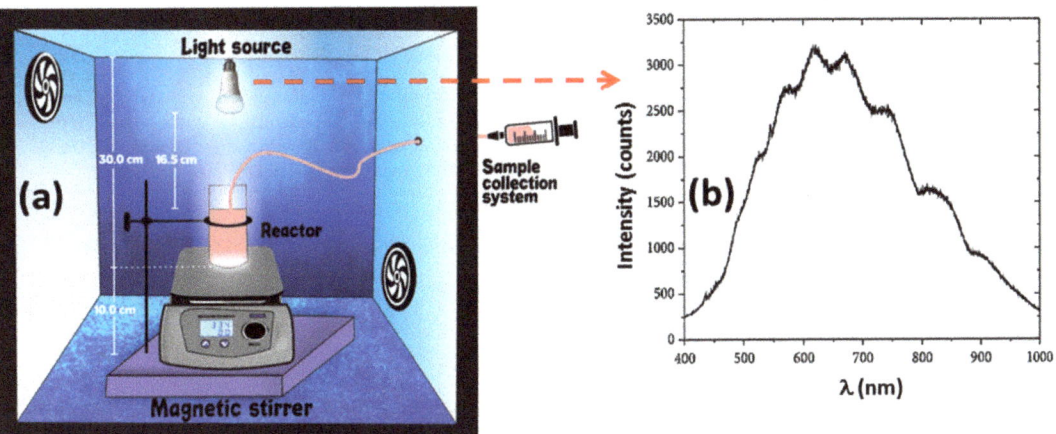

Figure 11. (**a**) Experimental apparatus for photocatalysis tests; (**b**) emission spectrum of white light applied during the photocatalysis experiments.

3.5. Statistical Analysis

The particles size, fibers diameter, and temporal dynamics of the dye photodegradation were given as an average of three independent measurements for each tested MON configuration. The normality and equality of means tests (95% confidence level, comparing the significance level (α) to the p-value returned by statistical tests) were performed as population statistics. The sample sizes (n) were obtained from SEM images using ImageJ software (National Institutes of Health, Bethesda, MD, USA).

4. Conclusions

The main aim of the current research was to investigate the preparation of vanadium-doped TiO_2/WO_3 nanostructures supported on electrospun membranes, their microstructural characterization, and their potential application as photocatalysts. The planned MON-loaded fibers were successfully produced from the sintering and electrospinning processes. All composites showed photocatalytic activity in visible light, with better performance for the sample with 5 wt% of V_2O_5.

The saturation in the unit cell volume and crystallite size of TiO_2, from the TWV3 to TWV5 sample, and the doping of TiO_2 (observed in the microstructural characterization) confirm that the best results of dye photodegradation, anotated for the TWV5-loaded fibers, are directly influenced by the higher doping level of V^{5+} ions in this system.

In addition, the use of polymer matrix as a support for metal oxide nanostructures allowed the production of flexible devices, provided better photocatalytic performance of oxides and ease of removal of oxides from the aqueous environment, without significant loss of photocatalyst for reuse.

Based on the excellent results of photocatalysis using TWV5 nanostructures, the continuity of research involves the execution of other important experiments to quantify parameters of the photodegradation mechanism of RhB (and other dyes) using these metal oxide nanocomposites, such as active species trapping and transient photocurrent test (evaluating the separation efficiency of charge carriers over the photocatalyst surface). Other perspectives involve the use of selective oxides as dopants (for example, Nb_2O_5, MoO_3 and In_2O_3); the improvement of the mechanical strength properties of the polymeric membrane that supports the MON nanostructures; development of the mechanical support apparatus for tests with the photocatalyst in an environment with larger proportions; and use of these nanostructured devices in other chemical decontamination processes.

Author Contributions: Sample preparation and characterization, M.F.G.P., M.M.N., P.H.N.C., C.Y.B.O., G.F.T. and E.S.A. Writing, editing, and review, M.F.G.P., C.Y.B.O. and E.S.A. All authors have read and agreed to the published version of the manuscript.

Funding: This research was funded by Conselho Nacional de Desenvolvimento Científico e Tecnológico (CNPq, Project 202451/2015-1), Fundação de Amparo à Pesquisa do Estado da Bahia (FAPESB, Project 1252/2018) and Fundação de Amparo a Ciência e Tecnologia do Estado de Pernambuco (FACEPE, project IBPG-0849-3.03/2020) brazilian research agencies.

Institutional Review Board Statement: Not applicable.

Informed Consent Statement: Not applicable.

Data Availability Statement: Not applicable.

Conflicts of Interest: The authors declare no conflict of interest.

References

1. Handojo, L.; Ikhsan, N.A.; Mukti, R.R.; Indarto, A. Nanotechnology for remediations of agrochemicals. In *Agrochemicals Detection, Treatment and Remediation*; Narasimha, M., Prasad, V., Eds.; Butterworth-Heinemann: London, UK, 2020; pp. 535–567.
2. Li, X.; Yu, J.; Jiang, C. Principle and surface science of photocatalysis. *Interface Sci. Technol.* **2020**, *31*, 1–38.
3. Kwaadsteniet, M.; Dobrowsky, P.; Deventer, A.V.; Khan, W.; Cloete, T. Domestic rainwater harvesting: Microbial and chemical water quality and point-of-use treatment systems. *Water Air Soil Pollut.* **2013**, *224*, 1629–1637. [CrossRef]
4. Presura, E.; Robescu, L. Energy use and carbon footprint for potable water and wastewater treatment. *Proc. Int. Conf. Bus. Excell.* **2017**, *11*, 191–198. [CrossRef]
5. Chen, J.; Luo, J.; Luo, Q.; Pang, Z. *Wastewater Treatment*; De Gruyter: Berlin, Germany; Boston, MA, USA, 2018.
6. Moga, I.C.; Ardelean, I.; Donțu, O.G.; Moisescu, C.; Băran, N.; Petrescu, G.; Voicea, I. Materials and Technologies Used in Wastewater Treatment. *IOP Conf. Ser. Mater. Sci. Eng.* **2018**, *374*, 012079. [CrossRef]
7. Carolin, C.F.; Kumar, P.S.; Saravanan, A.; Joshiba, G.J.; Naushad, M. Efficient techniques for the removal of toxic heavy metals from aquatic environment: A review. *J. Environ. Chem. Eng.* **2017**, *5*, 2782–2799. [CrossRef]
8. Yaqoob, A.A.; Parveen, T.; Umar, K.; Ibrahim, M.N.M. Role of Nanomaterials in the Treatment of Wastewater: A Review. *Water Sci. Eng.* **2020**, *12*, 495. [CrossRef]
9. Ganachari, S.V.; Hublikar, L.; Yaradoddi, J.S.; Math, S.S. Metal oxide nanomaterials for environmental applications. In *Handbook of Ecomaterials*; Martínez, L., Kharissova, O., Kharisov, B., Eds.; Springer: Singapore, 2019; pp. 2357–2368.
10. Yang, Y.; Niu, S.; Han, D.; Liu, T.; Wang, G.; Li, Y. Progress in Developing Metal Oxide Nanomaterials for Photoelectrochemical Water Splitting. *Adv. Energy Mater.* **2017**, *7*, 170–555. [CrossRef]
11. Choi, T.; Kim, J.S.; Kim, J.H. Transparent nitrogen doped TiO_2/WO_3 composite films for self-cleaning glass applications with improved photodegradation activity. *Adv. Powder Technol.* **2016**, *27*, 347–353. [CrossRef]
12. Baia, L.; Orbán, E.; Fodor, S.; Hampel, B.; Kedves, E.Z.; Székely, I.; Pap, Z. Preparation of TiO_2/WO_3 composite photocatalysts by the adjustment of the semiconductors' surface charge. *Mater. Sci. Semicond. Processing* **2016**, *42*, 66–71. [CrossRef]
13. Ibhadon, A.O.; Fitzpatrick, P. Heterogeneous Photocatalysis: Recent Advances and Applications. *Catalysts* **2013**, *3*, 189–218. [CrossRef]
14. Wang, H.; Li, X.; Zhao, X.; Li, C.; Song, X.; Zhang, P.; Huo, P. A review on heterogeneous photocatalysis for environmental remediation: From semiconductors to modification strategies. *Chin. J. Catal.* **2022**, *43*, 178–214. [CrossRef]

15. Upadhyay, G.K.; Rajput, J.K.; Pathak, T.K.; Kumar, V.; Purohit, L.P. Synthesis of ZnO: TiO$_2$ nanocomposites for photocatalyst application in visible light. *Vacuum* **2019**, *160*, 154–163. [CrossRef]
16. Caswell, T.; Dlamini, M.W.; Miedziak, P.J.; Pattisson, S.; Davies, P.R.; Taylor, S.H.; Hutchings, G.J. Enhancement in the rate of nitrate degradation on Au-and Ag-decorated TiO$_2$ photocatalysts. *Catal. Sci. Technol.* **2020**, *10*, 2082–2091. [CrossRef]
17. Xu, W.; Shu, G.; Zhang, S.; Song, L.; Ma, K.; Yue, H. In-Situ Fabricating V$_2$O$_5$/TiO$_2$-Carbon Heterojunction from Ti3C2 MXene as Highly Active Visible-Light Photocatalyst. *J. Nanomater.* **2022**, *12*, 1776. [CrossRef] [PubMed]
18. Julkapli, N.M.; Bagheri, S.; Yousefi, A.T. TiO$_2$ Hybrid Photocatalytic Systems: Impact of Adsorption and Photocatalytic Performance. *Rev. Inorg. Chem.* **2015**, *35*, 151–178.
19. Kadam, A.V. Propylene glycol-assisted seed layer-free hydrothermal synthesis of nanostructured WO$_3$ thin films for electrochromic applications. *J. Appl. Electrochem.* **2017**, *47*, 335–342. [CrossRef]
20. Qi, J.J.; Gao, S.; Chen, K.; Yang, J.; Zhao, H.W.; Guo, L.; Yang, S.H. Vertically aligned, double-sided, and self-supported 3D WO$_3$ nanocolumn bundles for low-temperature gas sensing. *J. Mater. Chem.* **2015**, *7*, 10108–10114. [CrossRef]
21. Zhang, Q.; Wu, Y.; Li, L.; Zuo, T. Sustainable Approach for Spent V$_2$O$_5$-WO$_3$/TiO$_2$ Catalysts Management: Selective Recovery of Heavy Metal Vanadium and Production of Value-Added WO$_3$-TiO$_2$ Photocatalysts. *ACS Sustain. Chem. Eng.* **2018**, *6*, 12502–12510. [CrossRef]
22. Araújo, E.S.; Leão, V.N.S. TiO$_2$:WO$_3$ heterogeneous structures prepared by electrospinning and sintering steps: Characterization and analysis of the impedance variation to humidity. *J. Adv. Ceram.* **2019**, *8*, 238–246. [CrossRef]
23. Yan, J.; Wu, G.; Guan, N.; Li, L. Nb2O5/TiO$_2$ Heterojunctions: Synthesis Strategy and Photocatalytic Activity. *Appl. Catal.* **2014**, *152*, 280–288. [CrossRef]
24. Mandal, R.K.; Kundu, S.; Sain, S.; Pradhan, S.K. Enhanced photocatalytic performance of V$_2$O$_5$-TiO$_2$ nanocomposites synthesized by mechanical alloying with morphological hierarchy. *New J. Chem.* **2019**, *43*, 2804–2816. [CrossRef]
25. Chen, M.; Wei, X.; Liang, J.; Li, S.; Zhang, Z.; Tang, F. Effects of CrOx species doping on V$_2$O$_5$-WO$_3$/TiO$_2$ catalysts on selective catalytic reduction of NOx by NH3 at low temperature. *Reac. Kinet. Mech. Cat.* **2022**, *135*, 1767–1783. [CrossRef]
26. Wu, Z.; Chen, H.; Wan, Z.; Zhang, S.; Zeng, Y.; Guo, H.; Zhong, Q.; Li, X.; Han, J.; Rong, W. Promotional Effect of S Doping on V$_2$O$_5$-WO$_3$/TiO$_2$ Catalysts for Low-Temperature NOx Reduction with NH3. *Ind. Eng. Chem. Res.* **2020**, *59*, 15478–15488. [CrossRef]
27. Hiemstra, T. Formation, stability, and solubility of metal oxide nanoparticles: Surface entropy, enthalpy, and free energy of ferrihydrite. *Geochim. Cosmochim. Acta* **2015**, *158*, 179–198. [CrossRef]
28. Tanev, P.; Chibwe, M.; Pinnavaia, T. Titanium-containing mesoporous molecular sieves for catalytic oxidation of aromatic compounds. *Nature* **1994**, *368*, 321–323. [CrossRef]
29. Lee, D.W.; Yoo, B.R. Advanced metal oxide (supported) catalysts: Synthesis and applications. *Ind. Eng. Chem. Res.* **2014**, *20*, 3947–3959. [CrossRef]
30. Fatimah, I.; Fadillah, G.; Yanti, I.; Doong, R. Clay-Supported Metal Oxide Nanoparticles in Catalytic Advanced Oxidation Processes: A Review. *J. Nanomater.* **2022**, *12*, 825. [CrossRef]
31. Araújo, E.S.; da Costa, B.P.; Oliveira, R.A.P.; Libardi, J.; Faia, P.M.; de Oliveira, H.P. TiO$_2$/ZnO hierarchical heteronanostructures: Synthesis, characterization and application as photocatalysts. *J. Environ. Chem. Eng.* **2016**, *4*, 2820–2829. [CrossRef]
32. Treacy, J.P.; Hussain, H.; Torrelles, X.; Grinter, D.C.; Cabailh, G.; Bikondoa, O.; Thornton, G. Geometric structure of anatase TiO$_2$(101). *Phys. Rev. B* **2017**, *95*, 075–416. [CrossRef]
33. Leng, X.; Pereiro, J.; Strle, J.; Bollinger, A.T.; Božović, I. Epitaxial growth of high quality WO$_3$ thin films. *APL Mater.* **2015**, *3*, 096–102. [CrossRef]
34. Faia, P.M.; Libardi, J.; Louro, C.S. Effect of V$_2$O$_5$ doping on P- to N-conduction type transition of TiO$_2$:WO$_3$ composite humidity sensors. *Sens. Actuators B Chem.* **2016**, *222*, 952–964. [CrossRef]
35. Chen, A.; Li, C.; Zhang, C.; Li, W.; Yang, Q. The mechanical hybrid of V$_2$O$_5$ microspheres/graphene as an excellent cathode for lithium-ion batteries. *J. Solid. State Electrochem.* **2022**, *26*, 729–738. [CrossRef]
36. Kniec, K.; Ledwa, K.; Marciniak, L. Enhancing the Relative Sensitivity of V5+, V4+ and V3+ Based Luminescent Thermometer by the Optimization of the Stoichiometry of Y3Al5−xGaxO12 Nanocrystals. *Nanomaterials* **2019**, *9*, 1375. [CrossRef] [PubMed]
37. Lima, F.M.; Martins, F.M.; Maia Júnior, P.H.F.; Almeida, A.F.L.; Freire, F.N. Nanostructured titanium dioxide average size from alternative analysis of Scherrer's Equation. *Matéria* **2018**, *23*, 1–9. [CrossRef]
38. Miranda, M.A.R.; Sasaki, J.M. The limit of application of the Scherrer equation. *Acta Crystallogr. A Found. Adv.* **2018**, *74*, 54–65. [CrossRef]
39. Sahoo, S.; Arora, A.K.; Sridharan, V. Raman line shapes of optical phonons of different symmetries in anatase TiO$_2$ nanocrystals. *J. Phys. Chem. C* **2009**, *113*, 16927–16933. [CrossRef]
40. Su, C.Y.; Lin, H.C.; Lin, C.K. Fabrication and optical properties of Ti-doped W18O49 nanorods using a modified plasma-arc gas-condensation technique. *J. Vac. Sci. Technol. B* **2009**, *27*, 2170–2174. [CrossRef]
41. Yang, C.C.; Li, S. Size-dependent Raman red shifts of semiconductor nanocrystals. *J. Phys. Chem. B* **2008**, *112*, 14193–14197. [CrossRef]
42. Sangeetha, P.; Jayapandi, S.; Saranya, C.; Ramakrishnan, V. Phonon confinement and size effect in Raman spectra of TiO$_2$ nanocrystal towards Photocatalysis Application. *J. Aust. Ceram. Soc.* **2021**, *57*, 533–541. [CrossRef]

43. Tan, Z.; Sato, K.; Ohara, S. Synthesis of layered nanostructured TiO$_2$ by hydrothermal method. *Adv. Powder Technol.* **2015**, *26*, 296–302. [CrossRef]
44. Patrycja Makuła, P.; Pacia, M.; Macyk, W. How to correctly determine the band gap energy of modified semiconductor photocatalysts Based on UV–Vis Spectra. *J. Phys. Chem. Lett.* **2018**, *9*, 6814–6817. [CrossRef] [PubMed]
45. Illangakoon, U.E.; Nazir, T.; Williams, G.R.; Chatterton, N.P. Mebeverineloaded electrospun nanofibers: Physicochemical characterization and dissolution studies. *J. Pharm. Sci.* **2014**, *103*, 283–292. [CrossRef] [PubMed]
46. Franco, P.; De Marco, I. Eudragit: A Novel Carrier for Controlled Drug Delivery in Supercritical Antisolvent Coprecipitation. *Polymers* **2020**, *12*, 234. [CrossRef] [PubMed]
47. Kishori, L.D.; Nilima, A.T.; Gide, P.S. Formulation and development of tinidazole microspheres for colon targeted drug delivery system. *J. Pharm. Res.* **2013**, *6*, 158–165.
48. Revellame, E.D.; Fortela, D.L.; Sharp, W.; Hernandez, R.; Zappi, M.E. Adsorption kinetic modeling using pseudo-first order and pseudo-second order rate laws: A review. *Clean. Eng. Technol.* **2020**, *1*, 100032. [CrossRef]
49. Luo, X.; Liu, F.; Li, X.; Gao, H.; Liu, G. WO$_3$/TiO$_2$ nanocomposites: Salt–ultrasonic assisted hydrothermal synthesis and enhanced photocatalytic activity. *Mater. Sci. Semicond. Processing* **2013**, *16*, 1613–1618. [CrossRef]
50. Bledowski, M.; Wang, L.; Ramakrishnan, A.; Khavryuchenko, O.V.; Khavryuchenko, V.D.; Ricci, P.C.; Beranek, R. Visible-light photocurrent response of TiO$_2$–polyheptazine hybrids: Evidence for interfacial charge-transfer absorption. *Phys. Chem. Chem. Phys.* **2011**, *13*, 21511–21519. [CrossRef]
51. Lee, W.H.; Lai, C.W.; Abd Hamid, S.B. One-step formation of WO$_3$-loaded TiO$_2$ nanotubes composite film for high photocatalytic performance. *Materials* **2015**, *8*, 2139–2153. [CrossRef]
52. Gao, L.; Gan, W.; Qiu, S.; Zhan, X.; Qiang, T.; Li, J. Preparation of heterostructured WO$_3$/TiO$_2$ cat-alysts from wood fibers and its versatile photodegradation abilities. *Sci. Rep.* **2017**, *7*, 1102. [CrossRef]
53. Enesca, A. The Influence of Photocatalytic Reactors Design and Operating Parameters on the Wastewater Organic Pollutants Removal—A Mini-Review. *Catalysts* **2021**, *11*, 556. [CrossRef]
54. Huang, Z.F.; Zou, J.J.; Pan, L.; Wang, S.; Zhang, X.; Wang, L. Synergetic promotion on photoactivity and stability of W$_{18}$O$_{49}$/TiO$_2$ hybrid. *Appl. Catal. B-Environ.* **2014**, *147*, 167–174. [CrossRef]
55. Luo, X.; He, G.; Fang, Y.; Xu, Y. Nickel sulfide/graphitic carbon nitride/strontium titanate (NiS/g-C3N4/SrTiO3) composites with significantly enhanced photocatalytic hydrogen production activity. *J. Colloid. Interface Sci.* **2018**, *518*, 184–191. [CrossRef] [PubMed]

Article

Application of Biobased Substances in the Synthesis of Nanostructured Magnetic Core-Shell Materials

Marcos E. Peralta [1,*], Alejandro Koffman-Frischknecht [1], M. Sergio Moreno [2], Daniel O. Mártire [3] and Luciano Carlos [1,*]

[1] Instituto de Investigación y Desarrollo en Ingeniería de Procesos, Biotecnología y Energías Alternativas, PROBIEN (CONICET-UNCo), Universidad Nacional Del Comahue, Neuquén 8300, Argentina
[2] Instituto de Nanociencia y Nanotecnología, INN (CNEA-CONICET), Centro Atómico Bariloche, Av. Bustillo 9500, San Carlos de Bariloche 8400, Argentina
[3] Instituto de Investigaciones Fisicoquímicas Teóricas y Aplicadas (INIFTA), CCT-La Plata-CONICET, Universidad Nacional de La Plata, La Plata 1900, Argentina
* Correspondence: marcos.peralta@probien.gob.ar (M.E.P.); luciano.carlos@probien.gob.ar (L.C.)

Abstract: We propose here a novel green synthesis route of core-shell magnetic nanomaterials based on the polyol method, which uses bio-based substances (BBS) derived from biowaste, as stabilizer and directing agent. First, we studied the effect of BBS concentration on the size, morphology, and composition of magnetic iron oxides nanoparticles obtained in the presence of BBS via the polyol synthesis method (MBBS). Then, as a proof of concept, we further coated MBBS with mesoporous silica (MBBS@mSiO$_2$) or titanium dioxide (MBBS@TiO$_2$) to obtain magnetic nanostructured core-shell materials. All the materials were deeply characterized with diverse physicochemical techniques. Results showed that both the size of the nanocrystals and their aggregation strongly depend on the BBS concentration used in the synthesis: the higher the concentration of BBS, the smaller the sizes of the iron oxide nanoparticles. On the other hand, the as-prepared magnetic core-shell nanomaterials were applied with good performance in different systems. In particular, MBBS@SiO$_2$ showed to be an excellent nanocarrier of ibuprofen and successful adsorbent of methylene blue (MB) from aqueous solution. MBBS@TiO$_2$ was capable of degrading MB with the same efficiency of pristine TiO$_2$. These excellent results encourage the use of bio-based substances in different types of synthesis methods since they could reduce the fabrication costs and the environmental impact.

Keywords: magnetic nanoparticles; magnetite; solvothermal synthesis; waste valorization; green chemistry

1. Introduction

Nowadays, magnetic nanoparticles and nanostructured magnetic materials with core-shell type structure are attracting widespread interest in material science due to their potential applications in many fields, including environmental remediation, drug delivery, magnetic resonance imaging, electronics, sensor developments, etc. [1]. Additionally, there are several new applications of magnetic nanoparticles and nanomagnetic fluids such as temperature sensors, heat exchangers, magnetic actuators, and energy harvesters that open up new possibilities for the development of these fields [2–4]. Among the magnetic nanomaterials, magnetic iron oxide nanoparticles (IONPs) (e.g., magnetite,Fe$_3$O$_4$, and maghemite,g-Fe$_2$O$_3$), have been widely studied due to their superparamagnetic behavior and because they are more environmentally friendly compared to other metals such as the Co and the Ni. [5,6]. Different methodologies have been implemented to prepare IONPs, such as coprecipitation, micelle synthesis, sol-gel method, thermal decomposition, and polyol method [7]. Among these synthetic methods, the polyol method is a well-suited technique to obtain pure and monodispersed nanoparticles with a narrow and controllable size distribution [8]. In this method, ferric salts mixed with a stabilizer are dissolved in a solvent, such as glycols (e.g., ethylene glycol or diethylene glycol), and heated to the boiling point of the solvent. The stabilizers are normally surfactants or polymers, whose main function is to regulate the size of the primary nanocrystals and secondary aggregates [9]. Various substances were studied as stabilizer including sodium citrate [10–12], polyethylene glycol (PEG) [13], polyacrylic acid (PAA) [14–16], polyvinylpyrrolidone (PVP) [17], cetyltrimethylammonium bromide (CTAB) [9], and sodium dodecyl benzene sulfonic acid (SDBS) [15]. However, a major drawback of using some synthetic surfactants is their lack of biodegradability or biocompatibility (e.g., PAA), or their toxicity (e.g., CTAB and SDBS), what discourages their applications.

The replacement of synthetic stabilizer with substances deriving from the treatment of biowaste is considered as a green process since it values waste as a source of renewable raw material for the synthesis of new technological materials, which leads to a reduction in the manufacturing costs and environmental impact. Aerobic biodegradation of the wet organic fraction of municipal waste has been shown to yield polymeric bio-based substances (BBS), which have chemical similarities to humic substances and possess surfactant properties which make them suitable as green auxiliary synthesis reactants [18,19]. In particular, BBS were used in the synthesis process of different materials as a structure directing agent in the preparation of mesoporous silica nanoparticles due to their ability to form micelles [20], as sacrificial carbon in the formation of zero-valent iron nanoparticles from carbothermal synthesis [21], and as stabilizer for the generation of silver nanoparticles [22]. On the other hand, BBS-coated magnetic nanoparticles and magnetic nanocomposites prepared using BBS have been successfully employed as adsorbents [23–25], photo-Fenton catalysts [26], and photocatalysts [27,28] for water treatment processes. However, the role of BBS as synthesis aids in the polyol method for the preparation of IONPs has not been addressed. Furthermore, since BBS could act as both nucleation directing agents and structure directing agents, they may play a critical role in the synthesis of nanostructured magnetic core-shell materials.

This paper investigates the possible application of BBS as green aids to act as stabilizers in the preparation of IONPs via polyol synthesis method and as nucleation directing agents in the synthesis of nanostructured core-shell magnetic oxide particles. For the latter, the as-prepared BBS-modified IONP nanoparticles were coated with mesoporous SiO_2 and TiO_2, and two types of magnetic core-shell structured nanoparticles, MBBS@mSiO$_2$ and MBBS@TiO$_2$, were prepared. Finally, the performance of nanostructured magnetic core-shell nanoparticles in specific applications was evaluated. The potential application of BBS in the preparation of nanomaterials opens up new ways for the development of novel synthesis strategies and represents an economically sustainable method of waste valorization.

2. Results and Discussion
2.1. Syntehsis of IONPs Using BBS

First, IONPs without BBS (Fe_3O_4) and with different amounts of BBS (35 mg, 70 mg, 200 mg, 500 mg and 1000 mg) were prepared via the polyol method. The obtained nanoparticles were called MBBS-X, where X stand for the mass of BBS used in the synthesis. The morphology of MBBS was studied by TEM images. Bare magnetite (Figure 1A) consists of spherical nanoparticles with a mean diameter of circa 80 nm with wide particle size distribution and poor dispersity. Figure 1B–D shows typical TEM images of the samples obtained with different BBS loads. It can be observed that the size of the nanocrystals and the secondary aggregates (or clusters) depend on the BBS concentration used in the synthesis. A particle size distribution obtained from the TEM images for each material is shown in Figure S1, Supplementary Materials. In particular, for MBBS-35 particles (Figure 1B), well-defined roughly spherical clusters with diameters of about 70–200 nm are observed. The clusters are composed of self-assembled small primary nanocrystals and an outer layer of organic matter less than 10 nm thick, as estimated from the lighter shell surrounding the cluster (Figure S2, Supplementary Materials). In sample MBBS-200, clusters with less geometrical shape are observed (Figure 1C). As the concentration of BBS increases, large aggregates of small IONPs are observed without a defined shape (e.g., sample MBBS-1000,

Figure 1D). Additionally, sample MBBS-1000 showed IONPs with the smallest particle size, which were around 5 nm (Figure S3, Supplementary Materials). These results suggest that BBS regulates the size of IONPs by limiting the growth of nanoparticles, the higher the concentration of BBS, the smaller the sizes of the IONPs. Furthermore, at high BBS concentration, BBS coated the majority of small IONPs, which avoids the formation of spherical cluster.

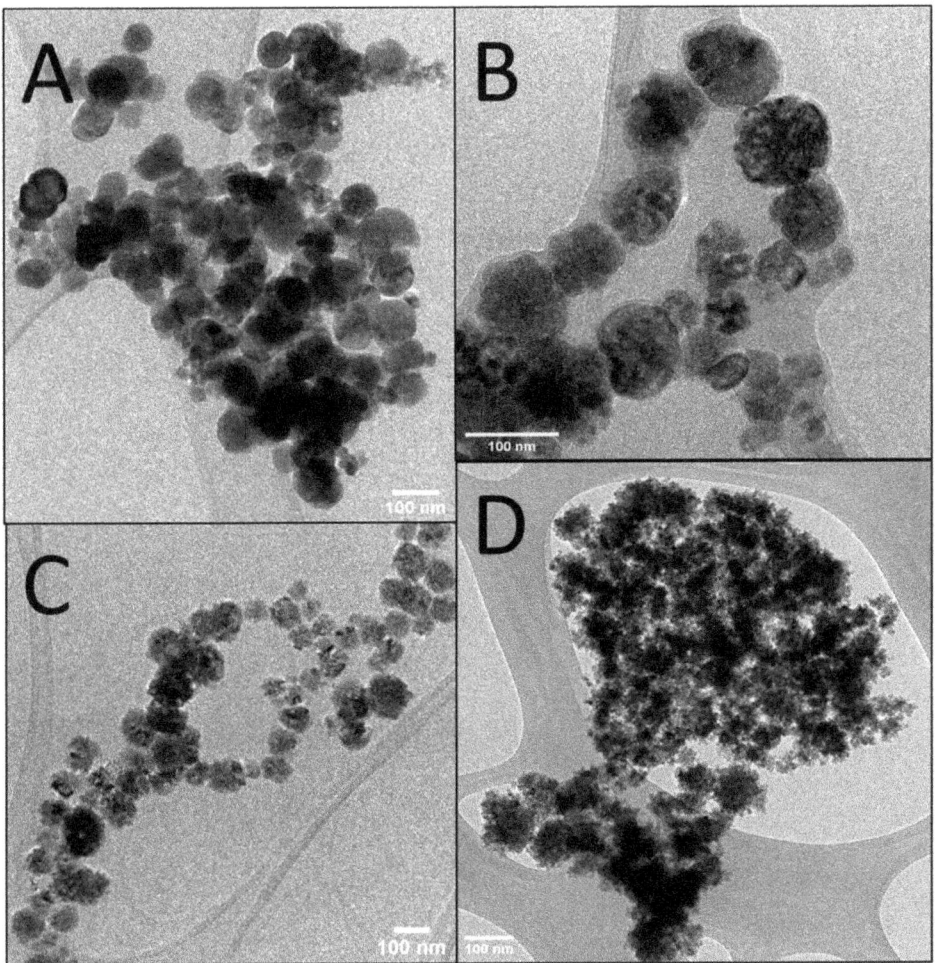

Figure 1. TEM images of Fe_3O_4 (**A**), MBBS-35 (**B**), MBBS-200 (**C**) y MBBS-1000 (**D**).

In order to characterize the presence of BBS on the surface of the obtained IONPs, FTIR spectra were performed (Figure 2A). The band at 1080 cm^{-1} (v_{C-O}), associated with organic matter, increases in intensity with the amount of BBS. The carboxylate group band (1620 cm^{-1}) and the carboxylate-iron stretching signal (1400 cm^{-1}) [29] are observed in all spectra, which suggests that carboxylate groups indeed play an important role in the bonding of the BBS to the IONPs surface. This result is in agreement with previous reports obtained for BBS-coated Fe_3O_4 nanoparticles prepared via the coprecipitation method [23]. The presence of carboxylate bands in the Fe_3O_4 spectrum could be due to the fact that sodium acetate was not completely removed in the washing steps. In the lower frequency FTIR region, the samples showed characteristic signals assigned to Fe-O bonds. The broad band in the range of 500–800 cm^{-1} is due to the stretching vibration of the Fe-O in tetrahedral sites (ν_1) [30]. The ν_1 frequency for Fe$_3$O$_4$ is observed at 567 cm^{-1}. It is important to note that this vibration mode was broadened after BBS coating and at high BBS amounts a splitting is observed, which might be due to symmetry lowering [31]. The broadening is due to the statistical distribution of cations over the octahedral and tetrahedral sites, while the local symmetry is disturbed [32]. Taking this into account, we propose that the interaction of BBS with surface Fe atoms induces a change in the local symmetry of the outer tetrahedral sites likely to result in peak broadening.

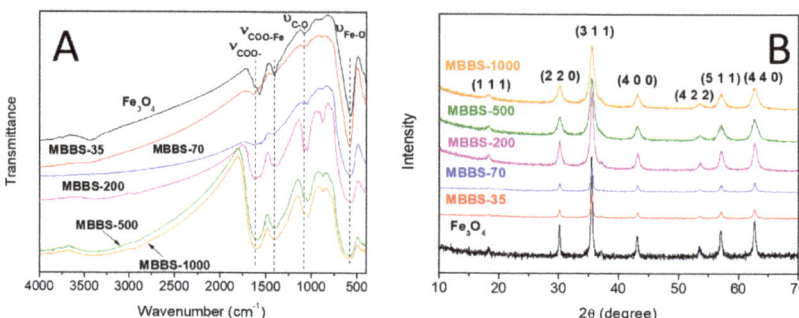

Figure 2. (**A**) FTIR spectra of MBBS nanoparticles. The main relevant peaks are labeled. (**B**) XRD diffraction patterns of MBBS nanoparticles. The indexed labels identify the miller indices of the Fe$_3$O$_4$ phases.

The XRD patterns of MBBS (Figure 2B) were recorded to determine the crystalline phases present in the nanoparticles. All diffractograms are very similar and show main diffraction peaks in good agreement with a reference pattern of magnetite. In particular, the XRD peak positions at 30.1 (220), 35.4 (311), 43.1 (400), 53.4 (422) 56.9 (511), and 62.5 (440) perfectly match with the lattice planes of magnetite (Card number 01-075-1610, ICDD Database). The phase identification of magnetite and maghemite by the conventional X-ray diffraction method is not a simple matter because both have the same cubic structure and their lattice parameters are almost identical [33]. However, the presence of doublets at the high-angle peaks as (5 1 1) and (4 4 0) is indicative of the presence of magnetite and maghemite in the samples [34]. Furthermore, a decrease in the lattice constant for the peaks (5 1 1) and (4 4 0) of magnetite is also indicative of the formation of maghemite [35]. For MBBS samples, a two-peak convolution was not noticed for the (5 1 1) and (4 4 0) planes and we did not observe any decrease in the lattice constant for the (5 1 1) and (4 4 0) peaks with respect to the Fe$_3$O$_4$ nanoparticles prepared without BBS. Therefore, this could indicate that in our samples Fe$_3$O$_4$ is the main phase of magnetic iron oxide, however, the presence of a small amount of maghemite cannot be ruled out. No other crystalline materials are detected, which indicates that BBS act as good stabilizer to obtain IONPs via the polyol method.

In order to determine the quantity of organic matter in MBBS samples, a thermogravimetric analysis (TGA) was performed. The amount of organic matter was calculated by measuring the total mass loss up to 1000 °C (Table S1, Supplementary Materials). As can be seen in Figure 3A, the BBS loading on the IONPs increases with the amount of BBS used in the synthesis, reaching the maximum value of near 47.2 wt.%.

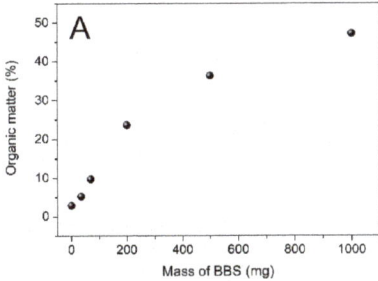 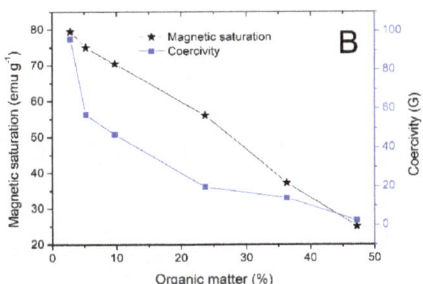

Figure 3. (**A**) Total organic matter (%) determined from the TGA vs. Mass of BBS used during synthesis (mg) for MBBS samples; (**B**) magnetization of saturation (emu g^{-1}) and Coercivity as a function of organic matter (%) determined from the TGA for MBBS samples.

The magnetic properties of MBBS samples and bare magnetite were evaluated by magnetization curves (Figure S4 and Table S2, Supplementary Materials). Magnetic saturation values (M_s) decrease with an almost linear relationship with the BBS loading (Figure 3B). The decrease in the saturation magnetization may be due to the non-magnetic fraction as well as the reduction in particle size. It is important to note that the larger the BBS loading, the smaller the coercivity (Figure 3B). This is due to the particles approaching superparamagnetism with size reduction [36,37]. For spherical magnetite nanoparticles, the critical superparamagnetic size is 26 nm [36], thus samples with high BBS content lead to superparamagnetic behavior. These results evidence the critical role of the adsorbed BBS on the magnetic properties of the IONPs.

Zeta potential values of bare magnetite and BBS coated IONPs measured in 10^{-2} M KCl aqueous solutions at pH 3, 6, and 10 are shown in Table S3, Supplementary Materials. The isoelectric point of Fe_3O_4 nanoparticles is 7.2 [38], which indicates that at lower pH values it presents positive charge. Because BBS is negatively charged in the range of pH 3–10, the positive charge observed for MBB-35 and MBB-70 at pH 3 could be the consequence of the limited covering of the IONPs surface by BBS. However, when higher amounts of BBS are used (i.e., MBB-200, MBB-500, and MBB-1000), a negative charge on the surface of the nanoparticles is observed over all the three pH values (3, 6, 10). Additionally, zeta potential becomes more negative as the pH increases because of the dissociation of carboxylic and phenolic groups of BBS [23].

2.2. MBBS@mSiO$_2$ and MBBS@TiO$_2$ Nanoparticles

In order to evaluate the role of BBS as a nucleation directing agent in the synthesis of nanostructured core-shell magnetic oxide particles, IONPs were coated directly by two types of common inorganic materials, mesoporous silica (mSiO$_2$) and titanium dioxide (TiO$_2$). Due to its magnetic properties, cluster size and particle size distribution we selected MBBS-35 for the synthesis of core-shell nanomaterials, which were named MBBS@mSiO$_2$ and MBBS@TiO$_2$. As control experiments, the synthesis of core@shell magnetic nanoparticles was performed with Fe_3O_4 nanoparticles prepared in the absence of BBS. No core@shell structure was formed, either for mSiO$_2$ or for TiO$_2$, when the IONP nanoparticles without BBS were used in the synthesis (data not shown). On the contrary, TEM images of MBBS@mSiO$_2$ show that the material consists of nanoparticles of spherical morphology with a dark core (iron oxide) covered by a quite uniform lighter porous shell (mesoporous silica) of c.a. 20 nm in thickness (Figure 4). This result evidences that BBS plays an important role in the growth of the mSiO$_2$ layer on the surface of IONPs. It is likely that the negatively charged carboxylate groups on MBBS surface interact with the positively charged CTAB molecules leading to the growth of the mesoporous silica layer onto the BBS-coated IONPs. FTIR spectra (Figure S5, Supplementary Materials) also support the presence of silica on the surface of IONPs. The wide band with a peak at around 1084 cm^{-1}

is associated with the stretching vibration of the Si–O–Si bond from the silica framework. The peak at 571 cm^{-1} corresponds to Fe-O stretching, whereas the bands at 3430 cm^{-1} and 1632 cm^{-1} can be associated with O–H stretching and H–O–H bending vibrations from absorbed water. The XRD pattern of MBBS@mSiO$_2$ (Figure 5A) indicates that the IONPs did not undergo any crystalline transformation to non-magnetic iron oxide phases during the mSiO$_2$ synthesis step and its subsequent calcination. The broad peak at around 2θ = 24° is assigned to the halo pattern of amorphous SiO$_2$ [39].

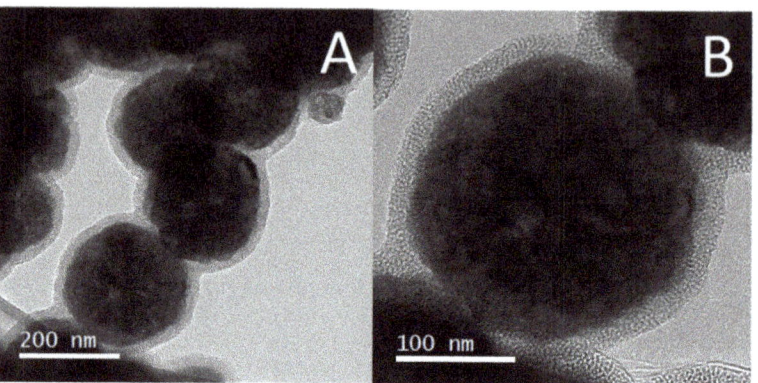

Figure 4. (**A**,**B**) TEM images of MBBS@SiO$_2$.

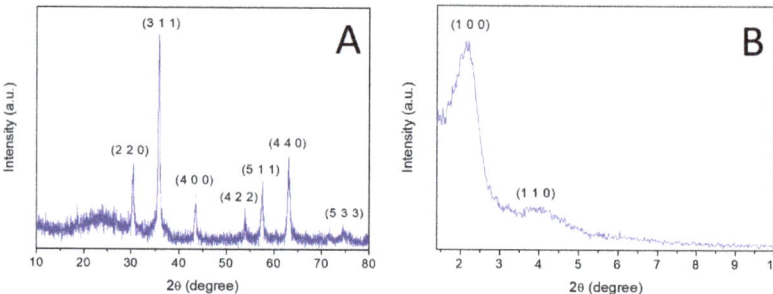

Figure 5. Wide-angle XRD patterns (**A**) and small-angle XRD (**B**) patterns of MBBS@SiO$_2$. The indexed labels identify the miller indices of the Fe$_3$O$_4$ phases (**A**) and Bragg reflections of MCM-41 (**B**).

The small-angle XRD patterns of MBBS@SiO$_2$ (Figure 5B) exhibit an intense peak at about 2θ = 2.2° and a weak peak at 2θ = 3.9°, which correspond to (100) and (110) Bragg reflections of MCM-41 according to previous reports [40,41]. This suggests a well-ordered hexagonal array of mesopores. The porous structure of silica shell was further investigated by the N$_2$ adsorption–desorption technique (Figure S6A, Supplementary Materials). The N$_2$ adsorbed quantity, the specific surface area (S$_{BET}$), and the pore volume (P$_V$) resulted to be higher for MBBS@SiO$_2$ (S$_{BET}$ = 136 m^2 g^{-1}, P$_v$ = 0.135 cm^3 g^{-1}) compared to MBBS-35 (S$_{BET}$ = 23 m^2 g^{-1}, P$_v$ = 0.056 cm^3 g^{-1}). This result is consistent with a mesoporous shell covering MBBS-35. MBBS@mSiO$_2$ displays type IV isotherms and H4 hysteresis loop according to IUPAC classification [42]. This behaviour can be related to the presence of pores in the mesopore range, as well as other pores with bigger diameters [43]. However, BJH pore size distribution (Figure S6B, Supplementary Materials) only shows a uniform mesoporous size of 2.9 nm, thus the larger pore size could be due to particle aggregation.

Magnetic hysteresis curve at room temperature of MBB@mSiO$_2$ is shown in Figure S7, Supplementary Materials. MBBS@mSiO$_2$ still displays a superparamagnetic behavior with low coercivity and remanence, similar to MBBS-35, and a magnetic saturation (M_s) value of 48.6 emu g^{-1} (Table S2, Supplementary Materials). It can be observed a reduction in M_s of MBBS@mSiO$_2$ compared to uncovered MBBS-35 (M_s = 75.4 emu g^{-1}), which is often attributed to presence of the diamagnetic silica [44].

The preparation of MBBS@TiO$_2$ was carried out via the sol-gel method using titanium (IV) butoxide (TBOT) as precursor of TiO$_2$. Noteworthily, no additional stabilizers/templates were used in the synthesis processes. TEM images of MBBS@TiO$_2$ showed the successful formation of a homogeneous TiO$_2$ layer on the surface of the IONPs with a thickness ranging between 15 and 33 nm (Figure 6 and Figure S8, Supplementary Materials). In this case, the hydroxyl and carboxylic groups of MBBS can promote intermolecular interactions with TBOT, leading to a uniform deposition of the TiO$_2$ layer onto the BBS-coated IONPs. The XRD pattern of MBBS@TiO$_2$ (Figure S9, Supplementary Materials) evidenced the formation of anatase as the only TiO$_2$ crystalline phase in the nanoparticles. Additionally, hematite (α-Fe$_2$O$_3$) and Fe$_3$O$_4$ were detected in the XRD diffraction pattern, which indicates that some oxidation degree of IONPs occurred in the calcination process at 500 °C. Magnetic curves obtained at 300 K clearly revealed a superparamagnetic behavior for MBBS@TiO$_2$ with a M_s value of 19.7 emu g^{-1} (Figure S7, Supplementary Materials). This lower M_s value compared to MBBS-35 can be explained by the presence of non-magnetic phases, such as the TiO$_2$ layer and hematite. Despite this, MBBS@TiO$_2$ nanoparticles possess a strong magnetic response and can be easily recovered by applying an external magnetic field.

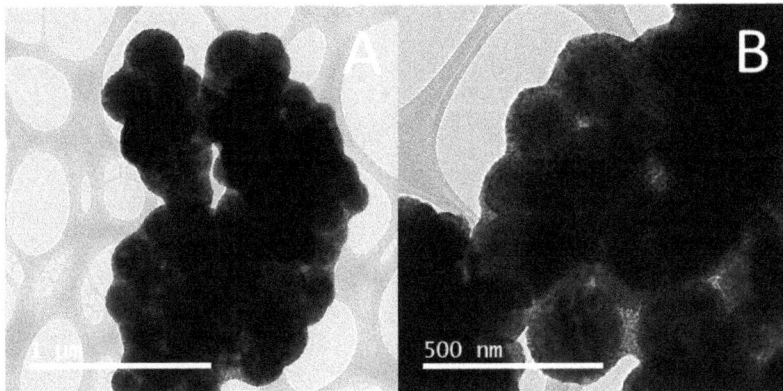

Figure 6. (A,B) TEM images of MBBS@TiO$_2$.

Magnetic core@shell nanoparticles have a wide range of applications in different research fields. In particular, due to its remarkable surface and magnetic features, MBBS@mSiO$_2$ raw or functionalized is a promising material for technological applications, such as adsorption [45,46], chromatography [47], and drug delivery [39], whereas MBBS@TiO$_2$ is a promising photocatalyst for environmental remediation [48]. In this study, we tested two potential applications of MBBS@mSiO$_2$, as a drug carrier and as adsorbent of pollutants from aqueous media.

The drug cargo capacity of MBBS@mSiO$_2$ by using ibuprofen (IBU), a well-known anti-inflammatory drug, was performed. The drug loading capacity, DLC (%) was defined as follows:

$$\text{DLC (\%)} = (\text{mass of IBU released}/\text{mass of nanoparticles}) \times 100 \quad (1)$$

Figure 7A shows the gradual release of IBU from MBBS@mSiO$_2$ at different times. The DLC of MBBS@mSiO$_2$ results to be around 13%, making them a good candidate as a drug delivery system. To better control the IBU release, the surface of MBBS@mSiO$_2$ should be modified with appropriate release triggers that specifically react with response to stimuli [49]. Thus, MBBS@mSiO$_2$ could be a potential starting material for a subsequent modification that results in a controlled drug delivery system.

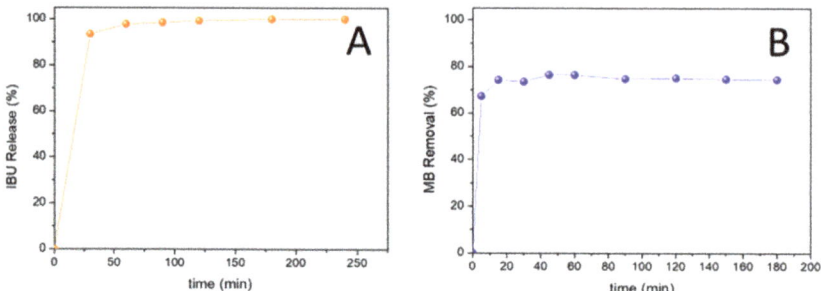

Figure 7. (**A**) Ibuprofen release from MBBS@mSiO$_2$ (T = 25 °C pH = 6.0) (**B**) MB removal from water using MBBS@mSiO$_2$ as adsorbent ([MB]$_0$ = 10 mg L^{-1}; MBBS@mSiO$_2$ dosage = 500 mg L^{-1}, T = 25 °C pH = 6.0).

On the other hand, MBBS@mSiO$_2$ was tested as adsorbent for the removal of methylene blue (MB), a toxic cationic dye that is used in textile industry. Figure 7B shows the % MB removal at different times. A very fast adsorption of MB can be observed, achieving a 76% dye removal in the first 15 min. Figure S10, Supplementary Materials, shows the evolution of UV-Vis spectra of MB at different times. The adsorption capacity of MB found for MBBS@mSiO$_2$ was similar to those of other magnetic mesoporous silicas reported in the literature [45].

MBBS@TiO$_2$ was studied as photocatalyst for the degradation of MB. Figure S11, Supplementary Materials, shows the evolution of the UV-Vis spectra of MB solution when irradiated in the presence of MBBS@TiO$_2$. The absorbance values at 664 nm were used to follow the MB photobleaching kinetics. For comparison purpose, MB photodegradation experiments were also carried out with TiO$_2$ synthesized in the absence of IONPs nanoparticles, but keeping the rest of the synthesis parameters unchanged. In these photocatalytic experiments, the same load of TiO$_2$ was used, so it was considered that MBBS@TiO$_2$ have an approximately composition of 33 wt.% of TiO$_2$. This composition was roughly estimated from the average size of cores and shells determined by TEM and the densities of pristine Fe$_3$O$_4$ and TiO$_2$. Figure 8 compares the evolution of the normalized absorbance at 664 nm with the reaction time obtained for MB degradation performed with MBBS@TiO$_2$ and TiO$_2$. It can be seen that MBBS@TiO$_2$ is capable of degrading MB with the same efficiency of pristine TiO$_2$. This is an excellent result, since a direct contact between Fe$_3$O$_4$ and TiO$_2$ typically brings about an unfavorable heterojunction, which accelerates the recombination of the electron–hole pairs and weakens the photocatalytic activity of titanium-based catalysts [50]. It was reported that the addition of a silica layer between an iron oxide core and a titania shell promotes the photocatalytic activity by decreasing the charge transfer between the IONPs and TiO$_2$, which could otherwise result in the recombination of photogenerated species on the IONPs surface [51]. Therefore, in our case, it is probable that BBS avoid the formation of a Fe$_3$O$_4$/TiO$_2$ heterojunction.

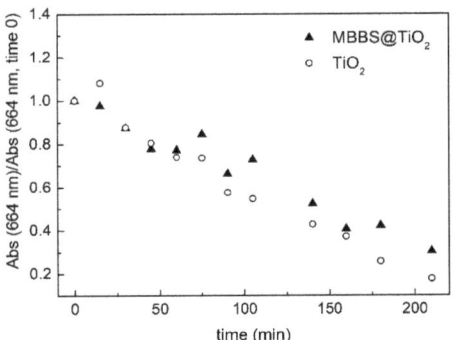

Figure 8. Photocatalytic activities of MBBS@TiO$_2$ and TiO$_2$. [MB]$_0$ = 5 mg L^{-1}; load MBBS@TiO$_2$ = 120 mg L^{-1}; load TiO$_2$ = 40 mg L^{-1}; pH = 6.

3. Materials and Methods

3.1. Reagents

FeCl$_3$x6H$_2$O (>99%) was purchased from Anedra (Buenos Aires, Argentina). Sodium acetate trihydrate (99%), ammonia solution (25–30), and absolute ethanol were obtained from Cicarelli (Santa Fe, Argentina). Cetyltrimethylammonium bromide (CTAB, >97%) was purchased from Merck (Burlington, MA, USA). Ibuprofen, Tetraethoxylsilane (TEOS, 98%) and titanium (IV) butoxide (TBOT, 97%) were from Sigma-Aldrich (Burlington, MA, USA). Acetonitrile (HPLC grade) was purchased from Biopack (Worcester, UK). All reactants were used without further purification. BBS were obtained following a previously reported protocol [22]; briefly, 50 g of green compost was treated with 1 L of 6 M NaOH aqueous solution under stirring at 60 °C for 4 h. The reaction mixture was then separated by centrifugation. The supernatant was concentrated, and different fractions were separated through a lab-scale ultrafiltration unit equipped with a membrane (molar mass cut-off 5 kDa). The retentate fraction was then dried at 60 °C for 24 h. The obtained BBS was about 20–30% in mass of the starting compost. Chemical composition of BBS is detailed in Table S4, in the Supplementary Material.

3.2. Syntehsis of IONPs and Core-Shell Magnetic Nanoparticles

The synthesis of magnetite nanoparticles was performed by a modification of the polyol method [52]. Briefly, 1.35 g of FeCl$_3$x6H$_2$O, 6 g of sodium acetate trihydrate and a defined mass of BBS (35 mg, 70 mg, 200 mg, 500 mg and 1000 mg) were dissolved in a Teflon vessel containing 50 mL of ethylene glycol. Thereafter, the Teflon vessel was placed into a stainless-steel autoclave reactor and heated at 200 °C for 8 h. Once cooled, the obtained black material was washed with ethanol and water. The solid was separated from the supernatant with the assistance of a Neodymium magnet in all washing steps. The obtained materials were named as MBBS-35, MBBS-70, MBBS-200, MBBS-500, and MBBS-1000 depending on the amount of BBS used in the synthesis process. Additionally, Fe$_3$O$_4$ without BBS was prepared with the same method as a reference material.

The covering of MBBS-35 with a mesoporous silica shell (MBBS@mSiO$_2$) was performed following our previous report [53]. In brief, 400 mg of MBBS-35, 500 mg of CTAB and 1.75 mL of ammonia solution were mixed in 250 mL of water. The suspension was kept under vigorous stirring and heating for 30 min. When the temperature reached 80 °C, 2.5 mL of TEOS was added drop by drop and the reaction was kept at 80 °C for 2 h. After cooling, the obtained material was magnetically separated, washed with distilled water and dried in an oven at 70 °C overnight. Finally, CTAB and BBS were removed by calcination in a furnace (500 °C for 1 h in air) and the final product was named as MBBS@mSiO$_2$.

MBBS-35 coating with titanium dioxide shell (MBBS@TiO$_2$) was performed via a sol-gel method [54]. For this, 50 mg of MBBS-35 were dispersed by sonication in a mixture of 90 mL absolute ethanol/30 mL acetonitrile. Then, 0.5 mL of ammonia solution and 1.0 mL of TBOT was added and the dispersion was kept under stirring at 25 °C for 1.5 h. Later, the sample was washed with ethanol, magnetically separated and dried at 70 °C. Then, the obtained material was calcined in argon atmosphere at 500 °C for 9 h. Pure TiO_2 nanoparticles were also prepared for comparison.

3.3. Application of Core-Shell Magnetic Nanoparticles

MBBS@mSiO$_2$ was evaluated as potential drug carrier. For this end, the drug loading capacity (DLC) using ibuprofen as model drug was measured according to our previous report [53]. First, MBBS@SiO$_2$ were loaded with IBU. For this, 20 mg of nanoparticles were dispersed in a hexane solution of IBU (0.16 M) and the system was kept under magnetic stirring in a closed flask for 24 h. Afterwards, the nanoparticles were magnetically separated from the supernatant and washed once with hexane. Later, a release of ibuprofen assay was conducted. For this purpose, fresh 0.9% NaCl was added to the flask containing the loaded nanoparticles. The system was stirred and at defined times an aliquot was magnetically separated and measured in a UV spectrophotometer (UV-T60, PG Instruments, Leicestershire, UK) to quantify the released amount of IBU (λ = 264 nm).

The photocatalytic activity of MBBS@TiO$_2$ was tested via the degradation of aqueous solutions of methylene blue (MB). The photochemical experiments were performed in a 200 mL cylindrical Pyrex vessel at room temperature and under continuous stirring. A solar simulator (SunLite™, ABET Technologies, Milford, CT, USA) equipped with a 100 W Xenon Short Arc Lamp was used as irradiation source. In all the experiments, the initial concentration of MB was 5 mg L^{-1}, and the load of the photocatalyst were 120 mg L^{-1} and 40 mg L^{-1} for MBBS@TiO$_2$ and TiO$_2$, respectively. The aqueous suspensions were irradiated for 4 h, and 3 mL aliquots were taken at different times. The separation of the supernatant was achieved by magnetic separation. Finally, the residual MB concentration was determined by UV-Vis Spectrophotometry.

3.4. Characterization Tecnhiques

Transmission electron microscopy (TEM) images were obtained using a Tecnai F20 (G2) UT microscope (ThermoFisher, Waltham, MA, USA), operated at 200 kV. X-ray diffraction (XRD) patterns were obtained on a SmartLab SE 3 KW (Rigaku, Tokyo, Japan) equipped with Cu anode (45 kV, 40 mA) and a graphite monochromador. Fourier transform infrared (FTIR) spectra were recorded by using an FT-08 spectrophotometer (Lumex, Wakendorf II, Germany) on KBr pellets (1/300 wt.) in transmission mode with 128 scans at 4 cm^{-1} resolution. Thermogravimetric analysis (TGA) was performed with a TGA-DSC Q600 Thermogravimetric Analyzer (TA Instruments, New Castle, DE, USA). The samples were pre-dried for 30 min at 100 °C before the analysis (heating rate of 10 °C min^{-1} from room temperature to 1000 °C in air). The analysis of textural properties was conducted with N$_2$ adsorption/desorption isotherms on previously degassed samples (105 °C for 12 h) at 77 K using an ASAP 2000 sortometer (Micromeritics, Norcross, GA, USA). The specific surface area was calculated via the Brunauer-Emmett-Teller (BET) method, the pore volume was calculated using Gurvich's rule at relative pressure p/p_0 = 0.98 and the pore size distribution was calculated according to the Barrett–Joyner–Halenda (BJH) method. Magnetization curves were registered at 300 K by using 7300 vibrating sample magnetometer (LakeShore, Westerville, OH, USA). The zeta potential measurements were carried out by using a zetasizer nano (Malvern Instruments, Malvern, UK).

4. Conclusions

A novel procedure based on the polyol method was applied for the synthesis of IONPs stabilized with BBS. TEM images show that both the size of the nanocrystals and their aggregation strongly depend on the BBS concentration used in the synthesis. In particular, under the conditions employed for the preparation of MBBS-35 particles, well-defined roughly spherical clusters with diameters of about 70–200 nm are observed.

BBS were also employed as nucleation directing agents in the synthesis of nanostructured core-shell MBBS@SiO$_2$ and MBBS@TiO$_2$ nanoparticles. Our results indicate that the obtained nanostructured magnetic core-shell nanomaterials using BBS can successfully be applied with good performance in different systems. In particular, MBBS@SiO$_2$ showed to be an excellent nanocarrier of ibuprofen. The material achieves a 13% of drug loading capacity and around 90% of the cargo was released within 90 min. MBBS@SiO$_2$ also were successfully employed as adsorbents for the quantitative removal of methylene blue from aqueous solution. On the other hand, MBBS@TiO$_2$ was capable of degrading MB with the same efficiency of pristine TiO$_2$. This excellent result shows that the BBS layer that covers the IONPs avoids formation of the unfavorable heterojunction between Fe$_3$O$_4$ and TiO$_2$.

Supplementary Materials: The following supporting information can be downloaded at: https://www.mdpi.com/article/10.3390/inorganics11010046/s1, Figure S1. Particle size distribution obtained from TEM images; Figure S2: High resolution TEM image of MBBS-35; Figure S3: High resolution TEM image of MBBS-1000; Figure S4: Magnetization curves (300 K) of Fe$_3$O$_4$ and covered magnetite with different amount of BBS; Figure S5: FTIR spectra of MBBS@mSiO$_2$ material; Figure S6: (A) N2 adsorption/desorption isotherms of MBBS-35 and of MBBS@SiO$_2$ (B) pore-size distribution of MBBS@SiO$_2$; Figure S7: Magnetization curves at 300 K of MBBS@mSiO$_2$ and MBBS@TiO$_2$; Figure S8. High resolution TEM image of MBBS@TiO$_2$. Red lines indicate the thickness of TiO$_2$ layer; Figure S9: XRD diffraction pattern of MBBS@TiO$_2$. M (Magnetite), A (Anatase), and H (Hematite).; Figure S10. UV-Vis absorption spectra of aqueous solutions of MB at different contact times using MBBS@mSiO$_2$ as adsorbent. ([MB]0 = 10 mg L^{-1}; MBBS@mSiO$_2$ dosage = 500 mg L^{-1}, T = 25 °C pH = 6.0); Figure S11. UV-Vis absorption spectra of aqueous solutions of MB at different irradiation times using MBBS@TiO$_2$ as photocatalyst. ([MB]0 = 5 mg L^{-1}; load MBBS@TiO$_2$ = 120 mg L^{-1}; load TiO$_2$ = 40 mg L^{-1}; pH = 6; Table S1: Total mass loss calculated by TGA for MBBS-X samples; Table S2: Magnetic properties of nanostructured core-shell prepared materials; Table S3: Zeta potential measurements of BBS coated magnetic iron oxide nanoparticles. Table S4: Chemical composition of BBS.

Author Contributions: Conceptualization, M.E.P. and L.C.; methodology, M.E.P. and A.K.-F.; investigation, M.E.P., A.K.-F. and M.S.M.; writing—original draft preparation, M.E.P., D.O.M. and L.C.; writing—review and editing, M.E.P., D.O.M. and L.C.; project administration, D.O.M., M.S.M. and L.C.; funding acquisition, D.O.M., M.S.M. and LC. All authors have read and agreed to the published version of the manuscript.

Funding: This research was funded by ANPCyT (PICT-1847-2017), UNCo (PIP 04/L009), and CONICET (PIP 11220200100768CO).

Data Availability Statement: Not applicable.

Acknowledgments: Authors would like to thank Giuliana Magnacca (UNITO, Italy) for providing BBS samples. M.E. Peralta thank CONICET for his postdoc grant. L. Carlos and M.S. Moreno are research members of CONICET (Argentina). D. Mártire is a research member of CIC (Buenos Aires, Argentina).

Conflicts of Interest: The authors declare no conflict of interest. The funders had no role in the design of the study; in the collection, analyses, or interpretation of data; in the writing of the manuscript; or in the decision to publish the results.

References

1. Nisticò, R.; Cesano, F.; Garello, F. Magnetic Materials and Systems: Domain Structure Visualization and Other Characterization Techniques for the Application in the Materials Science and Biomedicine. *Inorganics* **2020**, *8*, 6. [CrossRef]
2. Pathak, S.; Jain, K.; Kumar, P.; Wang, X.; Pant, R.P. Improved Thermal Performance of Annular Fin-Shell Tube Storage System Using Magnetic Fluid. *Appl. Energy* **2019**, *239*, 1524–1535. [CrossRef]
3. Pathak, S.; Jain, K.; Kumar, V.; Pant, R.P. Magnetic Fluid Based High Precision Temperature Sensor. *IEEE Sens. J.* **2017**, *17*, 2670–2675. [CrossRef]
4. Noh, B.-I.; Yang, S.C. Ferromagnetic, Ferroelectric, and Magnetoelectric Properties in Individual Nanotube-Based Magnetoelectric Films of CoFe$_2$O$_4$/BaTiO$_3$ Using Electrically Resistive Core-Shell Magnetostrictive Nanoparticles. *J. Alloys Compd.* **2022**, *891*, 161861. [CrossRef]

5. Peralta, M.E.; Ocampo, S.; Funes, I.G.; Medina, F.O.; Parolo, M.E.; Carlos, L. Nanomaterials with Tailored Magnetic Properties as Adsorbents of Organic Pollutants from Wastewaters. *Inorganics* **2020**, *8*, 24. [CrossRef]
6. Nisticò, R. Magnetic Materials and Water Treatments for a Sustainable Future. *Res. Chem. Intermed.* **2017**, *43*, 6911–6949. [CrossRef]
7. Carlos, L.; Garcia Einschlag, F.S.; González, M.C.; Mártire, D.O. Applications of Magnetite Nanoparticles for Heavy Metal Removal from Wastewater. In *Waste Water—Treatment Technologies and Recent Analytical Developments*; IntechOpen: London, UK, 2013; pp. 1–16. [CrossRef]
8. Fotukian, S.M.; Barati, A.; Soleymani, M.; Alizadeh, A.M. Solvothermal Synthesis of $CuFe_2O_4$ and Fe_3O_4 Nanoparticles with High Heating Efficiency for Magnetic Hyperthermia Application. *J. Alloys Compd.* **2020**, *816*, 152548. [CrossRef]
9. Liu, H.; Ji, S.; Zheng, Y.; Li, M.; Yang, H. Modified Solvothermal Synthesis of Magnetic Microspheres with Multifunctional Surfactant Cetyltrimethyl Ammonium Bromide and Directly Coated Mesoporous Shell. *Powder Technol.* **2013**, *246*, 520–529. [CrossRef]
10. Cheng, C.; Wen, Y.; Xu, X.; Gu, H. Tunable Synthesis of Carboxyl-Functionalized Magnetite Nanocrystal Clusters with Uniform Size. *J. Mater. Chem.* **2009**, *19*, 8782. [CrossRef]
11. Liu, J.; Sun, Z.; Deng, Y.; Zou, Y.; Li, C.; Guo, X.; Xiong, L.; Gao, Y.; Li, F.; Zhao, D. Highly Water-Dispersible Biocompatible Magnetite Particles with Low Cytotoxicity Stabilized by Citrate Groups. *Angew. Chemie Int. Ed.* **2009**, *48*, 5875–5879. [CrossRef]
12. Wang, W.; Tang, B.; Ju, B.; Zhang, S. Size-Controlled Synthesis of Water-Dispersible Superparamagnetic Fe_3O_4 Nanoclusters and Their Magnetic Responsiveness. *RSC Adv.* **2015**, *5*, 75292–75299. [CrossRef]
13. Deng, H.; Li, X.; Peng, Q.; Wang, X.; Chen, J.; Li, Y. Monodisperse Magnetic Single-Crystal Ferrite Microspheres. *Angew. Chem.* **2005**, *117*, 2842–2845. [CrossRef]
14. Ge, J.; Hu, Y.; Biasini, M.; Beyermann, W.P.; Yin, Y. Superparamagnetic Magnetite Colloidal Nanocrystal Clusters. *Angew. Chemie Int. Ed.* **2007**, *46*, 4342–4345. [CrossRef]
15. Liang, J.; Ma, H.; Luo, W.; Wang, S. Synthesis of Magnetite Submicrospheres with Tunable Size and Superparamagnetism by a Facile Polyol Process. *Mater. Chem. Phys.* **2013**, *139*, 383–388. [CrossRef]
16. Liu, Y.; Cui, T.; Li, Y.; Zhao, Y.; Ye, Y.; Wu, W.; Tong, G. Effects of Crystal Size and Sphere Diameter on Static Magnetic and Electromagnetic Properties of Monodisperse Fe_3O_4 Microspheres. *Mater. Chem. Phys.* **2016**, *173*, 152–160. [CrossRef]
17. Li, S.; Zhang, T.; Tang, R.; Qiu, H.; Wang, C.; Zhou, Z. Solvothermal Synthesis and Characterization of Monodisperse Superparamagnetic Iron Oxide Nanoparticles. *J. Magn. Magn. Mater.* **2015**, *379*, 226–231. [CrossRef]
18. Montoneri, E.; Prevot, A.B.; Avetta, P.; Arques, A.; Carlos, L.; Magnacca, G.; Laurenti, E.; Tabasso, S. CHAPTER 4:Food Wastes Conversion to Products for Use in Chemical and Environmental Technology, Material Science and Agriculture. In *RSC Green Chemistry*; Royal Society of Chemistry: London, UK, 2013; pp. 64–109. ISBN 9781849736152.
19. Quagliotto, P.; Montoneri, E.; Tambone, F.; Adani, F.; Gobetto, R.; Viscardi, G. Chemicals from Wastes: Compost-Derived Humic Acid-like Matter as Surfactant. *Environ. Sci. Technol.* **2006**, *40*, 1686–1692. [CrossRef]
20. Boffa, V.; Perrone, D.G.; Montoneri, E.; Magnacca, G.; Bertinetti, L.; Garlasco, L.; Mendichi, R. A Waste-Derived Biosurfactant for the Preparation of Templated Silica Powders. *ChemSusChem* **2010**, *3*, 445–452. [CrossRef]
21. Nisticò, R.; Carlos, L. High Yield of Nano Zero-Valent Iron (NZVI)from Carbothermal Synthesis Using Lignin-Derived Substances from Municipal Biowaste. *J. Anal. Appl. Pyrolysis* **2019**, *140*, 239–244. [CrossRef]
22. Nisticò, R.; Barrasso, M.; Carrillo Le Roux, G.A.; Seckler, M.M.; Sousa, W.; Malandrino, M.; Magnacca, G. Biopolymers from Composted Biowaste as Stabilizers for the Synthesis of Spherical and Homogeneously Sized Silver Nanoparticles for Textile Applications on Natural Fibers. *ChemPhysChem* **2015**, *16*, 3902–3909. [CrossRef] [PubMed]
23. Magnacca, G.; Allera, A.; Montoneri, E.; Celi, L.; Benito, D.E.; Gagliardi, L.G.; Gonzalez, M.C.; Mártire, D.O.; Carlos, L. Novel Magnetite Nanoparticles Coated with Waste-Sourced Biobased Substances as Sustainable and Renewable Adsorbing Materials. *ACS Sustain. Chem. Eng.* **2014**, *2*, 1518–1524. [CrossRef]
24. Nisticò, R.; Cesano, F.; Franzoso, F.; Magnacca, G.; Scarano, D.; Funes, I.G.; Carlos, L.; Parolo, M.E. From Biowaste to Magnet-Responsive Materials for Water Remediation from Polycyclic Aromatic Hydrocarbons. *Chemosphere* **2018**, *202*, 686–693. [CrossRef]
25. Nisticò, R.; Celi, L.R.; Bianco Prevot, A.; Carlos, L.; Magnacca, G.; Zanzo, E.; Martin, M. Sustainable Magnet-Responsive Nanomaterials for the Removal of Arsenic from Contaminated Water. *J. Hazard. Mater.* **2018**, *342*, 260–269. [CrossRef]
26. Franzoso, F.; Nisticò, R.; Cesano, F.; Corazzari, I.; Turci, F.; Scarano, D.; Bianco Prevot, A.; Magnacca, G.; Carlos, L.; Mártire, D.O. Biowaste-Derived Substances as a Tool for Obtaining Magnet-Sensitive Materials for Environmental Applications in Wastewater Treatments. *Chem. Eng. J.* **2017**, *310*, 307–316. [CrossRef]
27. Aparicio, F.; Mizrahi, M.; Ramallo-López, J.M.; Laurenti, E.; Magnacca, G.; Carlos, L.; Mártire, D.O. Novel Bimetallic Magnetic Nanocomposites Obtained from Waste-Sourced Bio-Based Substances as Sustainable Photocatalysts. *Mater. Res. Bull.* **2022**, *152*, 111846. [CrossRef]
28. Ayala, L.I.M.; Aparicio, F.; Boffa, V.; Magnacca, G.; Carlos, L.; Bosio, G.N.; Mártire, D.O. Removal of As(III) via Adsorption and Photocatalytic Oxidation with Magnetic Fe–Cu Nanocomposites. *Photochem. Photobiol. Sci.* **2022**, *1*, 1–10. [CrossRef] [PubMed]
29. Ou, X.; Chen, S.; Quan, X.; Zhao, H. Photochemical Activity and Characterization of the Complex of Humic Acids with Iron(III). *J. Geochem. Explor.* **2009**, *102*, 49–55. [CrossRef]
30. Yadav, R.S.; Havlica, J.; Kuřitka, I.; Kozakova, Z.; Masilko, J.; Hajdúchová, M.; Enev, V.; Wasserbauer, J. Effect of Pr^{3+} Substitution on Structural and Magnetic Properties of $CoFe_2O_4$ Spinel Ferrite Nanoparticles. *J. Supercond. Nov. Magn.* **2015**, *28*, 241–248. [CrossRef]

31. Arun, T.; Prabakaran, K.; Udayabhaskar, R.; Mangalaraja, R.V.; Akbari-Fakhrabadi, A. Carbon Decorated Octahedral Shaped Fe_3O_4 and α-Fe_2O_3 Magnetic Hybrid Nanomaterials for next Generation Supercapacitor Applications. *Appl. Surf. Sci.* **2019**, *485*, 147–157. [CrossRef]
32. Shirsath, S.E.; Kadam, R.H.; Mane, M.L.; Ghasemi, A.; Yasukawa, Y.; Liu, X.; Morisako, A. Permeability and Magnetic Interactions in Co^{2+} Substituted $Li_{0.5}Fe_{2.5}O_4$ Alloys. *J. Alloys Compd.* **2013**, *575*, 145–151. [CrossRef]
33. Vargas, J.M.; Shukla, D.K.; Meneses, C.T.; Mendoza Zélis, P.; Singh, M.; Sharma, S.K. Synthesis, Phase Composition, Mössbauer and Magnetic Characterization of Iron Oxide Nanoparticles. *Phys. Chem. Chem. Phys.* **2016**, *18*, 9561–9568. [CrossRef]
34. Kim, W.; Suh, C.Y.; Cho, S.W.; Roh, K.M.; Kwon, H.; Song, K.; Shon, I.J. A New Method for the Identification and Quantification of Magnetite-Maghemite Mixture Using Conventional X-Ray Diffraction Technique. *Talanta* **2012**, *94*, 348–352. [CrossRef] [PubMed]
35. Cuenca, J.A.; Bugler, K.; Taylor, S.; Morgan, D.; Williams, P.; Bauer, J.; Porch, A. Study of the Magnetite to Maghemite Transition Using Microwave Permittivity and Permeability Measurements. *J. Phys. Condens. Matter* **2016**, *28*, 106002. [CrossRef]
36. Arun, T.; Prakash, K.; Kuppusamy, R.; Joseyphus, R.J. Magnetic Properties of Prussian Blue Modified Fe_3O_4 Nanocubes. *J. Phys. Chem. Solids* **2013**, *74*, 1761–1768. [CrossRef]
37. Dutta, P.; Pal, S.; Seehra, M.S.; Shah, N.; Huffman, G.P. Size Dependence of Magnetic Parameters and Surface Disorder in Magnetite Nanoparticles. *J. Appl. Phys.* **2009**, *105*, 10–13. [CrossRef]
38. Peralta, M.E.; Nisticò, R.; Franzoso, F.; Magnacca, G.; Fernandez, L.; Parolo, M.E.; León, E.G.; Carlos, L. Highly Efficient Removal of Heavy Metals from Waters by Magnetic Chitosan-Based Composite. *Adsorption* **2019**, *25*, 1337–1347. [CrossRef]
39. Zhao, Q.; Xie, P.; Li, X.; Wang, Y.; Zhang, Y.; Wang, S. Magnetic Mesoporous Silica Nanoparticles Mediated Redox and PH Dual-Responsive Target Drug Delivery for Combined Magnetothermal Therapy and Chemotherapy. *Colloids Surf. A Physicochem. Eng. Asp.* **2022**, *648*, 129359. [CrossRef]
40. Kamarudin, N.H.N.; Jalil, A.A.; Triwahyono, S.; Sazegar, M.R.; Hamdan, S.; Baba, S.; Ahmad, A. Elucidation of Acid Strength Effect on Ibuprofen Adsorption and Release by Aluminated Mesoporous Silica Nanoparticles. *RSC Adv.* **2015**, *5*, 30023–30031. [CrossRef]
41. Bian, Q.; Xue, Z.; Sun, P.; Shen, K.; Wang, S.; Jia, J. Visible-Light-Triggered Supramolecular Valves Based on β-Cyclodextrin-Modified Mesoporous Silica Nanoparticles for Controlled Drug Release. *RSC Adv.* **2019**, *9*, 17179–17182. [CrossRef]
42. Thommes, M.; Kaneko, K.; Neimark, A.V.; Olivier, J.P.; Rodriguez-Reinoso, F.; Rouquerol, J.; Sing, K.S.W. Physisorption of Gases, with Special Reference to the Evaluation of Surface Area and Pore Size Distribution (IUPAC Technical Report). *Pure Appl. Chem.* **2015**, *87*, 1051–1069. [CrossRef]
43. Jin, C.; Wang, Y.; Wei, H.; Tang, H.; Liu, X.; Lu, T.; Wang, J. Magnetic Iron Oxide Nanoparticles Coated by Hierarchically Structured Silica: A Highly Stable Nanocomposite System and Ideal Catalyst Support. *J. Mater. Chem. A* **2014**, *2*, 11202–11208. [CrossRef]
44. Hou, X.; Xu, H.; Pan, L.; Tian, Y.; Zhang, X.; Ma, L.; Li, Y.; Zhao, J. Adsorption of Bovine Serum Albumin on Superparamagnetic Composite Microspheres with a Fe_3O_4/SiO_2 Core and Mesoporous SiO_2 Shell. *RSC Adv.* **2015**, *5*, 103760–103766. [CrossRef]
45. Nicola, R.; Muntean, S.G.; Nistor, M.A.; Putz, A.M.; Almásy, L.; Săcărescu, L. Highly Efficient and Fast Removal of Colored Pollutants from Single and Binary Systems, Using Magnetic Mesoporous Silica. *Chemosphere* **2020**, *261*, 127737. [CrossRef]
46. Nicola, R.; Costişor, O.; Muntean, S.G.; Nistor, M.A.; Putz, A.M.; Ianăşi, C.; Lazău, R.; Almásy, L.; Săcărescu, L. Mesoporous Magnetic Nanocomposites: A Promising Adsorbent for the Removal of Dyes from Aqueous Solutions. *J. Porous Mater.* **2020**, *27*, 413–428. [CrossRef]
47. Jiang, Y.; Liang, W.; Wang, B.; Feng, Q.; Xia, C.; Wang, Q.; Yan, Y.; Zhao, L.; Cui, W.; Liang, H. Magnetic Mesoporous Silica Nanoparticles Modified by Phosphonate Functionalized Ionic Liquid for Selective Enrichment of Phosphopeptides. *RSC Adv.* **2022**, *12*, 26859–26865. [CrossRef]
48. Gómez-Pastora, J.; Dominguez, S.; Bringas, E.; Rivero, M.J.; Ortiz, I.; Dionysiou, D.D. Review and Perspectives on the Use of Magnetic Nanophotocatalysts (MNPCs) in Water Treatment. *Chem. Eng. J.* **2017**, *310*, 407–427. [CrossRef]
49. Salve, R.; Kumar, P.; Ngamcherdtrakul, W.; Gajbhiye, V.; Yantasee, W. Stimuli-Responsive Mesoporous Silica Nanoparticles: A Custom-Tailored next Generation Approach in Cargo Delivery. *Mater. Sci. Eng. C* **2021**, *124*, 112084. [CrossRef]
50. Wang, D.; Yang, J.; Li, X.; Wang, J.; Zhai, H.; Lang, J.; Song, H. Effect of Thickness and Microstructure of TiO_2 Shell on Photocatalytic Performance of Magnetic Separable $Fe_3O_4/SiO_2/MTiO_2$ Core-Shell Composites. *Phys. Status Solidi Appl. Mater. Sci.* **2017**, *214*, 1600665. [CrossRef]
51. Mamba, G.; Mishra, A. Advances in Magnetically Separable Photocatalysts: Smart, Recyclable Materials for Water Pollution Mitigation. *Catalysts* **2016**, *6*, 79. [CrossRef]
52. Deng, Y.; Qi, D.; Deng, C.; Zhang, X.; Zhao, D. Superparamagnetic High-Magnetization Microspheres with an $Fe_3O_4@SiO_2$ Core and Perpendicularly Aligned Mesoporous SiO_2 Shell for Removal of Microcystins. *J. Am. Chem. Soc.* **2008**, *130*, 28–29. [CrossRef]
53. Peralta, M.E.; Jadhav, S.A.; Magnacca, G.; Scalarone, D.; Mártire, D.O.; Parolo, M.E.; Carlos, L. Synthesis and in Vitro Testing of Thermoresponsive Polymer-Grafted Core-Shell Magnetic Mesoporous Silica Nanoparticles for Efficient Controlled and Targeted Drug Delivery. *J. Colloid Interface Sci.* **2019**, *544*, 198–205. [CrossRef] [PubMed]
54. Zhang, Q.; Meng, G.; Wu, J.; Li, D.; Liu, Z. Study on Enhanced Photocatalytic Activity of Magnetically Recoverable $Fe_3O_4@C@TiO_2$ Nanocomposites with Core-Shell Nanostructure. *Opt. Mater.* **2015**, *46*, 52–58. [CrossRef]

Disclaimer/Publisher's Note: The statements, opinions and data contained in all publications are solely those of the individual author(s) and contributor(s) and not of MDPI and/or the editor(s). MDPI and/or the editor(s) disclaim responsibility for any injury to people or property resulting from any ideas, methods, instructions or products referred to in the content.

Article

Sacrificial Zinc Oxide Strategy-Enhanced Mesoporosity in MIL-53-Derived Iron–Carbon Composite for Methylene Blue Adsorption

Sander Dekyvere [1,2,†], Mohamed Elhousseini Hilal [1,2,3,†], Somboon Chaemchuen [1,2], Serge Zhuiykov [4] and Francis Verpoort [1,2,5,*]

1. State Key Laboratory of Advanced Technology for Materials Synthesis and Processing, Wuhan University of Technology, Wuhan 430070, China; sander4a@hotmail.com (S.D.); melhousseinihilal@hkcoche.org (M.E.H.); sama_che@hotmail.com (S.C.)
2. School of Materials Science and Engineering, Wuhan University of Technology, Wuhan 430070, China
3. Hong Kong Centre for Cerebro-Cardiovascular Health Engineering (COCHE), Building 17W, Hong Kong Science Park, Hong Kong 999077, China
4. Department of Solid State Science, Faculty of Science, Ghent University Global Campus, Songdo, 119 Songdomunhwa-Ro, Yeonsu-Gu, Incheon 404-840, Korea; serge.zhuiykov@ghent.ac.kr
5. National Research Tomsk Polytechnic University, Lenin Avenue 30, 634034 Tomsk, Russia
* Correspondence: francis@whut.edu.cn
† These authors contributed equally to this work.

Abstract: MOF-derived carbon-based materials have attracted widespread attention due to their relatively large surface area, morphology, and their stability in water. Considering these advantages, these materials present themselves as excellent adsorbents. In this work, a novel method was designed for the fabrication of a nano zero-valent-iron (nZVI) carbon composite. The utilization of zinc oxide nanorods (ZnONRs) in the role of sacrificial consumable nuclei for the synthesis of MIL-53 sacrificial zinc oxide nanorods (MIL-53-SNR) and the subsequent pyrolysis at 700 °C in the inert atmosphere led to a graphitic-supported nZVI material (Fe-C-SNR). Fe-C-SNR was compared with a commercial zinc oxide bulk (MIL-53-SB) and with a pristine MIL-53. By virtue of the ZnONRs, Fe-C-SNR exhibited a greatly improved mesoporous structure. Consequently, the pyrolyzed materials were applied as adsorbents for methylene blue. Fe-C-SNR's performance increased to more than double of the pyrolyzed MIL-53 (Fe-C), with a remarkably fast adsorption time (10 min) for a concentration of 10 mg L^{-1} with only 200 mg L^{-1} adsorbent required. This functional composite also displayed exceptional recyclability; after ten complete cycles, Fe-C-SNR was still capable of completely adsorbing the methylene blue. The utilization of ZnONRs proves itself advantageous and could further be extended to other MOFs for a wide range of applications.

Keywords: mesoporous material; zero-valent-iron nanoparticles; methylene blue adsorption; metal–organic framework

Citation: Dekyvere, S.; Hilal, M.E.; Chaemchuen, S.; Zhuiykov, S.; Verpoort, F. Sacrificial Zinc Oxide Strategy-Enhanced Mesoporosity in MIL-53-Derived Iron–Carbon Composite for Methylene Blue Adsorption. *Inorganics* **2022**, *10*, 59. https://doi.org/10.3390/inorganics10050059

Academic Editor: Roberto Nisticò

Received: 2 April 2022
Accepted: 22 April 2022
Published: 25 April 2022

Publisher's Note: MDPI stays neutral with regard to jurisdictional claims in published maps and institutional affiliations.

Copyright: © 2022 by the authors. Licensee MDPI, Basel, Switzerland. This article is an open access article distributed under the terms and conditions of the Creative Commons Attribution (CC BY) license (https://creativecommons.org/licenses/by/4.0/).

1. Introduction

Clean water is an indisputable necessity for humanity and the ever-increasing pollutant levels present threats to human health and our environment. Dye-containing wastewaters are some of the main contributors to the pollution of natural water, as they are toxic, carcinogenic, and are not biodegradable [1]. The primary sources of dye-containing wastewater are industrial sectors such as textile, pharmaceutical, food, mining, paper, printing, and leather industries. Of the main three categories of dyes, cationic dyes are of particular interest. Specifically, methylene blue (MB) is stable under light and heat and is difficult to biodegrade because of its complex structure [2]. Thus, developing effective and economical methods for removing MB dye from wastewater is a necessity. Many treatment techniques have been studied over the previous decades, such as photodegradation, biological and chemical degradation, chemical and physical adsorption, and combinations of these treatments [3]. Among these methods, adsorption shows many practical advantages, such as low cost, facile operation, and quick removal [4]. A wide variety of adsorbents have been applied to remove dyes. Conventional porous materials that have been widely studied include zeolites, metal oxide nanoparticles, and activated carbons [5]. However, all these materials are associated with similar drawbacks, including their low surface area and recyclability. Over the last few years, zero-valent-iron nanoparticles (nZVI) have gained increasing interest in wastewater treatment due to their large surface area, small particle size, high dispersibility in water, and magnetic separation [6–11]. The main challenge for the practical application of nZVIs is the synthesis and preservation of the material. nZVIs are readily oxidized in the presence of other oxidants, ref. [12], forming oxide layers on the surface of the nZVI, which unavoidably reduces their practical performance for this application [13]. Therefore, porous materials such as bentonite [14], chitosan [15], diatomite [16], kaolinite [17], polyphenols [18], resin [19], sepiolite [20], and activated carbon [21] have been used to support or modify nZVIs to prevent agglomeration and oxidation. This approach is widely regarded as an efficient and facile pathway, as the Fe^0 particles become more stable and the surface area and density of the functional groups can be increased [22,23]. Porous carbons are particularly regarded as exceptional hosts for nZVIs, endowing an enhanced surface area, highly mesoporous structure, good adsorption capacity, and providing the necessary protection from excessive detrimental surface oxidations [24–26]. Therefore, they present a higher activity, stability, and mobility towards removing various pollutants in wastewater compared to bare nZVIs. The composite of iron–carbon can be synthesized with generally facile methods using iron salts and porous carbon [27]. Even so, the dispersion and loading content of Fe^0 would still need to be improved [28]. In this context, metal–organic frameworks (MOFs) are materials constructed by organic ligands and metal clusters. Recently, they have been utilized as precursors for the synthesis of metal/carbon composites, which provides the needed loading and dispersion for Fe^0. Iron–carbon composites have been obtained by the pyrolysis of Fe-MOFs, e.g., MIL-53, by adjusting the temperature depending on the required iron phase [23,29,30]. However, the challenge of creating suitable pore sizes in the carbonized material remains.

This work reports a novel strategy for fabricating an iron–carbon-based mesoporous material derived from an iron-containing MOF. The first stage, represented in Scheme 1, is a modified synthesis of MIL-53 using zinc oxide nanorods as sacrificial nuclei named 'MIL-53-SNR'. The terephthalic acid binds with the outer surface of the ZnO, thereby etching zinc metal from the zinc oxide nanorods to form a MOF-5 layer. Simultaneously, the Zn in the MOF-5 is replaced by Fe, which synthesizes MIL-53. During these processes, the zinc gradually diffuses outward of the nuclei and thereby creates a highly defective MIL-53-SNR structure. The second step is the pyrolysis of the material under a nitrogen atmosphere at 700 °C. The as-prepared novel material iron–carbon sacrificial zinc oxide nanorods 'Fe-C-SNR' are characterized by XRD, FT-IR, SEM, and BET. A Fe-C-SNR is used as an adsorbent for methylene blue, displaying an excellent maximum adsorption capacity of 257 mg g^{-1} compared to the maximum of 125 mg g^{-1} by iron–carbon 'Fe-C' synthesized by the pyrolysis of a pristine MIL-53. To further assess the effect of the ZnO nanorods on performance, the third sample, iron–carbon sacrificial ZnO bulk 'Fe-C-SB', was prepared to evaluate the relationship between the ZnO morphology and the MB adsorption.

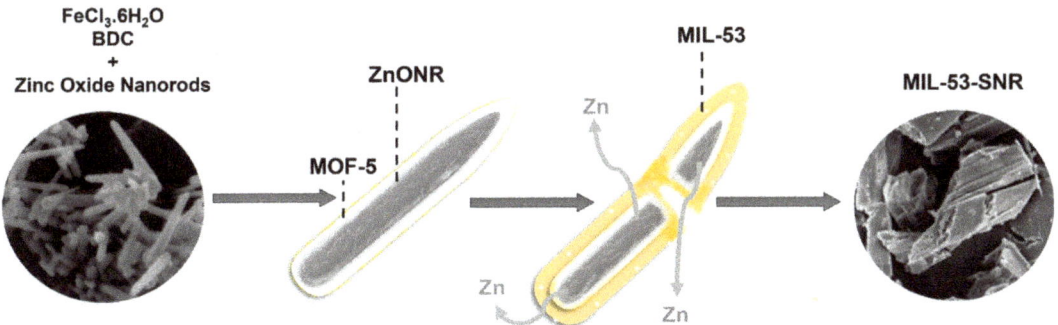

Scheme 1. Graphical representation of the sacrificial utilization of zinc oxide nanorods during the synthesis of MIL-53-SNR.

2. Results and Discussion

2.1. Characterizations

In this study, ZnO bulk and ZnONRs were implemented in the synthesis process to modify MIL-53. Commercial ZnO bulk possesses agglomerated particles of a few hundred micrometers in each direction. Scanning electron microscopy (SEM) was performed on as-synthesized zinc oxide nanorods, as illustrated in Figure S1. The nanorods displayed a smooth surface with an average length of 650 nm and a width of 60 nm; the XRD pattern is depicted in Figure S2. The X-ray diffraction patterns of MIL-53, MIL-53-SB, and MIL-53-SNR are displayed in Figure 1a and Figure S3; the peaks at 9.2°, 9.7°, 12.6°, 17.6°, 18.6°, 18.8°, and 25.4° were present in all three samples. The doublet peak at 9.2° and 9.7° corresponds to the crystalline plane (001) of MIL-53. The peaks at 12.6° and 25.4° are assigned to the (100) and (010) crystal planes. These results confirm the successful synthesis of MIL-53 in all three samples [31,32]. Furthermore, the peaks at 17.6°, 18.6°, and 18.8° match with the reported XRD patterns for MIL-53 [31,32]. The relatively small peak at 21.9°, present in MIL-53 and MIL-53-SB, is ascribed to the formation of MIL-88-B, which is synthesized through a similar procedure as MIL-53 but with some minor adaptations [33–35]. However, MIL-88-B can be a side product during the MIL-53 synthesis [36]. It has been reported that MIL-53 forms through homogenous nucleation and MIL-88-B forms through heterogeneous nucleation [34,35]; however, the peak matching with MIL-88-B was absent in the XRD pattern of MIL-53-SNR, which indicates that no heterogeneous nucleation occurred during the synthesis [34,35]. This difference between the zinc oxide bulk and the zinc oxide nanorods is attributed to the smaller size of the nanorods, which are quickly etched compared to the zinc oxide bulk. The successful preparation of MIL-53 was further confirmed by Fourier Transform Infrared (FTIR). The FTIR spectra of MIL-53, MIL-53-SB, and MIL-53-SNR are shown in Figure 1b. The broad peaks around 3300 cm^{-1} are ascribed to the -OH stretching vibrations of the water molecules adsorbed on the surface [37]. The absorption peaks in the region of 1400–1700 cm^{-1} exhibit the typical vibrational bands of the carboxylic acid function from the 1,4-BDC ligand [31,37]. A couple of sharp peaks around 1540 cm^{-1} and 1390 cm^{-1}, present in all three samples, correspond to the carboxyl groups' asymmetric and symmetric vibrations [37,38]. Typically, pure 1,4-BDC shows a peak at 1693 cm^{-1}; however, none of the samples exhibited this peak, confirming the absence of free BDC molecules. Instead, the band at 530 cm^{-1} represents the metal-oxo bound between the carboxylic group of BDC and Fe [31]. These results further confirm the successful formation of MIL-53 in all three samples. The remaining peaks at 750 cm^{-1}, 1665 cm^{-1}, and 1015 cm^{-1} are assigned to the C-H bending vibrations of the benzene rings and the carboxylate groups' vibrations, respectively [31].

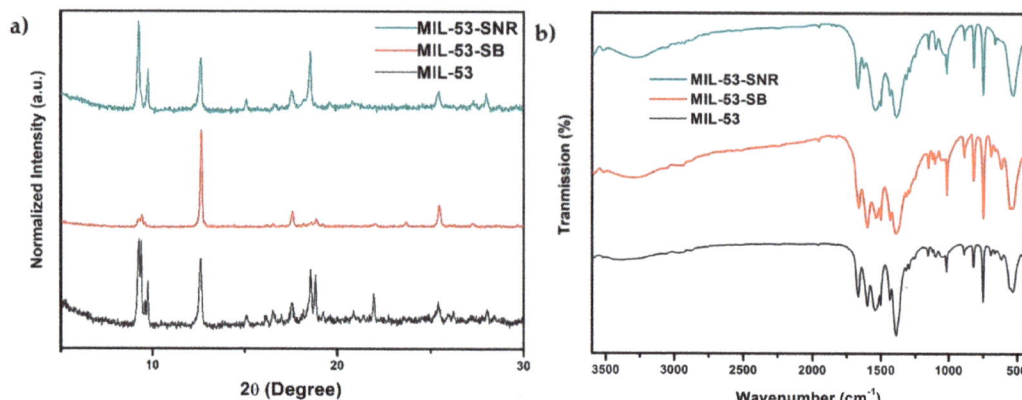

Figure 1. Structural analysis of MIL-53, MIL-53SB, and MIL-53-SNR, (**a**) XRD pattern, (**b**) FTIR spectra.

The influence of the sacrificial zinc oxide precursors on the morphology of the samples was further examined by SEM. The SEM micrographs of MIL-53, MIL-53-SB, and MIL-53-SNR are portrayed in Figure 2. The pristine MIL-53, Figure 2a, showed crystalline microrods of approximately 63.5 μm and 7.5 μm in length and width, respectively. These microrods had smooth surfaces and displayed a small amount of debris. Figure 2b represents the micrograph of MIL-53-SB. It is apparent that the MOF particles were less uniform, varying from small microrods to fragments of micro blocks; their texture was slightly rougher with a disordered structure.

In contrast, MIL-53-SNR in Figure 2c displayed fragmented micro shapes with a length of 12.54 μm (±2 μm) and a width of 4.84 μm (±1.3 μm), and hardly any microrods were apparent. The surface was extremely rough, and that comes in accordance with its lower XRD diffraction intensities as compared to the ZnO bulk-modified sample. This further indicates the lack of crystallinity in MIL-53-SNR due to the ZnO NRs. For these sacrificial ZnO materials, terephthalic acid is highly likely to bond with the outer surface of the ZnO, severely etching the zinc metal from the ZnO materials and creating MOF-5 layers, similar to previous research with ZIF-8 [39]. This etching process is supposedly continuous until the zinc source is consumed. Nevertheless, MOF-5 does not remain due to the favorable simultaneous process of metal exchange between Zn and Fe, which concurrently creates MIL-53. Therefore, ZnO bulk and ZnO nanorods serve as sacrificial nuclei. Due to their difference in morphology, they exhibit different outcomes. Massive and agglomerated ZnO bulk gets coated and slowly consumed by the 1,4-BDC ligand, serving as nuclei that eventually form MIL-53-SB as fragmented microrods in irregular shapes and sizes instead of the standard MIL-53 microrods. In contrast, ZnO nanorods are thin and well dispersed in the starting solution; thus, they engender more apparent defects as each adjacent nanorod will separately serve as a nuclei for the growth of MIL-53. Assumedly, the closely growing MOFs fuse, forming fragmented microrods with a disordered internal structure, causing the decrease in crystallinity shown in the XRD pattern. Figure 2d shows that Fe-C-SNR maintained the MIL-53-SNR morphology throughout the pyrolysis procedure.

The inductively coupled plasma spectroscopy (ICP) results are presented in Table S1, describing the content of the iron and zinc metal in the as-prepared samples MIL-53-SNR and MIL-53-SB and their pyrolyzed equivalents Fe-C-SNR and Fe-C-SB. Before pyrolysis, both samples showed the expected contents of Fe (21%), matching the reported value of Fe in MIL-53 [38]. In MIL-53-SNR and in MIL-53-SB, the amount of Zn metal detected was extremely low compared to the amount added during the synthesis, which further confirms the sacrificial role of the ZnO in the construction of MIL-53. The zinc metal gets etched to form MOF-5 and gradually diffuses outward of the nuclei during the metathesis replacement with iron. The relatively higher amount of the ZnO (13 times more) present in

MIL-53-SB than in MIL-53-SNR shows that this process of etching, replacing, and diffusing the Zn is more effective when using thin nanorods than bulk zinc oxide due to their lower concentration of localized zinc.

Figure 2. SEM micrographs of: (a) MIL-53; (b) MIL-53-SB; (c) MIL-53-SNR; (d) Fe-C-SNR.

The XRD patterns of the pyrolyzed samples, Fe-C, Fe-C-SB, and Fe-C-SNR, are displayed in Figure 3a. The prominent characteristic diffraction peak at 44.6°, present in all three samples, corresponds to the (110) plane of α-Fe, matching the reported XRDs for ZVI [1]. No other peaks were present, confirming iron is only present as a zero valent. However, Fe-C-SNR displayed a significant decrease in peak width and intensity compared to Fe-C and Fe-C-SB, affirming its smaller particle sizes and less crystalline ZVI phase. The nitrogen adsorption and desorption isotherms, depicted in Figure 3b, were measured to assess the porous structure of the pyrolyzed samples. The BET surface area results for Fe-C, Fe-C-SB, and Fe-C-SNR were 270 m^2 g^{-1}, 220 m^2 g^{-1}, and 185 m^2 g^{-1}, respectively. The volume and surface area of the micropores and mesopores for Fe-C, Fe-C-SB, and Fe-C-SNR are displayed in Table S2 and Figure 3c. The results for the pore distribution of Fe-C, Fe-C-SB, and Fe-C-SNR displayed normal differences in the microporous region below 5 nm, with Fe-C showing a superior pore volume. On the contrary, in the mesoporous region below 20 nm, Fe-C-SNR showed a remarkable improvement, reaching a volume of 0.169 cm^3 g^{-1} and a corresponding surface area of 92 m^2 g^{-1}. This increase in mesopores is linked to the disordered internal structure and the respectively formed defects caused after the pyrolysis of MIL-53-SNR due to the sacrificial zinc oxide nanorods. However, similar progress of the mesopores is not noticeable in Fe-C-SB, as the bulk zinc oxide did not create a similar disordered structure, therefore not favoring the generation of small-scale mesopores. The increase in small mesopores greatly enhanced Fe-C-SNR's MB removal capabilities.

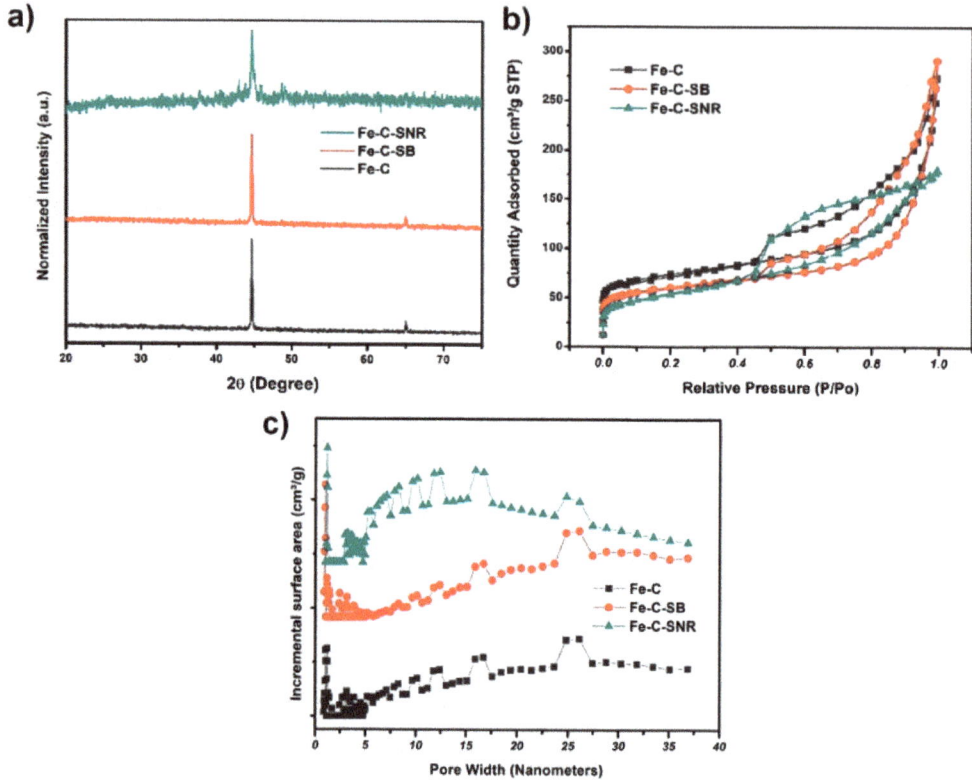

Figure 3. Crystallinity and adsorption analysis (**a**) XRD pattern, (**b**) N2 sorption isotherms, (**c**) pore distribution of Fe-C, Fe-C-SB, Fe-C-SNR.

2.2. Adsorption Properties

Fe-C, Fe-C-SB, and Fe-C-SNR were tested as methylene blue adsorbents, and their performances are presented in Figure 4. The UV-absorption patterns are shown in Figure S4. The concentration of the adsorbents used for the experiments was 200 mg L^{-1}, and the concentration of the methylene blue solution was 10 mg L^{-1}. The times selected were 2, 6, 10, and 30 min. The remaining dye in the solution was measured with UV spectroscopy; thus, the adsorbed MB could be calculated. Fe-C adsorbed 77% of MB after 6 min and reached its maximum adsorption of 85% in 30 min. The addition of the sacrificial ZnO bulk in the synthesis procedure of Fe-C-SB had a negative effect on the removal efficiency, which only reached 63% after 6 min, with a maximum adsorption of 73% after 30 min. This was expected as the N2 adsorption showed a noticeable decrease in early mesopores due to the use of ZnO bulk.

In contrast, the modification with the sacrificial ZnO nanorods stimulated an excellent improvement in the removal efficiency. Figure 4 displays the astonishing adsorption of 78% in the first 2 min, while after 6 min, a closely total adsorption of 99% was achieved. Further continuing the processing time until 30 min increased the adsorption of MB to full adsorption (100%). Although the surface area of Fe-C-SNR was the lowest of all three samples, the adsorption was greatly enhanced in comparison with Fe-C and Fe-C-SB. This sharp enhancement of MB adsorption was ascribed to the as-confirmed increase in the mesopores, which are superior to micropores for MB adsorption [40,41]. Moreover, due to the smaller particle size of Fe-C-SNR, the pores were easier and accessible faster.

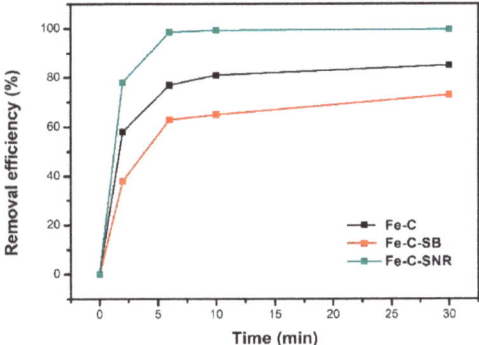

Figure 4. Comparison between the removal efficiencies of Fe-C, Fe-C-SB, Fe-C-SNR for methylene blue.

The adsorption kinetics were investigated using pseudo-first-order and pseudo-second-order kinetic models. Respectively, the experimental data were fitted into the Equations (1) and (2):

$$\ln(q_e - q_t) = \ln q_e - k_1 t \quad (1)$$

$$\frac{t}{q_t} = \frac{1}{k_2 q_e^2} + \frac{t}{q_e} \quad (2)$$

where q_e and q_t represent the amount of methylene blue adsorbed on the adsorbents at equilibrium, and time t. k_1 (min^{-1}) and k_2 (g mg^{-1} min^{-1}) are the rate constants of the first-order and second-order adsorption kinetics. The values of q_e and k_1 were calculated for the pseudo-first-order model. The values of q_e and k_2 were calculated for the pseudo-second-order model. Their linear plots are shown in Figure S5, and Table S3 presents the calculated values and the correlation coefficient R^2. It is seen by the $R^2 = 0.9997$ that the pseudo-second-order model describes the adsorption kinetics for MB on Fe-C-SNR well.

The effect of the different initial concentrations (10, 20, 40, 60, 80, 100, and 200 mg L^{-1}) was evaluated to study the maximum MB adsorption. The set treatment time was 60 min for each sample to ensure the full adsorption of MB onto Fe-C-SNR. The adsorption isotherm is shown in Figure S6. These results are an excellent match for the Langmuir model, which is a widely used model based on the assumption of monolayer adsorption on a homogenous surface. It is represented as follows:

$$= \frac{q_m b C_e}{1 + b C_e}$$

This equation can also be linearized:

$$\frac{C_e}{q_e} = \frac{1}{q_m b} + \frac{C_e}{q_m}$$

where C_e (mg L^{-1}) represents the equilibrium concentration of MB in the solution, Q_e and Q_m (mg g^{-1}) are the adsorption capacity at equilibrium and maximum, respectively, and b (L mg^{-1}) is the Langmuir constant related to the energy of adsorption. The Langmuir linear adsorption isotherm is shown in Figure 5a.

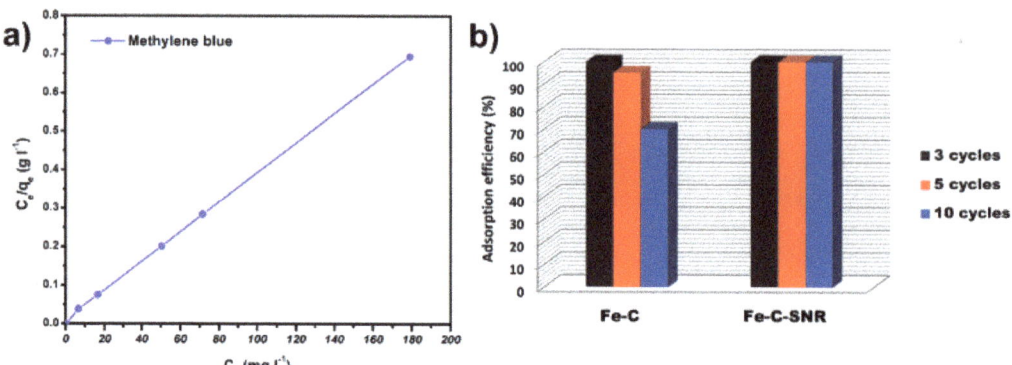

Figure 5. (**a**) Langmuir Linear adsorption isotherm, (**b**) recyclability of MB adsorption for Fe-C, Fe-C-SNR up to 10 cycles.

Figure 5a displays the linear equation of the Fe-C-SNR correlation coefficient (R^2) of 0.9995. Therefore, we can conclude that the adsorption of MB on Fe-C-SNR is a typical monomolecular-layer adsorption. Figure S7 presents the results fitted into the linear equation of the Freundlich model, presenting a correlation coefficient of 0.8479. Following the Langmuir model, the calculated maximum adsorption capacities of Fe-C, Fe-C-SB, and Fe-C-SNR were 125 mg g^{-1}, 139 mg g^{-1}, and 257 mg g^{-1}, respectively. A comparison with other materials reported so far in the literature is available in Table S4. The reuse of an adsorbent is of crucial importance for commercial applications. Furthermore, the methylene blue concentrations that need to be adsorbed in commercial applications are very low. Dependent on the exact location, the permitted methylene blue concentration is as low as 0.2 mg L^{-1}. Hence, it is of high importance that the adsorbent can almost entirely adsorb the MB in a solution even at low concentrations. The recyclability of Fe-C-SNR and Fe-C was determined by a series of MB adsorption experiments. The adsorbent (concentration: 1 g L^{-1}) was added to a 10 mg L^{-1} MB solution, and the treatment time was set to 60 min to reach the equilibrium time for both samples. Afterwards, the dye-adsorbed composites were collected using an external magnet and desorption was performed by using ethanol. Subsequently, the samples were washed with water to remove any adsorbed ethanol. Finally, the adsorbent was reused for up to 10 cycles. The results are presented in Figure 5b. After five cycles, the Fe-C still had a removal efficiency of 94.9%, whereas, after 10 cycles, the efficiency dropped to 70%. The Fe-C-SNR gave an excellent recyclability of 99.8% after five cycles and 99% after 10 cycles. The superior recyclability of Fe-C-SNR in contrast with Fe-C is expected to be due to the already fragmented microrods, which are less likely to be broken during the shaking, desorption, and washing process. To confirm the stability of the material, the XRD pattern before adsorption, with MB adsorbed, and after desorption were measured and are shown in Figure S8. No significant difference was observed, confirming the stability of the material and further confirming the remarkable stability. The highly enhanced maximum adsorption capacity and excellent recyclability makes Fe-SNR a remarkably outstanding adsorbent.

3. Materials and Methods

3.1. Methods

All chemicals were purchased from commercial products and used without further purification. Zinc oxide nanorods (ZnO NR) were fabricated by heating zinc acetate (Aladdin Ltd., Shanghai, China) at 300 °C for 12 h in a muffle furnace (TL 1200, Nanjing Bo Yun Tong Instrument Technology Co., Ltd., Nanjing, China). The ZnO NR were cooled and stored for further processing [42]. The MIL-53 (Fe), MIL-53-SNR, and MIL-53-SB were prepared by a solvothermal method. Firstly, 1.4 mmol ZnO NR/ZnO bulk (Sinopharm Chemical Reagent

Co., Ltd., Shanghai, China) were transferred in 20 mL N,N-dimethylformamide (DMF, Sinopharm Chemical Reagent Co., Ltd., Shanghai, China) for MIL-53-SNR and MIL-53-B, respectively. Then, 7 mmol of FeCl3.6H2O (Aladdin Ltd., Shanghai, China) and 7 mmol of terephthalic acid (BDC, Aladdin Ltd., Shanghai, China) were added to the solution together with 22 mL DMF. The mixture was stirred for 30 min before being transferred into a Teflon-lined autoclave and heated up at 150 °C for 24 h. The formed solid was collected by centrifugation and washed with DMF and ethanol (Sinopharm Chemical Reagent Co., Ltd., Shanghai, China) multiple times. Finally, the samples were dried at 70 °C in a vacuum oven for 24 h. In the second step, F-C, F-C-SNR, and F-C-SB were obtained using a pyrolysis system. Respectively, MIL-53, MIL-53-SNR, and MIL-53-SB (200 mg) were placed in an alumina boat (30 × 60 × 15 mm^3) and transferred into a quartz tube (OD:60 mm, length: 1000 mm) inside a muffle furnace (TL 1200, Nanjing Bo Yun Tong Instrument Technology Co., Ltd., Nanjing, China) and pyrolyzed at 700 °C with a heating rate of 5 °C min^{-1} for 3 h under a flowing Argon atmosphere (50 cm^3 min^{-1}) and cooled down naturally.

3.2. Characterizations

The material morphology was examined using scanning electron microscopy (FE-SEM, Zeiss Ultra Plus, Oberkochen, Germany). The Powder X-ray diffraction (PXRD) patterns of the samples were recorded on a Rigaku Ultima III diffract meter (Tokyo, Japan) in a wide-angle range (2θ = 3–80°) with Cu-Kα (λ = 0.15406 nm) 40 kV and 40 mA at a scan speed of 2° min^{-1}. The Fourier Transform Infrared Spectroscopy (FTIR) profiles of the samples were measured on a Bruker Tensor 2 apparatus (Ettlingen, Germany) using KBr as reference. The Brunauer–Emmett–Teller (BET) measurements and N2 adsorption–desorption isotherms were measured with a Porosity Analyzer ASAP 2020 Micromeritics apparatus (Norcross, United States of America) at 77.33 K. The inductively coupled plasma atomic emission spectroscopy (ICP-AES) was performed on an Optima 4300 DV, PerkinElmer Inc., Waltham, United States of America. The pH was measured using a Sartorius PB-10, Göttingen, Germany, at room temperature (25 °C). The UV–Vis absorption spectra were measured with a Shimadzu UV-1800 spectrophotometer, Canby, United States of America, using a quartz cuvette.

3.3. Adsorption Experiments

The methylene blue (MB) adsorption experiments were conducted at room temperature (25 °C). In a typical experiment, 2 mg of the adsorbent was brought into 10 mL of 10 mg L^{-1} methylene blue solution at its natural pH of 6.2. Then, the mixture was shaken for specific periods. At the required time intervals, a magnet was held next to the mixture to separate the adsorbent from the solution, then 2 mL of a liquid sample was taken, and the dye concentration was determined using spectroscopy by testing the absorbance at k$_{max}$ of 664 nm for MB. The MB concentrations were calculated using a calibration curve. The adsorbed MB concentrations were calculated using the initial MB concentration and the MB concentrations after the chosen time intervals. The removal efficiencies at various times were calculated by $C_t/C_0 \times 100$, where C_t presents the adsorbed MB concentration after time t and C_0 presents the initial MB concentration before adsorption. Blank experiments were also performed, using solutions without any adsorbent, to verify if there was any possible loss due to sorption onto the glass material/syringe surface. The results showed that the sorption onto surfaces was negligible. The adsorption kinetics were also tested using a range of initial MB concentrations (C_0) of 10 to 200 mg L^{-1}. Furthermore, recyclability tests were performed by the magnetic separation of our adsorbent, then desorption of the adsorbed MB was performed by shaking in 10 mL ethanol twice, subsequently washing the adsorbent with 10 mL water twice, and finally reusing the adsorbent in a new MB solution.

4. Conclusions

In summary, we successfully developed a new synthesis procedure to modify MOF-derived materials. Zinc oxide, shaped as nanorods, was added as sacrificial nuclei to form a modified MIL-53 precursor to synthesize an nZVI-mesoporous carbon nanocomposite. As expected, the nature of the zinc oxide was of crucial importance. The addition of commercial ZnO had no beneficial effect on the adsorption capacity of the pyrolyzed MIL-53. In contrast, sacrificially used zinc oxide nanorods induced mesopores in Fe-C-SNR, directly enhancing the adsorption capacity and recyclability. The maximum adsorption capacity was 257 mg g^{-1}, doubling the adsorption of the pyrolyzed MIL-53.

Furthermore, virtually 100% of MB was removed from water using the Fe-C-SNR material (concentration: 10 mg L^{-1}) after 60 min of treatment for 10 consecutive cycles. The zinc oxide nanorods assumedly enhanced the adsorption capability by creating an important number of mesopores, which increased the significant contact area between the ZVI and MB. As the tunability of pore size and the creation of defects in MOFs is of crucial importance, it is foreseen that this synthesis procedure could be beneficial for a wide variety of applications.

Supplementary Materials: The following supporting information can be downloaded at: https://www.mdpi.com/article/10.3390/inorganics10050059/s1, Figure S1: SEM micrograph of zinc oxide nanorods; Figure S2: X-ray diffraction pattern of zinc oxide nanorods; Figure S3: X-ray diffraction patterns of MIL-53-SNR, MIL-53-SB, MIL-53; Table S1: ICP results of MIL-53-SB and MIL-53-SNR; Table S2: Volume and surface area of micropores and mesopores of pyrolyzed samples; Figure S4: Adsorption performance of the three samples: (a) Fe-C, (b) Fe-C-SB, (c) Fe-C-SNR; Table S3: Kinetic parameters of pseudo first order and pseudo second order models for the adsorption of methylene blue onto Fe-C-SNR; Figure S5: Kinetic plots for the adsorption of methylene blue on Fe-C-SNR, (a) Plot of the pseudo first order kinetic model, (b) plot of the pseudo second order kinetic model; Figure S6: Adsorption isotherm of methylene blue on Fe-C-SNR; Figure S7: Freundlich Linear adsorption isotherm; Table S4 [1,38,43–49]: Comparative table of the maximum adsorbed methylene blue from literature; Figure S8: X-ray diffraction pattern of Fe-C-SNR before adsorption, with MB adsorbed, and after desorption

Author Contributions: S.D.: conceptualization, investigation, writing—original draft preparation; M.E.H.: conceptualization, investigation, writing—original draft preparation; S.C.: investigation, methodology; S.Z.: validation, writing—review and editing; F.V.: supervision, funding acquisition, writing—review and editing. All authors have read and agreed to the published version of the manuscript.

Funding: This research was funded by the National Natural Science Foundation of China (Contract number 21172027), and the Chinese Scholarship Council (CSC No. 2017GXZ006032).

Institutional Review Board Statement: Not applicable.

Informed Consent Statement: Not applicable.

Conflicts of Interest: The authors declare no conflict of interest. The funders had no role in the design of the study; in the collection, analyses, or interpretation of data; in the writing of the manuscript; or in the decision to publish the results.

References

1. Liu, J.; Wang, Y.; Fang, Y.; Mwamulima, T.; Song, S.; Peng, C. Removal of Crystal Violet and Methylene Blue from Aqueous Solutions Using the Fly Ash-Based Adsorbent Material-Supported Zero-Valent Iron. *J. Mol. Liq.* **2018**, *250*, 468–476. [CrossRef]
2. Ahmad, A.; Rafatullah, M.; Sulaiman, O.; Ibrahim, M.H.; Hashim, R. Scavenging Behaviour of Meranti Sawdust in the Removal of Methylene Blue from Aqueous Solution. *J. Hazard. Mater.* **2009**, *170*, 357–365. [CrossRef] [PubMed]
3. Katheresan, V.; Kansedo, J.; Lau, S.Y. Journal of Environmental Chemical Engineering. *J. Environ. Chem. Eng.* **2018**, *6*, 4676–4697. [CrossRef]
4. Dang, T.-D.; Le, H.T.T.; Nguyen, D.A.; Duc, D.; Dinh, L.; Nguyen, D. A Magnetic Hierarchical Zero-Valent Iron Nanoflake-Decorated Graphene Nanoplate Composite for Simultaneous Adsorption and Reductive Degradation of Rhodamine B. *New J. Chem.* **2020**, *44*, 9083. [CrossRef]
5. Fan, Y.; Ida, S.; Staykov, A.; Akbay, T.; Hagiwara, H.; Matsuda, J.; Kaneko, K.; Ishihara, T. Ni-Fe Nitride Nanoplates on Nitrogen-Doped Graphene as a Synergistic Catalyst for Reversible Oxygen Evolution Reaction and Rechargeable Zn-Air Battery. *Small* **2017**, *13*, 1700099. [CrossRef] [PubMed]

6. Shao, Y.; Gao, Y.; Yue, Q.; Kong, W.; Gao, B.; Wang, W.; Jiang, W. Degradation of Chlortetracycline with Simultaneous Removal of Copper (II) from Aqueous Solution Using Wheat Straw-Supported Nanoscale Zero-Valent Iron. *Chem. Eng. J.* **2020**, *379*, 122384. [CrossRef]
7. Zhang, S.; Zhao, Y.; Yang, K.; Liu, W.; Xu, Y.; Liang, P.; Zhang, X.; Huang, X. Versatile Zero Valent Iron Applied in Anaerobic Membrane Reactor for Treating Municipal Wastewater: Performances and Mechanisms. *Chem. Eng. J.* **2020**, *382*, 123000. [CrossRef]
8. Sun, Y.; Li, J.; Huang, T.; Guan, X. The Influences of Iron Characteristics, Operating Conditions and Solution Chemistry on Contaminants Removal by Zero-Valent Iron: A Review. *Water Res.* **2016**, *100*, 277–295. [CrossRef]
9. Fu, F.; Dionysiou, D.D.; Liu, H. The Use of Zero-Valent Iron for Groundwater Remediation and Wastewater Treatment: A Review. *J. Hazard. Mater.* **2014**, *267*, 194–205. [CrossRef]
10. Shaibu, S.E.; Adekola, F.A.; Adegoke, H.I.; Ayanda, O.S. A Comparative Study of the Adsorption of Methylene Blue onto Synthesized Nanoscale Zero-Valent Iron-Bamboo and Manganese-Bamboo Composites. *Materials* **2014**, *7*, 4493–4507. [CrossRef]
11. Ezzatahmadi, N.; Ayoko, G.A.; Millar, G.J.; Speight, R.; Yan, C.; Li, J.; Li, S.; Zhu, J.; Xi, Y. Clay-Supported Nanoscale Zero-Valent Iron Composite Materials for the Remediation of Contaminated Aqueous Solutions: A Review. *Chem. Eng. J.* **2017**, *312*, 336–350. [CrossRef]
12. Woo, H.; Park, J.; Lee, S.; Lee, S. Effects of Washing Solution and Drying Condition on Reactivity of Nano-Scale Zero Valent Irons (NZVIs) Synthesized by Borohydride Reduction. *Chemosphere* **2014**, *97*, 146–152. [CrossRef] [PubMed]
13. Teng, W.; Fan, J.; Wang, W.; Bai, N.; Liu, R.; Liu, Y.; Deng, Y.; Kong, B.; Yang, J.; Zhao, D.; et al. Nanoscale Zero-Valent Iron in Mesoporous Carbon (NZVI@C): Stable Nanoparticles for Metal Extraction and Catalysis. *J. Mater. Chem. A* **2017**, *5*, 4478–4485. [CrossRef]
14. Bao, T.; Damtie, M.M.; Hosseinzadeh, A.; Wei, W.; Jin, J.; Phong Vo, H.N.; Ye, J.S.; Liu, Y.; Wang, X.F.; Yu, Z.M.; et al. Bentonite-Supported Nano Zero-Valent Iron Composite as a Green Catalyst for Bisphenol A Degradation: Preparation, Performance, and Mechanism of Action. *J. Environ. Manag.* **2020**, *260*, 110105. [CrossRef] [PubMed]
15. Liu, T.; Wang, Z.L.; Zhao, L.; Yang, X. Enhanced Chitosan/Fe 0-Nanoparticles Beads for Hexavalent Chromium Removal from Wastewater. *Chem. Eng. J.* **2012**, *189–190*, 196–202. [CrossRef]
16. Sun, Z.; Zheng, S.; Ayoko, G.A.; Frost, R.L.; Xi, Y. Degradation of Simazine from Aqueous Solutions by Diatomite-Supported Nanosized Zero-Valent Iron Composite Materials. *J. Hazard. Mater.* **2013**, *263*, 768–777. [CrossRef]
17. Wang, J.; Liu, G.; Li, T.; Zhou, C.; Qi, C. Zero-Valent Iron Nanoparticles (NZVI) Supported by Kaolinite for CuII and NiII Ion Removal by Adsorption: Kinetics, Thermodynamics, and Mechanism. *Aust. J. Chem.* **2015**, *68*, 1305. [CrossRef]
18. Mystrioti, C.; Sparis, D.; Papasiopi, N.; Xenidis, A.; Dermatas, D.; Chrysochoou, M. Assessment of Polyphenol Coated Nano Zero Valent Iron for Hexavalent Chromium Removal from Contaminated Waters. *Bull. Environ. Contam. Toxicol.* **2015**, *94*, 302–307. [CrossRef]
19. Shu, H.Y.; Chang, M.C.; Chen, C.C.; Chen, P.E. Using Resin Supported Nano Zero-Valent Iron Particles for Decoloration of Acid Blue 113 Azo Dye Solution. *J. Hazard. Mater.* **2010**, *184*, 499–505. [CrossRef]
20. Fu, R.; Yang, Y.; Xu, Z.; Zhang, X.; Guo, X.; Bi, D. The Removal of Chromium (VI) and Lead (II) from Groundwater Using Sepiolite-Supported Nanoscale Zero-Valent Iron (S-NZVI). *Chemosphere* **2015**, *138*, 726–734. [CrossRef]
21. Mortazavian, S.; An, H.; Chun, D.; Moon, J. Activated Carbon Impregnated by Zero-Valent Iron Nanoparticles (AC/NZVI) Optimized for Simultaneous Adsorption and Reduction of Aqueous Hexavalent Chromium: Material Characterizations and Kinetic Studies. *Chem. Eng. J.* **2018**, *353*, 781–795. [CrossRef]
22. Li, J.; Zhou, Q.; Liu, Y.; Lei, M. Recyclable Nanoscale Zero-Valent Iron-Based Magnetic Polydopamine Coated Nanomaterials for the Adsorption and Removal of Phenanthrene and Anthracene. *Sci. Technol. Adv. Mater.* **2017**, *18*, 3–16. [CrossRef]
23. Tristão, J.C.; de Mendonça, F.G.; Lago, R.M.; Ardisson, J.D. Controlled Formation of Reactive Fe Particles Dispersed in a Carbon Matrix Active for the Oxidation of Aqueous Contaminants with H_2O_2. *Environ. Sci. Pollut. Res.* **2015**, *22*, 856–863. [CrossRef]
24. Van Tran, T.; Bui, Q.T.P.; Nguyen, T.D.; Thanh Ho, V.T.; Bach, L.G. Application of Response Surface Methodology to Optimize the Fabrication of $ZnCl_2$-Activated Carbon from Sugarcane Bagasse for the Removal of Cu^{2+}. *Water Sci. Technol.* **2017**, *75*, 2047–2055. [CrossRef]
25. Van Thuan, T.; Quynh, B.T.P.; Nguyen, T.D.; Ho, V.T.T.; Bach, L.G. Response Surface Methodology Approach for Optimization of Cu^{2+}, Ni^{2+} and Pb^{2+} Adsorption Using KOH-Activated Carbon from Banana Peel. *Surf. Interfaces* **2017**, *6*, 209–217. [CrossRef]
26. Shi, X.; Ruan, W.; Hu, J.; Fan, M.; Cao, R.; Wei, X. Optimizing the Removal of Rhodamine B in Aqueous Solutions by Reduced Graphene Oxide-Supported Nanoscale Zerovalent Iron (NZVI/RGO) Using an Artificial Neural Network-Genetic Algorithm (ANN-GA). *Nanomaterials* **2017**, *7*, 134. [CrossRef]
27. Stefaniuk, M.; Oleszczuk, P.; Ok, Y.S. Review on Nano Zerovalent Iron (NZVI): From Synthesis to Environmental Applications. *Chem. Eng. J.* **2016**, *287*, 618–632. [CrossRef]
28. Wang, Z.; Yang, J.; Li, Y.; Zhuang, Q.; Gu, J. In Situ Carbothermal Synthesis of Nanoscale Zero-Valent Iron Functionalized Porous Carbon from Metal-Organic Frameworks for Efficient Detoxification of Chromium(VI). *Eur. J. Inorg. Chem.* **2018**, *2018*, 23–30. [CrossRef]
29. Zan, J.; Song, H.; Zuo, S.; Chen, X.; Xia, D.; Li, D. MIL-53(Fe)-Derived Fe_2O_3 with Oxygen Vacancy as Fenton-like Photocatalysts for the Elimination of Toxic Organics in Wastewater. *J. Clean. Prod.* **2020**, *246*, 118971. [CrossRef]

30. Liang, R.; Jing, F.; Shen, L.; Qin, N.; Wu, L. MIL-53(Fe) as a Highly Efficient Bifunctional Photocatalyst for the Simultaneous Reduction of Cr(VI) and Oxidation of Dyes. *J. Hazard. Mater.* **2015**, *287*, 364–372. [CrossRef]
31. Vu, T.A.; Le, G.H.; Dao, C.D.; Dang, L.Q.; Nguyen, K.T.; Nguyen, Q.K.; Dang, P.T.; Tran, H.T.K.; Duong, Q.T.; Nguyen, T.V.; et al. Arsenic Removal from Aqueous Solutions by Adsorption Using Novel MIL-53(Fe) as a Highly Efficient Adsorbent. *RSC Adv.* **2015**, *5*, 5261–5268. [CrossRef]
32. Serre, C.; Millange, F.; Thouvenot, C.; Noguès, M.; Marsolier, G.; Louër, D.; Férey, G. Very Large Breathing Effect in the First Nanoporous Chromium(III)-Based Solids: MIL-53 or CrIII(OH)·{O2C-C6H4- CO2}·{HO2C-C6H4 -CO2H}x·H2Oy. *J. Am. Chem. Soc.* **2002**, *124*, 13519–13526. [CrossRef] [PubMed]
33. Cho, W.; Park, S.; Oh, M. Coordination Polymer Nanorods of Fe-MIL-88B and Their Utilization for Selective Preparation of Hematite and Magnetite Nanorods. *Chem. Commun.* **2011**, *47*, 4138–4140. [CrossRef] [PubMed]
34. Scherb, C.; Schödel, A.; Bein, T. Metal-Organic Frameworks Directing the Structure of Metal-Organic Frameworks by Oriented Surface Growth on an Organic Monolayer. *Angew. Chem. Int. Ed.* **2008**, *47*, 5777–5779. [CrossRef]
35. Pham, H.; Ramos, K.; Sua, A.; Acuna, J.; Slowinska, K.; Nguyen, T.; Bui, A.; Weber, M.D.R.; Tian, F. Tuning Crystal Structures of Iron-Based Metal–Organic Frameworks for Drug Delivery Applications. *ACS Omega* **2020**, *5*, 3418–3427. [CrossRef]
36. Cai, X.; Lin, J.; Pang, M. Facile Synthesis of Highly Uniform Fe-MIL-88B Particles. *Cryst. Growth Des.* **2016**, *16*, 3565–3568. [CrossRef]
37. Naeimi, S.; Faghihian, H. Application of Novel Metal Organic Framework, MIL-53(Fe) and Its Magnetic Hybrid: For Removal of Pharmaceutical Pollutant, Doxycycline from Aqueous Solutions. *Environ. Toxicol. Pharmacol.* **2017**, *53*, 121–132. [CrossRef]
38. Banerjee, A.; Gokhale, R.; Bhatnagar, S.; Jog, J.; Bhardwaj, M.; Lefez, B.; Hannoyer, B.; Ogale, S. MOF Derived Porous Carbon-Fe$_3$O$_4$ Nanocomposite as a High Performance, Recyclable Environmental Superadsorbent. *J. Mater. Chem.* **2012**, *22*, 19694–19699. [CrossRef]
39. Stassen, I.; Campagnol, N.; Fransaer, J.; Vereecken, P.; De Vos, D.; Ameloot, R. Solvent-Free Synthesis of Supported ZIF-8 Films and Patterns through Transformation of Deposited Zinc Oxide Precursors. *CrystEngComm* **2013**, *15*, 9308–9311. [CrossRef]
40. Cao, X.; Wang, R.; Peng, Q.; Zhao, H.; Fan, H.; Liu, H.; Liu, Q. Effect of Pore Structure on the Adsorption Capacities to Different Sizes of Adsorbates by Ferrocene-Based Conjugated Microporous Polymers. *Polymer* **2021**, *233*, 124192. [CrossRef]
41. Yuan, X.; Zhuo, S.P.; Xing, W.; Cui, H.Y.; Dai, X.D.; Liu, X.M.; Yan, Z.F. Aqueous Dye Adsorption on Ordered Mesoporous Carbons. *J. Colloid Interface Sci.* **2007**, *310*, 83–89. [CrossRef] [PubMed]
42. Horzum, N.; Hilal, M.E.; Isık, T. Enhanced Bactericidal and Photocatalytic Activities of ZnO Nanostructures by Changing the Cooling Route. *New J. Chem.* **2018**, *42*, 11831–11838. [CrossRef]
43. Frost, R.L.; Xi, Y.; He, H. Synthesis, Characterization of Palygorskite Supported Zero-Valent Iron and Its Application for Methylene Blue Adsorption. *J. Colloid Interface Sci.* **2010**, *341*, 153–161. [CrossRef]
44. Shi, Z.; Wang, Y.; Sun, S.; Zhang, C.; Wang, H. Removal of Methylene Blue from Aqueous Solution Using Mg-Fe, Zn-Fe, Mn-Fe Layered Double Hydroxide. *Water Sci. Technol.* **2020**, *81*, 2522–2532. [CrossRef]
45. Hamdy, A.; Hamdy, M.K.; Nasr, M. Zero-Valent Iron Nanoparticles for Methylene Blue Removal from Aqueous Solutions and Textile Wastewater Treatment, with Cost Estimation. *Water Sci. Technol.* **2018**. [CrossRef]
46. Zhou, Y.; Gao, B.; Zimmerman, A.R.; Chen, H.; Zhang, M.; Cao, X. Biochar-Supported Zerovalent Iron for Removal of Various Contaminants from Aqueous Solutions. *Bioresour. Technol.* **2014**, *152*, 538–542. [CrossRef]
47. Arabi, S.; Sohrabi, M.R. Removal of Methylene Blue, a Basic Dye, from Aqueous Solutions Using Nano-Zerovalent Iron. *Water Sci. Technol.* **2014**, *70*, 24–31. [CrossRef]
48. Aslam, S.; Zeng, J.; Subhan, F.; Li, M.; Lyu, F.; Li, Y.; Yan, Z.; Aslam, S.; Zeng, J.; Subhan, F.; et al. Accepted Manuscript In Situ One-Step Synthesis of Fe$_3$O$_4$ @MIL-100(Fe) Core-Shells for Adsorption of Methylene Blue from Water In Situ One-Step Synthesis of Fe$_3$O$_4$ @MIL-100(Fe) Core-Shells for Adsorption of Methylene Blue from Water. *J. Colloid Interface Sci.* **2017**. [CrossRef]
49. Wu, R.; Liu, L.-H.; Zhao, L.; Zhang, X.; Xie, J.; Yu, B.; Ma, X.; Yang, S.-T.; Wang, H.; Liu, Y. Hydrothermal Preparation of Magnetic Fe$_3$O$_4$ @C Nanoparticles for Dye Adsorption. *Biochem. Pharmacol.* **2014**, *2*, 907–913. [CrossRef]

Article

Evaluating the Sorption Affinity of Low Specific Activity ^{99}Mo on Different Metal Oxide Nanoparticles

Mohamed F. Nawar [1,2,*], Alaa F. El-Daoushy [2], Ahmed Ashry [3], Mohamed A. Soliman [3,4] and Andreas Türler [1]

1. Department of Chemistry, Biochemistry, and Pharmaceutical Sciences, Faculty of Science, University of Bern, Freiestrasse 3, CH-3012 Bern, Switzerland
2. Radioactive Isotopes and Generators Department, Hot Laboratories Center, Egyptian Atomic Energy Authority, Cairo 13759, Egypt
3. Radiation Protection and Civil Defense Department, Nuclear Research Center, Egyptian Atomic Energy Authority, Cairo 13759, Egypt
4. Egypt Second Research Reactor, Nuclear Research Center, Egyptian Atomic Energy Authority, Cairo 13759, Egypt
* Correspondence: mohamed.nawar@unibe.ch

Abstract: ^{99}Mo/^{99m}Tc generators are mainly produced from ^{99}Mo of high specific activity generated from the fission of ^{235}U. Such a method raises proliferation concerns. Alternative methods suggested the use of low specific activity (LSA) ^{99}Mo to produce ^{99m}Tc generators. However, its applicability is limited due to the low adsorptive capacity of conventional adsorbent materials. This study attempts to investigate the effectiveness of some commercial metal oxides nanoparticles as adsorbents for LSA ^{99}Mo. In a batch equilibration system, we studied the influence of solution pH (from 1–8), contact time, initial Mo concentration (from 50–500 mg·L^{-1}), and temperature (from 298–333 K). Moreover, equilibrium isotherms and thermodynamic parameters (changes in free energy ΔG^0, enthalpy change ΔH^0, and entropy ΔS^0) were evaluated. The results showed that the optimum pH of adsorption ranges between 2 and 4, and that the equilibrium was attained within the first two minutes. In addition, the adsorption data fit well with the Freundlich isotherm model. The thermodynamic parameters prove that the adsorption of molybdate ions is spontaneous. Furthermore, some investigated adsorbents showed maximum adsorption capacity ranging from 40 ± 2 to 73 ± 1 mg Mo·g^{-1}. Therefore, this work demonstrates that the materials used exhibit rapid adsorption reactions with LSA ^{99}Mo and higher capacity than conventional alumina (2–20 mg Mo·g^{-1}).

Keywords: LSA ^{99}Mo; thermodynamic parameters; solid-phase extraction; isotherm; metal oxides NPs

Citation: Nawar, M.F.; El-Daoushy, A.F.; Ashry, A.; Soliman, M.A.; Türler, A. Evaluating the Sorption Affinity of Low Specific Activity ^{99}Mo on Different Metal Oxide Nanoparticles. *Inorganics* 2022, 10, 154. https://doi.org/10.3390/inorganics10100154

Academic Editor: Roberto Nisticò

Received: 7 August 2022
Accepted: 23 September 2022
Published: 26 September 2022

Publisher's Note: MDPI stays neutral with regard to jurisdictional claims in published maps and institutional affiliations.

Copyright: © 2022 by the authors. Licensee MDPI, Basel, Switzerland. This article is an open access article distributed under the terms and conditions of the Creative Commons Attribution (CC BY) license (https://creativecommons.org/licenses/by/4.0/).

1. Introduction

^{99}Mo/^{99m}Tc radioisotope generators have a growing importance in nuclear medicine investigations. They are the primary source of supplying ^{99m}Tc radionuclide for diagnostic purposes [1–3]. ^{99m}Tc is considered the workhorse of all nuclear medicine applications [4,5]. It is involved in more than 80% of all in vivo diagnostic procedures because of its ideal nuclear characteristics, such as the short half-life of 6 h, absence of beta particles, and emission of a mono-energetic photon with low energy at 140 keV [3,6]. Therefore, this leads to less radiation exposure dose to the patients, and it produces a high-quality image for better diagnosis aspects. Furthermore, its unique labeling chemistry allows the use of a wide range of ^{99m}Tc-labelled compounds to visualize different body organs [7,8]. For instance, ^{99m}Tc-DTPA and ^{99m}Tc-MAG3 are used to monitor renal functions [9]. In addition, ^{99m}Tc-tetrofosmin, ^{99m}Tc-sestamibi, and ^{99m}Tc-teboroxime are utilized for the diagnosis of cardiac disease [10]. Moreover, ^{99m}Tc-lidofenin is applied for liver diagnostics [11]. Furthermore, ^{99m}Tc-medronate, ^{99m}Tc-propyleneamineoxime, and ^{99m}Tc-MDP (methylene diphosphonate) are involved in skeletal imaging, cerebral perfusion, and diagnosis of bone metastases, respectively [12–15].

Among the developed ^{99}Mo/^{99m}Tc generators, the chromatographic column type is the most widely used system [3,16]. This system is based on adsorbing ^{99}Mo on a column filled with a suitable material from which ^{99m}TcO$_4^-$ can be easily eluted while ^{99}Mo remains adsorbed [3]. The differences between these generators include the column material and the origin of the parent, ^{99}Mo. The main practical difficulties linked to the preparation of ^{99}Mo/^{99m}Tc generators are the low sorption capacity of the bulk conventional inorganic sorbents usually used. These sorbents have low sorption capacity (2–20 mg Mo/g) due to the low availability of active sites and relatively limited surface area [3]. Consequently, such sorbents require a parent of high specific activity to prepare a useful generator of a proper radioactivity level. A high specific activity parent can be produced from the fission of ^{235}U. Fission-produced ^{99}Mo faces some critical difficulties. For example, sophisticated infrastructures and well-qualified personnel are needed to separate and purify ^{99}Mo from the irradiated ^{235}U target and other fission products. In addition, a considerable level of radioactive waste is generated during the manufacturing process, which increases the cost of production [17,18]. Alternatively, research studies focused on developing clinical-grade chromatographic ^{99}Mo/^{99m}Tc generators based on ^{99}Mo of low specific activity (LSA) [3,19,20]. However, this proposal demands using high-capacity sorbents to compensate for the LSA ^{99}Mo and make it more reliable from the economic point of view [17,21].

The use of advanced nanomaterials has generated a growing interest in developing diagnostic ^{99m}Tc generators [3]. Nawar and Türler [3] highlighted several nanomaterial adsorbents that have been developed for ^{99}Mo/^{99m}Tc generator application. This class of sorbents possesses appreciable adsorption capacity and unique performance [20]. In this regard, the utilization of advanced commercial metal-oxide nanoparticles is an exciting idea due to their improved properties. In contrast to traditional sorbents, these nano-adsorbents have large surface-to-volume ratios, enhanced porosity, improved surface reactivity, and significant radiation resistance and chemical stability [21,22]. Therefore, they show high adsorption efficiency and selectivity [23].

In this study, we intend to evaluate the sorption efficiency of some commercially available nano-metal oxides towards LSA ^{99}Mo. To achieve this goal, we investigated the adsorption behavior of the selected materials for LSA ^{99}Mo under different experimental conditions. These conditions include the pH, initial concentration of molybdate ions, contact time, and temperature. In addition, to better understand their sorption behavior, the sorption kinetics, equilibrium isotherms, and thermodynamic behavior were evaluated.

2. Results and Discussion

2.1. Effect of Solution pH

The solution pH has a profound impact on the efficiency of the adsorption process. The influence of pH can be clarified by understanding its role in varying the ionic state of the functional groups on the adsorbent surface. Moreover, it affects the ionization and/or the dissociation of the studied ions [24]. In this context, a batch equilibration experiment was conducted at a pH range from 1 to 8 to determine the optimum pH value that shows the maximum ^{99}Mo retention on each adsorbent. Figure 1a depicts the distribution coefficients (K_d) of CA-^{99}Mo at different pH values. The data presented in this figure show that higher K_d values are observed at pH values (2–4). Beyond this region, the K_d values decrease with increasing the solution pH, which agrees with previously published studies [25].

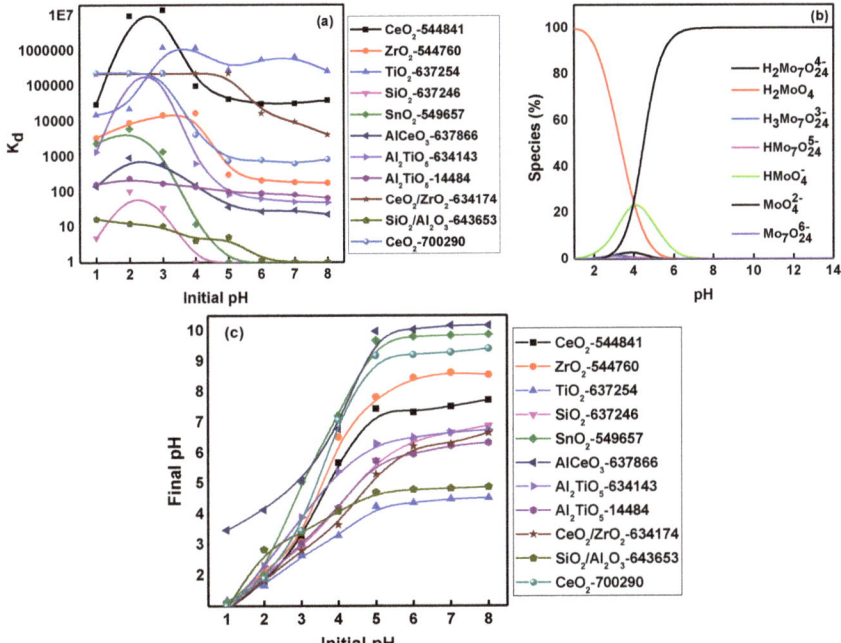

Figure 1. Effect of initial pH on (**a**) the distribution coefficients (K_d) of CA-^{99}Mo on different metal oxides NPs (C_0 = 50 mg·L^{-1}, V/m = 100 mL·g^{-1}, and temperature = 298 ± 1 K), (**b**) Speciation of molybdenum [22], and (**c**) variation of the final pH values.

Since the adsorbents are metal oxides, they might have similar surface chemistry. Moreover, since the adsorption process depends mainly on the aqueous phase's pH values and the adsorbent material's surface characteristics, we investigated the isoelectric point (pHI$_{EP}$) of each adsorbent (Table 1). The pHI$_{EP}$ measurements help to clarify the sorption mechanism. The sorbent surface carries a positive charge at pH < pHI$_{EP}$, zero charge at pH~pHI$_{EP}$, and is negatively charged at pH > pHI$_{EP}$. Consequently, there is a change in the pHI$_{EP}$ of the sorbent with the pH of an aqueous solution. Nawar et al. [22] reported that this behavior might occur because amphoteric hydroxyl groups cover the adsorbent surface. Hence, based on the pH of the medium, these groups develop different reactions in different pH media, resulting in positive or negative charges appearing on the adsorbent surface. Herein, at pH < pHI$_{EP}$, they are protonated, and the surface develops a positive charge as follows:

$$\text{Adsorbent} - \text{OH}_{\text{Surface}} + \text{H}^+_{\text{solution}} \rightleftharpoons \text{Adsorbent} - \text{OH}_2^+ \qquad (1)$$

The data presented in Figure 1a can be interpreted by considering the speciation diagram of molybdenum shown in Figure 1b [22]. The speciation data are generated using the PHREEQC software (version 3) to determine the predominant Mo species at different pHs for the following conditions: C_0 = 50 mg·L^{-1} at 298 1 K and using the built-in database of stability constants [22]. At acidic medium, the molybdate anionic species exist and polymerize, increasing the molybdenum content per unit charge as follows:

$$7\,^{99}\text{MoO}_4^{2-} + 8\,\text{H}^+ \rightleftharpoons\, ^{99}\text{Mo}_7\text{O}_{24}^{6-} + 4\,\text{H}_2\text{O} \qquad (2)$$

Consequently, this results in favorable interactions between negatively charged molybdenum polyanions and positively charged adsorbents surfaces [26]. At higher pH values, the speciation shifts to less negatively charged Mo species, and the density of hydroxyl groups (OH$^-$) increases in solution. These hydroxyl anions compete with less negatively charged molybdenum anions to retain the available active sites on adsorbents surfaces, explaining the low K_d distribution values at higher pH values [22,27].

Table 1. Description of the analyzed commercial metal oxides NPs *.

No.	Name	Description	Particle Size, (nm)	Surface Area, (m$^2\cdot$g^{-1})	Isoelectric Point (pH$_{IEP}$)
CeO$_2$-544841	Cerium oxide-SA-544841	Molecular formula: CeO$_2$ Molecular weight: 172.11 Density: 7.13 g·mL^{-1} at 298 K	<25	N.A	5
ZrO$_2$-544760	Zirconium oxide-SA-544760	Molecular formula: ZrO$_2$ Molecular weight: 123.22 Density: 5.89 g·mL^{-1} at 298 K	<100	≥25	6.1
TiO$_2$-637254	Titanium oxide-SA-637254	Molecular formula: TiO$_2$ Molecular weight: 79.87 Density: 3.9 g·mL^{-1} at 298 K	<25	45–55	6.6
SnO$_2$-549657	Tin oxide-SA-549657	Molecular formula: SnO$_2$ Molecular weight: 150.71 Density: 6.95 g·mL^{-1} at 298 K	≤100	20.1	3.8
SiO$_2$-637246	Silicon oxide-SA-637246	Molecular formula: SiO$_2$ Molecular weight: 60.08 Density: 2.2–2.6 g·mL^{-1} at 298 K	5–20	590–690	2.5
AlCeO$_3$-637866	Cerium aluminium oxide-SA-637866	Molecular formula: AlCeO$_3$ Molecular weight: 215.1	≤80	N.A	4.8
Al$_2$TiO$_5$-634143	Aluminium titanium oxide-SA-634143	Molecular formula: Al$_2$TiO$_5$ Molecular weight: 181.83	<25	N.A	6.4
Al$_2$TiO$_5$-14484	Aluminium titanium oxide-AA-14484	Molecular formula: Al$_2$TiO$_5$ Molecular weight: 181.86	100 mesh	N.A	6.5
CeO$_2$/ZrO$_2$-634174	Cerium zirconium oxide-SA-634174	Molecular formula: (CeO$_2$)·(ZrO$_2$) Molecular weight: 295.34 Density: 6.61 g·mL^{-1} at 298 K	<50	N.A	6.7
SiO$_2$/Al$_2$O$_3$-643653	Aluminosilicate-SA-643653	Molecular formula: (SiO$_2$)$_x$(Al$_2$O$_3$)$_y$ pore volume: 0.8–1.1 cm$^3\cdot$g^{-1} mesostructured, pore size: 2–4 nm	4.5–4.8	900–1100	6
CeO$_2$-700290	Cerium oxide-SA-700290	Molecular formula: CeO$_2$ Molecular weight: 172.11 Density: 7.13 g·mL^{-1} at 298 K	<50	30	4.5

* The information was provided by the supplier. Only the isoelectric point data were determined experimentally. Abbreviations: AA: Alfa Aesar (Kandel, Germany); N.A: Not Available; SA: Sigma-Aldrich (Buchs, Switzerland).

Moreover, based on the isoelectric point (pH$_{IEP}$) of each sorbent material (Table 1) and the measured final solution pH (Figure 1c), it can be observed that K_d values start to decrease when the final solution pH exceeds the sorbent's pH$_{IEP}$, which can be attributed to the expected change in the surface charge of the sorbent material. As previously mentioned, at solution pH values above the pH$_{IEP}$, the sorbent surface becomes predominately negatively charged. As a result, repulsion between the negatively charged sorbent surface and the negatively charged molybdenum polyanions takes place, leading to the observed decrease in K_d values [22,28].

It can also be observed that both silicon oxide and aluminosilicate nanoparticles possess small particle sizes (5–20 and 4.5–4.8 nm) and high surface area (590–690 and 900–1100 $m^2 \cdot g^{-1}$), respectively (Table 1). However, both adsorbents show a weak affinity for Mo species. This behavior may be attributed to their poor stability with increasing pH values. At high pH values, the dissolution of silica occurs, resulting in the formation of monomeric ortho-silicic acid (H_4SiO_4), which can be explained due to the presence of more hydroxyl groups. These hydroxyl groups are chemisorbed on the adsorbent surface, which increases the number of coordination bonds around the silicon atom to more than four bonds. Consequently, it may lead to Si-O bond rupture, and the silicon atom dissolves as $Si(OH)_4$ and ortho-silicic acid [29].

According to the obtained results, we selected six adsorbents that showed high distribution coefficient values towards CA-^{99}Mo for the subsequent investigations. These adsorbents are CeO_2-544841, ZrO_2-544760, TiO_2-637254, Al_2TiO_5-634143, CeO_2/ZrO_2-634174, and CeO_2-700290.

2.2. Adsorption Isotherm

Equilibrium isotherms are essential in describing the adsorption mechanisms for the interaction of Mo(VI) ions with the surfaces of the investigated metal oxides NPs. These mechanisms describe the adsorption process successfully. Here, we investigated equilibrium data obtained for adsorption of CA-^{99}Mo on CeO_2-544841, ZrO_2-544760, TiO_2-637254, Al_2TiO_5-634143, CeO_2/ZrO_2-634174, and CeO_2-700290 with various isotherm models to find out which one is the most suitable for describing the obtained adsorption equilibrium data.

2.2.1. Freundlich Isotherm

Many studies have utilized the Freundlich adsorption isotherm model proposed as a general power equation used to describe the adsorption of radionuclides in a large number of studies [30–32]. The Freundlich isotherm has the form shown as follows:

$$q_e = K_f (C_e)^{\frac{1}{n_f}} \quad (3)$$

where q_e ($mg \cdot g^{-1}$) is the concentration of CA-^{99}Mo adsorbed and C_e ($mg \cdot L^{-1}$) is the concentration of Mo remaining in the solution. K_f ($mg^{1-n}L^n \cdot g^{-1}$) and n_f (dimensionless) are constants unique to each combination of adsorbent and adsorbate.

2.2.2. Langmuir Isotherm

Langmuir (1918) developed an equation to describe the adsorption of gases on a solid surface that was subsequently adapted to describe the adsorption of solutes onto solids in aqueous solutions [31,33,34], as shown in Equation (4):

$$q_e = \frac{n_L K_L C_e}{1 + K_L C_e} \quad (4)$$

where q_e ($mg \cdot g^{-1}$) is the total concentration of solute adsorbed, K_L ($L \cdot mg^{-1}$) is an equilibrium constant, and n_L ($mg \cdot g^{-1}$) is the adsorption capacity.

Figure 2 presents the experimental adsorption equilibrium data obtained for Mo ions on the investigated metal oxide adsorbents as a plot of adsorption equilibrium capacity (q_e) against initial concentration (C_0). It is observed that there is an increase in the amount of Mo ions taken up with the increase in the initial metal ion concentration. This increase in the adsorbate uptake can be explained by the driving force for mass transfer [34].

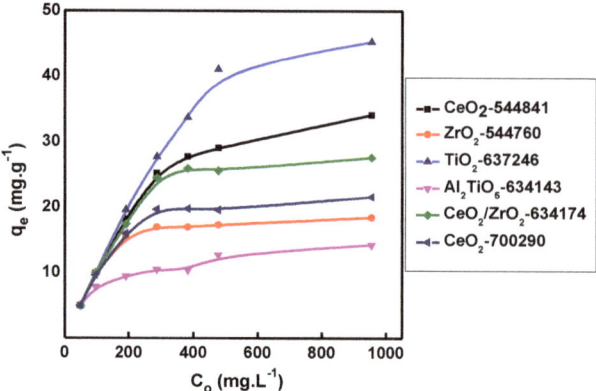

Figure 2. The influence of initial molybdate concentration on the equilibrium sorption capacity (q_e) of CA-^{99}Mo on different metal oxides NPs (pH = 3, V/m =100 mL·g^{-1}, t = 24 h, and temperature = 298 ± 1 K).

The non-linear forms of both isotherm models were applied to the measured adsorption data (C_e versus q_e), and the data were displayed in Figure 3. Adsorption parameters were optimized using the add-ins "Solver" function in Microsoft Excel. Table 2 gives the Freundlich parameters (K_f and n_f), Langmuir parameters (K_L and n_L), and the goodness of fit of the model lines to the experimental data (R^2). Based on the regression coefficient values reported in Table 2, it is observed that good to excellent correlations between the experimental results and the fitted data of the Freundlich isotherm model were obtained for all the investigated sorbents. In contrast, the Langmuir model failed to fit any equilibrium sorption isotherm of the CA-^{99}Mo on all tested adsorbents; lower R^2 values were obtained.

These findings suggest that CA-^{99}Mo adsorption on metal oxide nanomaterials under investigation mainly occurred through multilayer adsorption at heterogeneous surfaces [31,35]. The Freundlich adsorption constant (n_f) is usually used as a measure of adsorption intensity as follows; (i) $n_f < 1$ indicates that adsorption takes place via a chemical process, (ii) $n_f = 1$ shows linear adsorption, (iii) while $n_f > 1$ indicates physisorption [35]. The n_f values displayed in Table 2 were higher than 1, indicating that CA-^{99}Mo adsorption on the materials used in this study was physisorption and favorable under the investigated conditions. Furthermore, the closer the 1/n value to 0 than unity (ranging from 0.10 to 0.25), the more heterogeneous the surface is, implying a broad distribution of adsorption sites on the adsorbent surface [32,33].

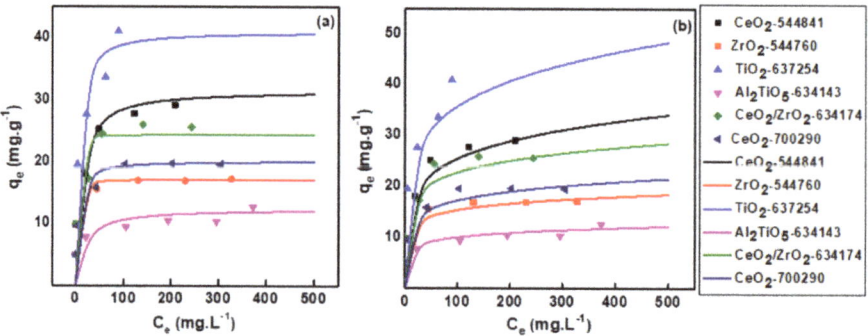

Figure 3. Adsorption isotherms: (**a**) Langmuir and (**b**) Freundlich of CA-^{99}Mo on different metal oxides NPs.

Table 2. Isotherm parameters calculations for the adsorption of CA-^{99}Mo on different metal oxides NPs.

Isotherm Model	Parameter	CeO$_2$-544841	ZrO$_2$-544760	TiO$_2$-637254	Al$_2$TiO$_5$-634143	CeO$_2$/ZrO$_2$-634174	CeO$_2$-700290
Langmuir	n_L (mg·g^{-1})	26.704	16.814	36.980	10.207	23.470	19.603
	K_L (L·mg^{-1})	0.407	0.993	0.311	0.0907	3.537	0.254
	R^2	0.911	0.957	0.930	0.836	0.870	0.954
Freundlich	K_F (mg^{1-n}L^n·g^{-1})	10.514	9.058	13.064	6.364	11.506	8.876
	n_f	5.010	8.294	4.079	10.325	6.346	6.644
	R^2	0.982	0.968	0.989	0.898	0.955	0.966

2.3. Thermodynamic Studies

We determined the amount of CA-^{99}Mo adsorbed on the surface of the materials investigated in the current study as a function of temperature (T) using adsorption thermodynamic parameters. These parameters include the Gibbs free energy ΔG^0 (kJ·mol^{-1}), the standard enthalpy change ΔH^0 (kJ·mol^{-1}), and the standard entropy change ΔS^0 (J·mol^{-1}·K^{-1}). They were investigated at different temperatures (298, 313, 323, and 333 K) using Equations (5) and (6) [34,36,37] and are tabulated in Table 3:

$$\Delta G^0 = -RT \ln K_d \quad (5)$$

$$\ln K_d = \frac{\Delta S^0}{R} - \frac{\Delta H^0}{RT} \quad (6)$$

where R is the universal gas constant (8.314 J·mol^{-1}·K^{-1}), T is the absolute temperature (K), and K_d (mL·g^{-1}) is the distribution coefficient.

Table 3. Thermodynamic parameters for the sorption of CA-^{99}Mo on different metal oxides NPs.

Adsorbent	Temperature (K)	ΔG^0 (kJ·mol^{-1})	ΔH^0 (kJ·mol^{-1})	ΔS^0 (J·mol^{-1}·K^{-1})
CeO$_2$-544841	298	−9.8 ± 2.9	−5.1 ± 1.5	16.0 ± 4.7
	313	−10.1 ± 3.0		
	323	−10.2 ± 3.0		
	333	−10.4 ± 3.1		
ZrO$_2$-544760	298	−7.8 ± 1.4	−1.4 ± 0.7	21.3 ± 2.2
	313	−8.1 ± 1.4		
	323	−8.3 ± 1.4		
	333	−8.5 ± 1.5		
TiO$_2$-637254	298	−10.9 ± 3.1	4.7 ± 1.5	52.0 ± 5.0
	313	−11.7 ± 3.1		
	323	−12.2 ± 3.2		
	333	−12.7 ± 3.2		
Al$_2$TiO$_5$-634143	298	−7.2 ± 2.0	−4.5 ± 1	9.1 ± 3.2
	313	−7.4 ± 2.0		
	323	−7.5 ± 2.0		
	333	−7.6 ± 2.0		

Table 3. Cont.

Adsorbent	Temperature (K)	ΔG^0 (kJ·mol^{-1})	ΔH^0 (kJ·mol^{-1})	ΔS^0 (J·mol^{-1}·K^{-1})
CeO$_2$/ZrO$_2$-634174	298	-9.0 ± 1.4	1.3 ± 0.7	34.6 ± 2.2
	313	-9.5 ± 1.4		
	323	-9.9 ± 1.4		
	333	-10.2 ± 1.4		
CeO$_2$-700290	298	-8.2 ± 1.9	-0.1 ± 0.9	27.2 ± 3.0
	313	-8.6 ± 1.9		
	323	-8.9 ± 2.0		
	333	-9.1 ± 2.0		

Figure 4 shows linear plots of ln K_d versus (1/T). The calculated ΔG^0 values at each temperature for all nano-adsorbents are $\Delta G^0 < 0$, which implies that the Mo(VI) adsorption process on the surfaces of all adsorbents is spontaneous and the reaction is feasible. Likewise, ΔG^0 values decrease with increasing temperature, indicating that the degree of spontaneity can be enhanced by increasing the temperature. Furthermore, the adsorption process is physisorption ($-20 < \Delta G^0 < 0$) [38]. The positive values of ΔS^0 ($\Delta S^0 > 0$) report random adsorption reactions of CA-^{99}Mo at all adsorbents surfaces. The values of ΔH^0 are positive ($\Delta H^0 > 0$) for both TiO$_2$-637254 and CeO$_2$/ZrO$_2$-634174, implying that CA-^{99}Mo adsorption at their surfaces is endothermic [39]. While for CeO$_2$-544841, ZrO$_2$-544760, Al$_2$TiO$_5$-634143, and CeO$_2$-700290, the change in enthalpy (ΔH^0) is negative ($\Delta H^0 < 0$), indicating that the adsorption of CA-^{99}Mo at their surfaces is exothermic [38,40].

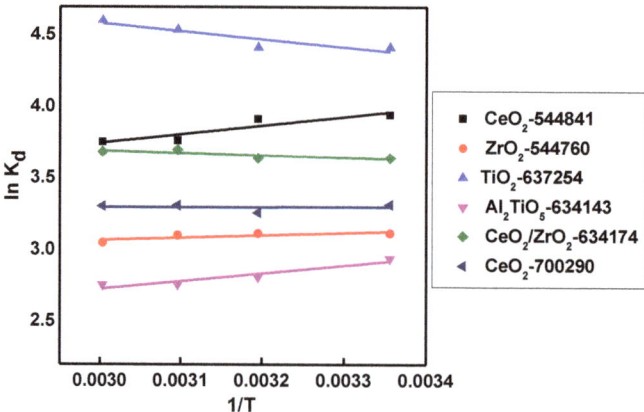

Figure 4. Van't Hoff plot for the sorption of CA-^{99}Mo on different metal oxides NPs.

2.4. Determining the Maximum Sorption Capacity

In order to evaluate the maximum sorption capacity of each adsorbent, the equilibrations of CA-^{99}Mo with each adsorbent were performed separately. Batch equilibrations were repeated until no further ^{99}Mo(IV) uptake was observed, and the adsorbents became fully saturated with ^{99}Mo. After each equilibration, 1 mL aliquot was decanted, centrifuged, and counted. Ultimately, the maximum sorption capacity (q_{max}) for each material was calculated by applying the following equation:

$$q_{max} = \frac{\sum U\%}{100} \times C_o \times \frac{V}{m} \quad (\text{mg·g}^{-1}) \tag{7}$$

where U% is the uptake percent of CA-^{99}Mo, C_0(mg·L^{-1}) is the starting Mo(IV) concentration, V (L) is the liquid phase volume, and m (g) is the adsorbent weight. Figure 5 shows CA-^{99}Mo maximum sorption capacity on different studied metal oxides NPs. It can be concluded that the studied metal oxide NPs show better sorption capacity than conventional alumina currently used in ^{99}Mo/^{99m}Tc generators. Nonetheless, the obtained capacities are insufficient for developing a clinical-grade ^{99m}Tc generator based on LSA ^{99}Mo.

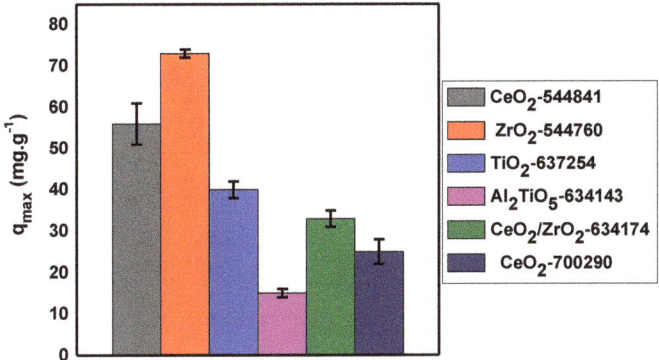

Figure 5. The maximum sorption capacity of different metal oxide NPs for CA-^{99}Mo.

2.5. Effect of Contact Time

The effect of contact time on the uptake percent of CA-^{99}Mo was monitored for an initial Mo(IV) concentration of 50 mg·L^{-1} (pH~3), using an adsorbent dose of 200 mg. The reaction temperature was adjusted to 298 ± 1 K. The results are shown in Figure 6. The results show that the Mo uptake sharply increased at the beginning of the adsorption process and reached a constant value (a plateau value) in the first two minutes. This behavior indicates a rapid and almost instantaneous removal of CA-^{99}Mo from the solution, and a dynamic equilibrium is established under the given experimental conditions. In order to design an effective adsorption process, determining the kinetic parameters is crucial. The kinetic data shown in Figure 6 revealed that the equilibrium for adsorption of Mo on metal oxide nano-adsorbents is already reached at the very beginning of the adsorption process. Consequently, using the current methodology, such data cannot be modeled with adsorption kinetic models.

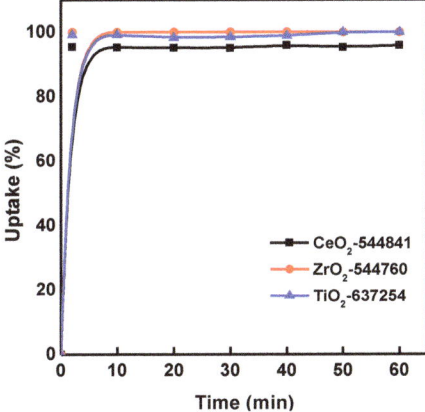

Figure 6. Effect of contact time on CA-^{99}Mo uptake on different metal oxide NPs (C_0 = 50 mg·L^{-1}, pH = 3, V/m = 100 mL·g^{-1}, and temperature = 298 ± 1 K).

3. Materials and Methods

3.1. Materials

All chemicals are of analytical grade purity (A. R. grade) and were used without further purification. Milli-Q water was used for the preparation of solutions and washings. Sodium hydroxide and nitric acid were purchased from Merck, Darmstadt, Germany. The metal oxide nanomaterials were purchased from different suppliers (Table 1).

^{99}Mo radiotracer solution was obtained by eluting a 40 GBq fission ^{99}Mo alumina-based ^{99}Mo/^{99m}Tc generator (Pertector, manufactured by National Centre for Nuclear Research, POLATOM, Otwock, Poland) with 5 mL of 1 M NaOH solution after ~7 d from the calibration date. The total ^{99}Mo radioactivity was measured with a Capintec Radioisotopes Calibrator (model CRC-55tR Capintec, Inc., Florham Park, NJ, USA). The ^{99}Mo eluate solution was passed through a 0.45 micro-Millipore filter to retain alumina particles. Then, the ^{99}Mo solution was treated with nitric acid to attain the desired pH value.

3.2. Batch Equilibrium Studies

A batch equilibration experiment was conducted to investigate the adsorption behavior of carrier-added (CA) ^{99}Mo (Mo(IV) treated with ^{99}Mo) on several commercial metal oxide nanoparticles (NPs) under different conditions. These conditions included the influence of pH, contact time, reaction temperature, and initial adsorbate concentration. In a series of clean glass bottles, we added 200 mg of each adsorbent to 20 mL of ^{99}Mo(IV) solution of a given concentration and pH value. Subsequently, the mixtures were shaken in a thermostatic shaker water bath (Julabo GmbH, Seelbach, Germany) at 298 ± 1 K for 24 h. Eventually, the supernatant solution was collected, centrifuged, and 1 mL was separated for radiometric measurements. For all radiometric identifications and γ-spectrometry, we used a multichannel analyzer (MCA) of Inspector 2000 model, Canberra Series, Mirion Technologies, Inc., Meriden, CT, USA, coupled with a high-purity germanium coaxial detector (HPGe). All samples have fixed geometry and were counted at a low dead time (<2%). The measurements were done by using an appropriate gamma-ray peak of 740 keV.

3.2.1. Distribution Ratio (K_d)

The distribution coefficient (K_d) values of CA-^{99}Mo were investigated at a wide range of pH (from 1–8). For adjusting the desired pH value of the solutions, few drops of 0.5 M nitric acid or 0.5 M sodium hydroxide were added. The pH values of the solutions were measured before and after reaching the equilibrium state. pH values were determined using a pH-meter with a microprocessor (Mettler Toledo, Seven Compact S210 model, Greifensee, Switzerland).

3.2.2. Adsorption Isotherm

In order to determine the sorption isotherms, we used different initial molybdate ion concentrations from 50 to 500 mg·L^{-1} while keeping the adsorbent amount constant. Moreover, the solution pH, equilibrium time, and reaction temperature were kept at pH~3, 24 h, and 298 ± 1 K, respectively. In addition, the equilibrium adsorption capacity (q_e) was calculated. Finally, we used the obtained results to determine the sorption isotherm model.

3.2.3. Thermodynamic Studies

The reaction temperature effect on the uptake of carrier-added ^{99}Mo was studied at four different reaction temperatures (298, 313, 323, and 333 K). At each temperature, we added 20 mL of CA-^{99}Mo solution (pH 3) in contact with 200 mg of the adsorbent material for 24 h. From the resulting data, we calculated different thermodynamic parameters, namely the standard enthalpy change (ΔH^0), standard entropy change (ΔS^0), and Gibbs free energy change (ΔG^0).

3.2.4. Effect of Contact Time

In order to investigate the ^{99}Mo adsorption rate on the studied metal oxides NPs, we monitored the progress of the uptake capacity of ^{99}MoO$_4^{2-}$ ions (50 mg·L^{-1} and pH~3) at different time slots. The adsorption of CA-^{99}Mo was followed with time until the equilibrium was established. Finally, we calculated the ^{99}Mo capacity (q_t) in mg·g^{-1} at each time (t).

3.3. Calculations

The adsorption data of CA-^{99}Mo include uptake percent (U%), distribution coefficient (K_d), equilibrium capacity (q_e), and equilibrium concentration (C_e). These data were calculated according to the following equations:

$$U\% = \frac{(A_i - A_f)}{A_i} \times 100 \tag{8}$$

$$q_e = \frac{U\%}{100} \times C_0 \times \frac{V}{m} \; (\text{mg·g}^{-1}) \tag{9}$$

$$C_e = A_i - \left(A_i \times \frac{U\%}{100}\right) \; (\text{mg·L}^{-1}) \tag{10}$$

$$K_d = \frac{A_i - A_f}{A_i} \times \frac{V'}{m} \; (\text{mL·g}^{-1}) \tag{11}$$

where A_i and A_f are the initial and final ^{99}Mo radioactivity in counts/min. C_0 (mg·L^{-1}) is the initial concentration of CA-^{99}Mo, V (L) and V' (mL) represent the volume of liquid phases, and m (g) is the weight of the solid phase.

4. Summary and Conclusions

The main objective of this study was to evaluate the adsorption affinity of different commercial metal oxides NPs purchased from different suppliers towards LSA ^{99}Mo. All experiments were conducted at static equilibrium conditions. We studied the distribution ratio of CA-^{99}Mo in a pH range of 1 to 8. The optimum adsorption pH was found to be in the range of pH 2 to 4. In addition, the Freundlich isotherm model fitted the experimental data of the CA-^{99}Mo on all adsorbent materials investigated in this study. Moreover, we determined the values of enthalpy change (ΔH^0), entropy change (ΔS^0), and free energy change (ΔG^0) at the different reaction temperatures. Furthermore, the maximum adsorption capacities were evaluated, and the best adsorbents showed a capacity of 40 ± 2 to 73 ± 1 mg Mo·g^{-1}. Summing up the results, it can be concluded that the adsorption behavior of the materials investigated depends on the solution pH, contact time, initial metal ion concentration, and temperature. Furthermore, the investigated materials showed higher static sorption capacities than conventional alumina (2–20 mg Mo·g^{-1}). Nonetheless, they are not suitable to build a useful ^{99}Mo/^{99m}Tc generator using LAS ^{99}Mo for radiopharmaceutical applications. Since the available specific activity of LAS ^{99}Mo is 2.5–5 Ci/g Mo, approximately 20–25 g of each material would be required to prepare a ^{99m}Tc generator of 37 GBq (1 Ci). Using such a massive amount of sorbent material per generator would deteriorate the elution performance and the radioactive concentration of the produced ^{99m}Tc.

Author Contributions: Conceptualization, M.F.N.; methodology, M.F.N. and A.F.E.-D.; software, M.F.N., A.F.E.-D. and A.A.; validation, M.F.N., A.F.E.-D., A.A., M.A.S. and A.T.; formal analysis, M.F.N., A.F.E.-D. and A.A.; investigation, M.F.N.; data curation, M.F.N., A.F.E.-D. and A.T.; writing—original draft preparation, M.F.N.; writing—review and editing, A.F.E.-D., A.A., M.A.S. and A.T.; visualization, M.F.N. and A.F.E.-D.; supervision, A.T.; project administration, M.F.N. and A.T.; funding acquisition, M.F.N. All authors have read and agreed to the published version of the manuscript.

Funding: The research was funded by the Swiss National Science Foundation (grant number CR-SII5_180352). Mohamed F. Nawar gratefully acknowledges the funding support of the Swiss Government Excellence fellowships program (fellowship No: 2017.1028).

Institutional Review Board Statement: Not applicable.

Informed Consent Statement: Not applicable.

Data Availability Statement: Data is contained within the article.

Acknowledgments: The authors would like to express their sincere thanks to Marcel Langensand, Managing Director, Medeo AG, CH-5040 Schöftland, Switzerland, for his valuable support in supplying ^{99}Mo/^{99m}Tc generators to conduct this research study.

Conflicts of Interest: The authors declare no conflict of interest.

References

1. Cutler, C.S. Supply of Mo-99: Focus on U.S. supply. Trends in Radiopharmaceuticals (ISTR-2019). In Proceedings of the International Symposium, Vienna, Austria, 28 October–1 November 2019; International Atomic Energy Agency (IAEA): Vienna, Austria, 2020.
2. Munir, M.; Sriyono; Abidin; Sarmini, E.; Saptiama, I.; Kadarisman; Marlina. Development of mesoporous γ-alumina from aluminium foil waste for 99Mo/99mTc generator. *J. Radioanal. Nucl. Chem. Artic.* **2020**, *326*, 87–96. [CrossRef]
3. Nawar, M.F.; Türler, A. New strategies for a sustainable ^{99m}Tc supply to meet increasing medical demands: Promising solutions for current problems. *Front. Chem.* **2022**, *10*, 926258. [CrossRef] [PubMed]
4. Capogni, M.; Pietropaolo, A.; Quintieri, L.; Angelone, M.; Boschi, A.; Capone, M.; Cherubini, N.; De Felice, P.; Dodaro, A.; Duatti, A.; et al. 14 MeV Neutrons for ^{99}Mo/^{99m}Tc Production: Experiments, Simulations and Perspectives. *Molecules* **2018**, *23*, 1872. [CrossRef] [PubMed]
5. Duatti, A. Review on ^{99m}Tc radiopharmaceuticals with emphasis on new advancements. *Nucl. Med. Biol.* **2021**, *92*, 202–216. [CrossRef] [PubMed]
6. Boschi, A.; Uccelli, L.; Martini, P. A Picture of Modern Tc-99m Radiopharmaceuticals: Production, Chemistry, and Applications in Molecular Imaging. *Appl. Sci.* **2019**, *9*, 2526. [CrossRef]
7. Mohan, A.-M.; Beindorff, N.; Brenner, W. Nuclear Medicine Imaging Procedures in Oncology. *Metastasis* **2021**, *2294*, 297–323. [CrossRef]
8. Kniess, T.; Laube, M.; Wüst, F.; Pietzsch, J. Technetium-99m based small molecule radiopharmaceuticals and radiotracers targeting inflammation and infection. *Dalton Trans.* **2017**, *46*, 14435–14451. [CrossRef] [PubMed]
9. Momin, M.; Abdullah, M.; Reza, M. Comparison of relative renal functions calculated with 99m Tc-DTPA and 99m Tc-DMSA for kidney patients of wide age ranges. *Phys. Med.* **2018**, *45*, 99–105. [CrossRef]
10. Fang, W.; Liu, S. New ^{99m}Tc Radiotracers for Myocardial Perfusion Imaging by SPECT. *Curr. Radiopharm.* **2019**, *12*, 171–186. [CrossRef]
11. Marlina, M.; Lestari, E.; Abidin, A.; Hambali, H.; Saptiama, I.; Febriana, S.; Kadarisman, K.; Awaludin, R.; Tanase, M.; Nishikata, K.; et al. Molybdenum-99 (^{99}Mo) Adsorption Profile of Zirconia-Based Materials for 99Mo/99mTc Generator Application. *At. Indones.* **2020**, *46*, 91–97. [CrossRef]
12. Papagiannopoulou, D. Technetium-99m radiochemistry for pharmaceutical applications. *J. Label. Compd. Radiopharm.* **2017**, *60*, 502–520. [CrossRef] [PubMed]
13. Sharma, S.; Jain, S.; Baldi, A.; Singh, R.K.; Sharma, R.K. Intricacies in the approval of radiopharmaceuticals-regulatory perspectives and the way forward. *Curr. Sci.* **2019**, *116*, 47–55. [CrossRef]
14. Lassen, A.; Stokely, E.; Vorstrup, S.; Goldman, T.; Henriksen, J.H. Neuro-SPECT: On the development and function of brain emission tomography in the Copenhagen area. *Clin. Physiol. Funct. Imaging* **2021**, *41*, 10–24. [CrossRef]
15. Chen, B.; Wei, P.; Macapinlac, H.A.; Lu, Y. Comparison of ^{18}F-Fluciclovine PET/CT and ^{99m}Tc-MDP bone scan in detection of bone metastasis in prostate cancer. *Nucl. Med. Commun.* **2019**, *40*, 940–946. [CrossRef] [PubMed]
16. Hasan, S.; Prelas, M.A. Molybdenum-99 production pathways and the sorbents for ^{99}Mo/^{99m}Tc generator systems using (n, γ) ^{99}Mo: A review. *SN Appl. Sci.* **2020**, *2*, 1782. [CrossRef]
17. Marlina; Ridwan, M.; Abdullah, I.; Yulizar, Y. Recent progress and future challenge of high-capacity adsorbent for non-fission molybdenum-99 (^{99}Mo) in application of ^{99}Mo/^{99m}Tc generator. *AIP Conf. Proc.* **2021**, *2346*, 030003. [CrossRef]
18. Munir, M.; Herlina; Sriyono; Sarmini, E.; Abidin; Lubis, H.; Marlina. Influence of GA Siwabessy Reactor Irradiation Period on The Molybdenum-99 (^{99}Mo) Production by Neutron Activation of Natural Molybdenum to Produce Technetium-99m (^{99m}Tc). *J. Phys. Conf. Ser.* **2019**, *1204*, 012021. [CrossRef]
19. Nawar, M.F.; El-Daoushy, A.F.; Ashry, A.; Türler, A. Developing a Chromatographic ^{99m}Tc Generator Based on Mesoporous Alumina for Industrial Radiotracer Applications: A Potential New Generation Sorbent for Using Low-Specific-Activity ^{99}Mo. *Molecules* **2022**, *27*, 5667. [CrossRef]

20. Nawar, M.F.; Türler, A. Development of New Generation of 99Mo/99mTc Radioisotope Generators to Meet the Continuing Clinical Demands. In Proceedings of the 2nd International Conference on Radioanalytical and Nuclear Chemistry (RANC 2019), Budapest, Hungary, 5–10 May 2019.
21. Moreno-Gil, N.; Badillo-Almaraz, V.E.; Pérez-Hernández, R.; López-Reyes, C.; Issac-Olivé, K. Comparison of the sorption behavior of ^{99}Mo by Ti-, Si-, Ti-Si-xerogels and commercial sorbents. *J. Radioanal. Nucl. Chem. Artic.* **2021**, *328*, 679–690. [CrossRef]
22. Nawar, M.F.; El-Daoushy, A.F.; Madkour, M.; Türler, A. Sorption Profile of Low Specific Activity ^{99}Mo on Nanoceria-Based Sorbents for the Development of ^{99m}Tc Generators: Kinetics, Equilibrium, and Thermodynamic Studies. *Nanomaterials* **2022**, *12*, 1587. [CrossRef]
23. Sakr, T.M.; Nawar, M.F.; Fasih, T.; El-Bayoumy, S.; Abd El-Rehim, H.A. Nano-technology contributions towards the development of high performance radioisotope generators: The future promise to meet the continuing clinical demand. *Appl. Radiat. Isot.* **2017**, *129*, 67–75. [CrossRef] [PubMed]
24. Monir, T.; El-Din, A.S.; El-Nadi, Y.; Ali, A. A novel ionic liquid-impregnated chitosan application for separation and purification of fission ^{99}Mo from alkaline solution. *Radiochim. Acta* **2020**, *108*, 649–659. [CrossRef]
25. Van Tran, T.; Nguyen, V.H.; Nong, L.X.; Nguyen, H.-T.T.; Nguyen, D.T.C.; Nguyen, H.T.T.; Nguyen, T.D. Hexagonal Fe-based MIL-88B nanocrystals with NH_2 functional groups accelerating oxytetracycline capture via hydrogen bonding. *Surf. Interfaces* **2020**, *20*, 100605. [CrossRef]
26. Dash, A.; Chakravarty, R.; Ram, R.; Pillai, K.; Yadav, Y.Y.; Wagh, D.; Verma, R.; Biswas, S.; Venkatesh, M. Development of a ^{99}Mo/^{99m}Tc generator using alumina microspheres for industrial radiotracer applications. *Appl. Radiat. Isot.* **2012**, *70*, 51–58. [CrossRef] [PubMed]
27. Moret, J.; Alkemade, J.; Upcraft, T.; Oehlke, E.; Wolterbeek, H.; van Ommen, J.; Denkova, A. The application of atomic layer deposition in the production of sorbents for ^{99}Mo/^{99m}Tc generator. *Appl. Radiat. Isot.* **2020**, *164*, 109266. [CrossRef]
28. Reedijk, J.; Poeppelmeier, K. *Comprehensive Inorganic Chemistry II: From Elements to Applications*, 2nd ed.; Elsevier Ltd.: Singapore, 2013; pp. 1–7196. [CrossRef]
29. Bernauer, U.; Bodin, L.; Chaudhry, Q.; Coenraads, P.J.; Dusinska, M.; Gaffet, E.; Panteri, E.; Rousselle, C.; Stepnik, M.; Wijnhoven, S.; et al. *SCCS Opinion on Solubility of Synthetic Amorphous Silica (SAS)-SCCS/1606/19*; Publications Office of the European Union: Luxembourg, 2020; Available online: https://hal.archives-ouvertes.fr/hal-03115473 (accessed on 23 September 2022).
30. Ashry, A.; Bailey, E.H.; Chenery, S.R.N.; Young, S.D. Kinetic study of time-dependent fixation of U(VI) on biochar. *J. Hazard Mater.* **2016**, *320*, 55–66. [CrossRef]
31. Ashry, A. Adsorption and Time Dependent Fixation of Uranium (VI) in Synthetic and Natural Matrices. Ph.D. Thesis, University of Nottingham, Nottingham, UK, 2017.
32. Mahmoud, M.R.; Soliman, M.A.; Allan, K.F. Removal of Thoron and Arsenazo III from radioactive liquid waste by sorption onto cetyltrimethylammonium-functionalized polyacrylonitrile. *J. Radioanal. Nucl. Chem. Artic.* **2014**, *300*, 1195–1207. [CrossRef]
33. Mahmoud, M.R.; Sharaf El-deen, G.E.; Soliman, M.A. Surfactant-impregnated activated carbon for enhanced adsorptive re-moval of Ce(IV) radionuclides from aqueous solutions. *Ann. Nucl. Energy* **2014**, *72*, 134–144. [CrossRef]
34. Langmuir, I. The adsorption of gases on plane surfaces of glass, mica and platinum. *J. Am. Chem. Soc.* **1918**, *40*, 1361–1403. [CrossRef]
35. Akbas, Y.A.; Yusan, S.; Sert, S.; Aytas, S. Sorption of Ce(III) on magnetic/olive pomace nanocomposite: Isotherm, kinetic and thermodynamic studies. *Environ. Sci. Pollut. Res.* **2021**, *28*, 56782–56794. [CrossRef]
36. Alothman, Z.A.; Naushad, M.; Ali, R. Kinetic, equilibrium isotherm and thermodynamic studies of Cr(VI) adsorption onto low-cost adsorbent developed from peanut shell activated with phosphoric acid. *Environ. Sci. Pollut. Res.* **2013**, *20*, 3351–3365. [CrossRef] [PubMed]
37. Zhang, J.; Deng, R.j.; Ren, B.Z.; Hou, B.; Hursthouse, A. Preparation of a novel Fe3O4/HCO composite adsorbent and the mechanism for the removal of antimony (III) from aqueous solution. *Sci. Rep.* **2019**, *9*, 13021. [CrossRef] [PubMed]
38. Zheng, H.; Wang, Y.; Zheng, Y.; Zhang, H.; Liang, S.; Long, M. Equilibrium, kinetic and thermodynamic studies on the sorption of 4-hydroxyphenol on Cr-bentonite. *Chem. Eng. J.* **2008**, *143*, 117–123. [CrossRef]
39. Naiya, T.K.; Bhattacharya, A.K.; Mandal, S.; Das, S.K. The sorption of lead(II) ions on rice husk ash. *J. Hazard. Mater.* **2009**, *163*, 1254–1264. [CrossRef] [PubMed]
40. Hamdaoui, O.; Naffrechoux, E. Modeling of adsorption isotherms of phenol and chlorophenols onto granular activated carbon. Part I. Two-parameter models and equations allowing determination of thermodynamic parameters. *J. Hazard Mater.* **2007**, *147*, 381–394. [CrossRef] [PubMed]

Article

Fabrication of a Potential Electrodeposited Nanocomposite for Dental Applications

Chun-Wei Chang [1,†], Chen-Han Tsou [2,†], Bai-Hung Huang [3,4], Kuo-Sheng Hung [5,6], Yung-Chieh Cho [7], Takashi Saito [8], Chi-Hsun Tsai [8], Chia-Chien Hsieh [9], Chung-Ming Liu [3,*] and Wen-Chien Lan [10,*]

1. Division of Endodontics, Department of Dentistry, Taipei Medical University Hospital, Taipei 110, Taiwan
2. Department of Dentistry, Zuoying Branch of Kaohsiung Armed Forces General Hospital, Kaohsiung 813, Taiwan
3. Department of Biomedical Engineering, College of Biomedical Engineering, China Medical University, Taichung 404, Taiwan
4. Graduate Institute of Dental Science, College of Dentistry, China Medical University, Taichung 404, Taiwan
5. Graduate Institute of Injury Prevention and Control, College of Public Health, Taipei Medical University, Taipei 110, Taiwan
6. Department of Neurosurgery, Taipei Medical University-Wan Fang Hospital, Taipei, Taipei 116, Taiwan
7. School of Dentistry, College of Oral Medicine, Taipei Medical University, Taipei 110, Taiwan
8. Division of Clinical Cariology and Endodontology, Department of Oral Rehabilitation, School of Dentistry, Health Sciences University of Hokkaido, Hokkaido 061-0293, Japan
9. Graduate Institute of Biomedical Optomechatronics, College of Biomedical Engineering, Taipei Medical University, Taipei 110, Taiwan
10. Department of Oral Hygiene Care, Ching Kuo Institute of Management and Health, Keelung 203, Taiwan
* Correspondence: liuc@mail.cmu.edu.tw (C.-M.L.); jameslan@ems.cku.edu.tw (W.-C.L.)
† These authors contributed equally to this work.

Abstract: In the present study, a nanocrystalline Ni-Fe matrix with reinforced TiO_2 nanoparticles as a functional nanocomposite material was fabricated by pulsed current electroforming in UV-LIGA (lithography, electroplating, and molding). The influences of TiO_2 nanoparticles on the Ni-Fe nanocomposite deposition were also investigated using scanning electron microscopy, transmission electron microscopy, and in vitro cytotoxicity assay. It was found that the Ni-Fe nanocomposite with 5 wt.% TiO_2 nanoparticles showed a smooth surface and better dispersion property. When the Ni-Fe nanocomposite is combined with 20 wt.% TiO_2, it resulted in congeries of TiO_2 nanoparticles. In addition, TiO_2 nanoparticles possessed better dispersion properties as performed in pulse current electrodeposition. The microstructure of the electrodeposited Ni-Fe-TiO_2 nanocomposite was a $FeNi_3$ phase containing anatase nano-TiO_2. Moreover, the electrodeposited Ni-Fe-5 wt.% TiO_2 nanocomposite exhibited a smooth surface and structural integrity. Cytotoxicity assay results also proved that the Ni-Fe nanocomposite with different concentrations of TiO_2 nanoparticles had good biocompatibility. Therefore, the optimization of pulse current electroforming parameters was successfully applied to fabricate the Ni-Fe-TiO_2 nanocomposite, and thus could be used as an endodontic file material for dental applications.

Keywords: Ni-Fe-TiO_2 nanocomposite; microstructure; biocompatibility; endodontic file

1. Introduction

Nickel-based alloys have been extensively used as endodontic files for root canal treatment in dental fields because of their unique advantages such as high strength and toughness, superior flexibility, good shape-memory ability, excellent corrosion resistance, and acceptable biocompatibility [1–5]. Despite the advantages of nickel-based alloys, nickel-based endodontic files fracture caused by torsional and cyclic fatigue in root canal treatment still is the main issue in clinical applications [6–8]. It is well known that metallic material with a nanocrystalline structure improves mechanical properties [9–12]. Accordingly, it would be desirable if the traditional nickel-based alloy could be developed as a nanocomposite with high torsional and cyclic fatigue resistance using potential fabrication methods or by adding reinforced nanostructured materials to avoid fracture formation in root canal without sacrificing desirable physicochemical and biological properties.

There are extensive studies on methods to produce nanocrystalline materials such as current electroplating, electroless plating, co-deposition processes, pulse plating, etc. [9,12–18]. In general, electrodeposition produced porosity-free products and no consolidation was required. In contrast, nanocrystalline precursor powders were considered a vital raw material in other processing methods. Many pure metals such as Pd, Co, Ni, Ni–Mo Ni–Zn, Ni–Fe, binary alloys, and Ni–Fe–Cr ternary alloys have been fabricated by the above-mentioned methods [19–24]. Electrodeposition can produce nanostructured materials with selected treatment parameters such as pH value, overpotential, bath composition, temperature, etc. [25–27], while electrocrystallization occurs through two competing processes (existing crystals buildup and new materials formation) and can be affected by several factors. The ratio of the two processes was determined by the diffusion status of anions on the surface of crystal and charge transfer rate on the surface of the electrode [28]. Low over potential and high surface diffusion rates could cause grain growth, while high over potential and low diffusion rates of surface promote the new nuclei formation. The current density allowed by pulse plating is much higher than the limited direct current density, which can improve the new nuclei formation [29]. As stated above, the goal of the present study aimed to fabricate the Ni-Fe with reinforced Ti dioxide (Ni-Fe-TiO_2) nanocomposite (nanocrystalline Ni-Fe matrix and TiO_2 nanoparticles) as a potential endodontic instrument by the co-deposition approach with pulse electroplating for dental applications. Properties of the Ni-Fe-TiO_2 nanocomposites were evaluated through material analyses and biocompatibility assays.

2. Results

2.1. Electrodeposition of the Ni-Fe-TiO_2 Nanocomposites

According to the used parameters of pulse plating, it was found that the deposited nanocomposites exhibited extremely shiny and smooth surface features. Moreover, the TiO_2 concentration of the samples changed from 0.1 ± 0.1 wt.% (minimum) to a concentration of 2.9 ± 0.1 wt.% (maximum) for the plating solution. Different parameters such as TiO_2 concentration in the plating solution, duty cycle, current density, and solution agitation affected the TiO_2 concentration in the electrodeposits. The TiO_2 concentration in the plating solution was affected most at peak current density. Enhancing the TiO_2 concentration of the solution could increase the TiO_2 amount in the electrodeposition. However, the amount of electrodeposited TiO_2 reduced as the peak current density increased.

2.2. Microstructural Characterization of the Ni-Fe-TiO_2 Nanocomposites

The microstructure of the Ni-Fe-5 wt.% TiO_2 nanocomposite is shown in Figure 1. The formation of a nanocrystalline Ni-Fe matrix with irregularly shaped TiO_2 particles (as indicated by black arrows) can be clearly seen. The TiO_2 particles were evenly dispersed in the Ni-Fe matrix. Chemical compositions of the Ni-Fe-5 wt.% TiO_2 nanocomposite are illustrated in Figure 2. Based on the analysis by EDS, only the presence of Ni, Fe, Ti, and O elements can be seen in the matrix. No other contaminants or impurity substances were formed in the matrix during the electrodeposition. Similar microstructural characteristics and chemical compositions could also be found in the other two nanocomposites. The average size of grain in the Ni-Fe matrix was measured using the Scherrer method [11]. It was found that the average size of the grain is between 15 nm~20 nm. One thing to consider during the analysis is that the Scherrer method used in this study might not provide accurate grain sizes because of some effects; for example, the presence of stacking faults and broadening of the diffracted beam caused by microstrains. Thus, TEM was also employed to determine the grain size for the investigated nanocomposites. Figure 3 shows the microstructures and grain sizes of the as-deposited Ni-Fe-5 wt.%

TiO$_2$. Grain size is calculated from a plan-view of TEM micrograph. Apparently, the as-deposited Ni-Fe matrix was a polycrystalline structure with a grain size of ~50 nm (Figure 3a). The Ni-Fe matrix belongs to the FeNi3 phase with a face-centered cubic crystalline structure. Moreover, a diffraction ring pattern (Figure 3a) and numerous white nanocrystalline structures (Figure 3b) were seen in the matrix indicating that the TiO$_2$ nanoparticles were successfully deposited in the nanocomposite matrix. According to the camera length and d-spacings from the selected-area diffraction pattern, the deposited TiO$_2$ belongs to the anatase with a tetragonal crystalline structure. Hence, the microstructure of the electrodeposited Ni-Fe-5 wt.% TiO$_2$ nanocomposite was FeNi3 phase containing anatase nano-TiO$_2$. These features were also discovered in the Ni-Fe-10 wt.% TiO$_2$ and Ni-Fe-20 wt.% TiO$_2$ nanocomposites. Figures 4 and 5 present a car-like shape model fabricated with Ni-Fe-5 wt.% TiO$_2$ nanocomposite through an optimal UV-LIGA method. Clearly, the micro-scale car-like shape model with a smooth surface and structural integrity could be fabricated via this potential method.

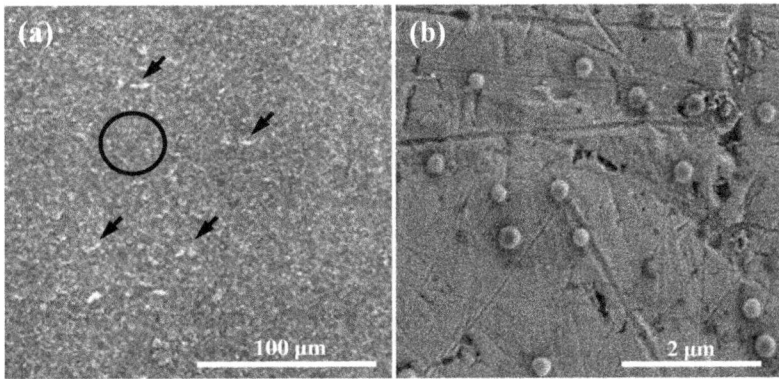

Figure 1. The FE-SEM micrographs of the Ni-Fe-5 wt.% TiO$_2$ nanocomposite: (**a**) irregular shaped TiO$_2$ particles are embedded in a nanocrystalline Ni-Fe matrix and (**b**) a higher magnification image taken from the Ni-Fe matrix (marked as the black circular area) in (**a**).

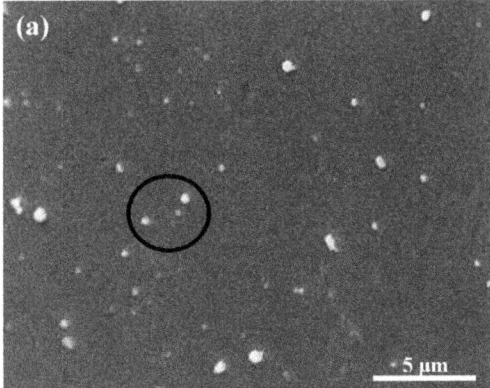

Figure 2. *Cont.*

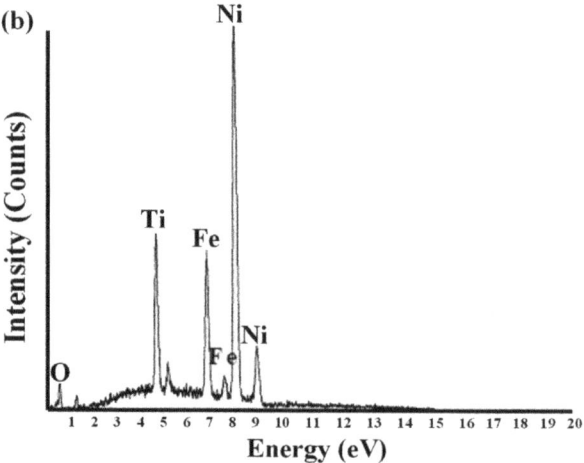

Figure 2. The chemical compositions of the Ni-Fe-5 wt.%TiO$_2$ nanocomposite: (**a**) a higher magnification FE-SEM micrograph for EDS analysis and (**b**) an EDS spectrum taken from the matrix with TiO$_2$ particles (marked as the black circular area) in (**a**).

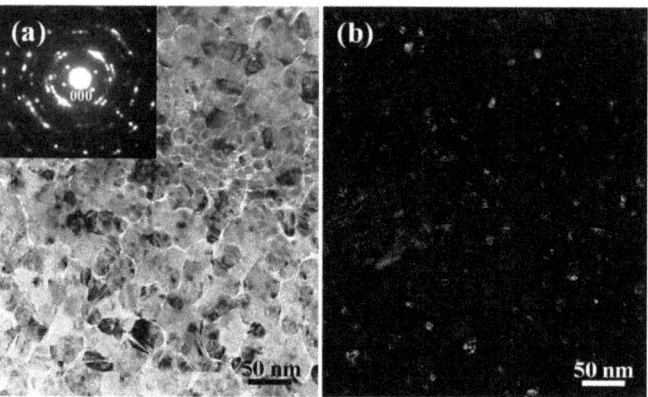

Figure 3. The TEM micrographs of the Ni-Fe-5 wt.%TiO$_2$ nanocomposite: (**a**) bright-field image showing the polycrystalline structure in the Ni-Fe matrix and (**b**) dark-field image indicating the TiO$_2$ nanocrystallization formation.

2.3. Cell Viability and Adhesion Behavior of the Ni-Fe-TiO$_2$ Nanocomposites

The cell viability of L929 of the Ni-Fe-TiO$_2$ nanocomposites for 24 h is shown in Figure 6a. Obviously, the Ni-Fe-TiO$_2$ nanocomposites exhibited a cell survival rate of more than 70%. It is considered an acute cytotoxic potential when the cell viability of the sample is less than <70% of the blank, according to ISO 109993-5. No statistically significant difference (n = 5) between investigated Ni-Fe-TiO$_2$ nanocomposites. Following cell seeding on the Ni-Fe-TiO$_2$ nanocomposites, morphology and cell adhesion in Ni-Fe-TiO$_2$ nanocomposites were observed through FE-SEM as illustrated in Figure 6b. It was found that all Ni-Fe-TiO$_2$ nanocomposites showed numerous elongated filopodia after 3 days of cell seeding. The filopodia of cells not only adhered flat, but also tightly grabbed the surface structure (as indicated by arrows). The cell viability and response features demonstrated all Ni-Fe-TiO$_2$ nanocomposites possessed good biocompatibility.

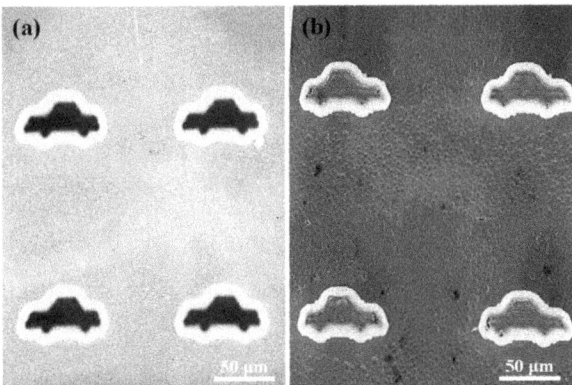

Figure 4. The FE-SEM micrographs of a car-like shape model of the Ni-Fe-5 wt.% TiO_2 nanocomposite fabricated through an optimal UV-LIGA method: (**a**) car-like cavity image and (**b**) car-like shape image.

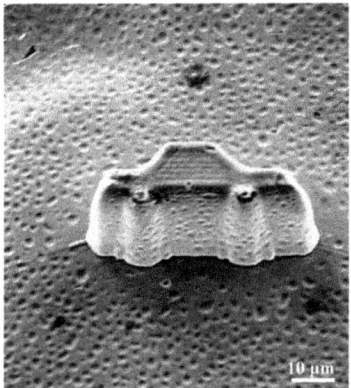

Figure 5. A higher magnification FE-SEM micrograph of the car-like shape model. The electrodeposited micro-scale car-like shape model exhibits a smooth surface and structural integrity.

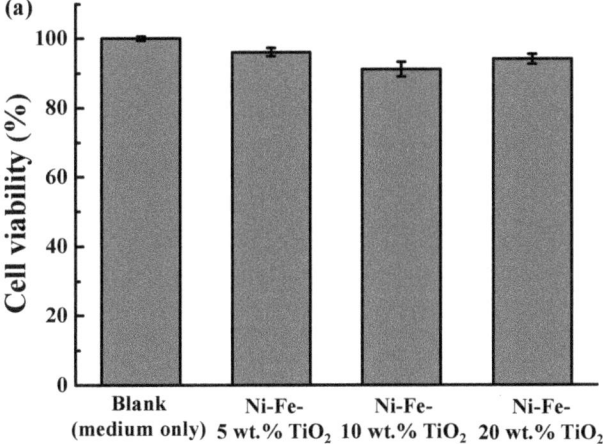

Figure 6. *Cont.*

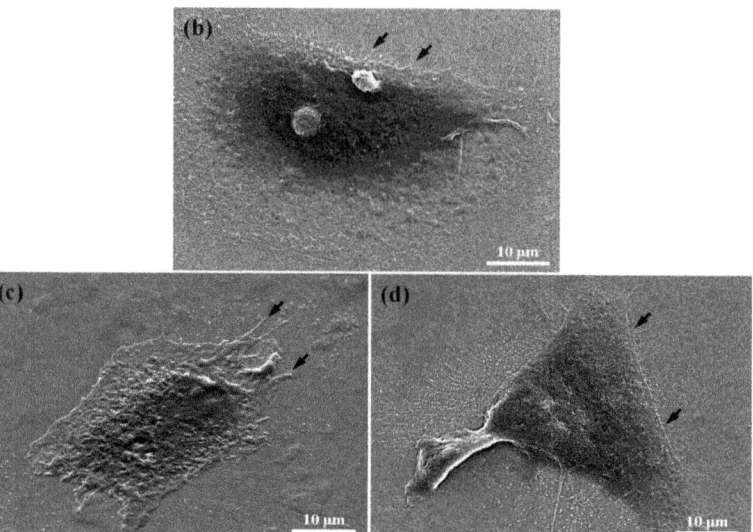

Figure 6. (a) Cell viability of L929 of the Ni-Fe-TiO$_2$ nanocomposites for 24 h and cell morphologies of the Ni-Fe-TiO$_2$ nanocomposites after culturing with L929 cells for 3 days: (b) Ni-Fe-5 wt.% TiO$_2$, (c) Ni-Fe-10 wt.% TiO$_2$, and (d) Ni-Fe-20 wt.% TiO$_2$. The filopodia (as indicated by arrows) of cells not only adhered flat, but also tightly grabbed the surface structure.

3. Discussion

In this study, it was found that the electrodeposited nanocomposite produced better smooth sidewall and surface as well as structural integrity, revealing that the Ni-Fe alloy with TiO$_2$ nanoparticles promoted formability. To explore possible reinforcement factors in nanocomposite materials, we compared results from different groups of nanocomposites including a traditional matrix of polycrystalline Ni containing reinforced nanoparticles of Al$_2$O$_3$ [30]. Oberle et al. [31] reported that with large hardness and mechanical strength enhancement in the composite matrix, the volume fraction of the co-deposited oxide reinforcements with 50 nm and 300 nm particle sizes is relatively low (i.e., 1–2% volume). Muller et al. [32] also indicated that there was a hardness and mechanical strength increased in the co-deposited Ni nanocomposite matrix containing at least 23 vol.% of Al$_2$O$_3$ (average 14 nm in diameter), and grain size of Ni is over 50 nm. However, the fabricated Ni-Fe with reinforced TiO$_2$ nanoparticles nanocomposite exhibited a similar microstructure texture with a grain size of ~50 nm. The grain size decreased when TiO$_2$ nanoparticles were added, which indicated that the nanocrystallization effect would occur due to the reactions of energetic ions in the solution during electrodeposition [22,33].

For a small matrix, the grain size was only meaningful within a limited range when used in nanocrystalline range dislocation models. The concept of the original dislocation model of the Hall-Petch relationship was based on that grain boundaries were played as barriers to dislocation movement, so a pile-up of dislocation formed at grain boundaries. As the length of pile-up with a range of 10~100 nm, the models of pile-up become doubtful. For example, the nanocrystalline range of dislocations in a pile-up rapidly reduced when the size of the grain decreased [34]. Likewise, the typical mechanism of the Orowan type was unlikely to work in the investigated nanocomposite material, since the particles of reinforcing were one order of magnitude larger than the average grain size of the matrix at least, leading to a structure in which one hard particle was surrounded by numerous differently oriented grains in the matrix of Ni-Fe. Accordingly, the nanocomposite materials with higher mechanical strength were mainly owing to the nanocrystalline Ni matrix with the presence of a reinforced second phase [35]. It is believed that the mechanical properties can be promoted for the Ni-Fe-TiO$_2$ nanocomposite.

Cytotoxicity testing is designed to assess the general toxicity of biomaterials and medical devices. The test consists of extracting the device in a cell culture medium and then exposing the extract to L929 mouse fibroblasts. ISO 10993-5 specification indicates reasonable cell viability of 70% under the MTT assay. Cell viability near or below 70% may be highly toxic to cells. In the present study, cell viability results showed that the investigated Ni-Fe-TiO$_2$ nanocomposites had high cell viability of over 90% after culturing with L929 cells for 24 h. A high cell viability rate reveals that the investigated material possesses better cell proliferation behavior and excellent biocompatibility [36,37]. In addition, it was found that there was no statistically significant difference between investigated Ni-Fe-TiO$_2$ nanocomposites. These findings demonstrated the electrodeposited Ni-Fe-TiO$_2$ nanocomposites with different concentrations of TiO$_2$ nanoparticles did not influence the proliferation and adhesion behaviors of L929 cells. Nanostructured TiO$_2$ has broad potential applications due to its nanosized features, low toxicity, and good biocompatibility [38]. Mohammadi et al. [39] also indicated that the mechanical strength of calcium phosphate cement can be enhanced with TiO$_2$ nanoparticles addition in the short-term. As a result, it is believed that the electrodeposited Ni-Fe-TiO$_2$ nanocomposites can not only enhance cell viability but can also improve mechanical properties to obtain the desired results. As discussed above, the optimal Ni-Fe-TiO$_2$ nanocomposite can be fabricated through the UV-LIGA approach. The Ni-Fe-TiO$_2$ is a potential nanocomposite material that could be unitized as an endodontic file for dental applications. However, further studies should be performed to provide additional information concerning the mechanical properties including micro-hardness, tensile strength, torsional strength, and cyclic fatigue in the presence of electrodeposited Ni-Fe-TiO$_2$ nanocomposites.

4. Materials and Methods

4.1. Materials Preparation

In this study, a modified Watts bath solution [40] was adopted as a plating solution and TiO$_2$ nanoparticles with an average diameter of 40 nm (Merk Taiwan, Taipei, Taiwan) were used as reinforced material. The pH value was controlled at a range of 2.5–3.5. Ni carbonate and hydrochloric/sulphuric acid (ratio 1:9) were used for pH value adjustment. Before fabrication, the TiO$_2$ nanoparticles powder was slowly added to the modified Watts bath solution with continuous mixing to avoid the TiO$_2$ nanoparticles agglomeration. Subsequently, the TiO$_2$ slurry with different concentrations of 5 wt.%, 15 wt.%, and 20 wt.% was added to the solution of bulk in the final bath, respectively. Hereafter, the pulsed electrodeposition process was performed with a galvanostat/potentiostat electrochemical instrument (EG & G, Princeton Applied Research 263A, Artisan Technology Group, Champaign, IL, USA), which could control the direct or pulsed electrodepositions. The parameter of pulse plating of duty cycle (pulse on time divided by pulse on time plus pulse off time) was set in direct current between 30% and 100%. The peak current densities were up to 10 A/dm^2. The stirring rate was calculated by means of a mechanical impeller to maintain the TiO$_2$ particulate in suspension. The time of plating was set constantly at 750 rpm for 3 h. The plating solution contained in the plating cell was kept in a water bath at 60 °C. The plating cell anode was made by the electrolytic Ni (purity 99.99%) containing Ti basket. The cathode was made by the Ti substrate (1 cm × 2 cm). Finally, the electrodeposits were mechanically removed from the Ti substrate (cathode side) to analyze surface and microstructural characterizations.

4.2. Surface Characterization

Surface morphology was studied by a JEOL JSM-6500F field emission scanning electron microscope (FE-SEM, Tokyo, Japan) equipped with a high-energy dispersive X-ray spectroscope (EDS; INCA, Oxford Instruments, Abingdon, UK). The operating voltage was kept at 20 kV. The samples were observed and analyzed under different magnifications.

4.3. Microstructure Identification

The model 657 dimple grinder (Gatan Inc., Pleasanton, CA, USA) and model 691 precision ion polishing system (Gatan Inc., Pleasanton, CA, USA) were used to prepare the TEM specimen with an electron transparent area. Hereafter, a model JEOL-2100 high-resolution transmission electron microscope (TEM; JEOL Ltd., Tokyo, Japan) was used to research crystallinity and phase identification with an accelerating voltage of 200 kV.

4.4. Cytotoxicity Assay

In this study, the L929 RM60091 mouse fibroblast cell line (Bioresource Collection, and Research Center, Hsinchu, Taiwan) was used for cytotoxicity evaluation. The cells were seeded in culture dishes at a density of 5×10^4 cells per 100 µL in α-Minimum Essential Medium (MEM; Level Biotechnology, New Taipei City, Taiwan). Cells from passage 2 were harvested at 80% confluence and used for further 3-(4,5-dimethylthiazol-2-yl)-2,5-diphenyltetrazolium bromide (MTT) assay. The extracts of the investigated samples were placed in an orbital shaker maintained at 37 °C for 24 h with a mass to volume extraction ratio of 0.2 g/mL, which was followed by filtering and sealing in sterile bottles. L929 cells at a density of 1×10^4 cells/well were cultured in MEM and seeded on the 24-well culture plates. After obtaining a confluent monolayer, the medium was replaced by 0.1 mL sample extracts and incubated for 24 h at 37 °C in an atmosphere of 5% CO_2 ($n = 5$). Afterward, a 10 µL MTT assay kit (R&D system, Minneapolis, MN, USA) was added to each well and incubated for 2 h. The optical density value of each plate was read at 570 nm through the ELx800 microplate reader (BioTek, Winooski, VT, USA). According to ISO 10993-5 specification, the cell viability (%) in the short-term culturing (24 h) experiment was adopted to assess the material's acute cytotoxicity response.

4.5. Cell Morphology Observation

The morphology of L929 cells was observed after 3 days of culture. The adhered L929 cells were washed with PBS, placed in a fixative consisting of 2.5% glutaraldehyde in 0.1 M sodium cacodylate buffer for 1 h at 4 °C, rinsed in deionized water, and dehydrated in serial of ethanol solutions for 15 min each concentration. Hereafter, dehydrated samples were soaked in hexamethyldisilazane, sputter coated with platinum, and analyzed with JEOL-6500F FE-SEM at 25 kV under different magnifications.

4.6. Statistical Analysis

The experimental results with multiple readings are presented as mean ± standard deviation. Data were analyzed through the variance from the Student's t-test (Excel 2016 version, Microsoft Corporation, Redmond, WA, USA). P values ≤ 0.05 were considered statistically significant.

5. Conclusions

The present work fabricated a potential nanocomposite material consisting of anatase TiO_2 nanoparticles contained in the nanocrystalline matrix of Ni-Fe using electrodeposition from a modified Watts bath. The size of grain in the nanocrystalline Ni-Fe matrix decreased with the addition of the TiO_2 nanoparticles. The Ni-Fe-TiO_2 nanocomposite exhibited a smooth surface and structural integrity. The TiO_2 nanoparticles doped in the Ni-Fe alloy can facilitate formability. The electrodeposited Ni-Fe-TiO_2 nanocomposites with different concentrations of TiO_2 nanoparticles did not influence the proliferation and adhesion behaviors of cells. Therefore, the electrodeposited Ni-Fe-TiO_2 nanocomposite is a promising endodontic file material for dental applications.

Author Contributions: Writing—original draft, C.-W.C.; Investigation, C.-W.C. and C.-H.T. (Chen-Han Tsou); Data curation, C.-C.H. and C.-H.T. (Chi-Hsun Tsai); Methodology, B.-H.H.; Supervision, T.S.; Resources, Y.-C.C.; Validation, K.-S.H.; Writing—review & editing, W.-C.L. and C.-M.L. All authors have read and agreed to the published version of the manuscript.

Funding: The authors would like to thank the Zuoying Branch of Kaohsiung Armed Forces General Hospital and Taipei Medical University Hospital for financially supporting this research under contract no. KAFGH-ZY-A-110015 and 111-D-TMUH-003.

Institutional Review Board Statement: Not applicable.

Informed Consent Statement: Not applicable.

Data Availability Statement: Data is contained within the article.

Conflicts of Interest: The authors declare no conflict of interest.

References

1. Bhagyashree, B.; Rao, D.; Panwar, S.; Kothari, N.; Gupta, S. An in vitro comparative evaluation of dentinal crack formation caused by three different nickel-titanium rotary file systems in primary anterior teeth. *J. Indian Soc. Pedod. Prev. Dent.* **2022**, *40*, 188–194.
2. Dhaimy, S.; Kim, H.C.; Bedida, L.; Benkiran, I. Efficacy of reciprocating and rotary retreatment nickel-titanium file systems for removing filling materials with a complementary cleaning method in oval canals. *Restor. Dent. Endod.* **2021**, *46*, e13. [CrossRef] [PubMed]
3. Hasheminia, S.M.; Farhad, A.; Davoudi, H.; Sarfaraz, D. Microleakage of five separated nickel-titanium rotary file systems in the apical portion of the root canal. *Dent. Res. J.* **2022**, *19*, 46.
4. Huang, Z.; Quan, J.; Liu, J.; Zhang, W.; Zhang, X.; Hu, X. A microcomputed tomography evaluation of the shaping ability of three thermally-treated nickel-titanium rotary file systems in curved canals. *J. Int. Med. Res.* **2019**, *47*, 325–334. [CrossRef]
5. Tabassum, S.; Zafar, K.; Umer, F. Nickel-titanium rotary file systems: What's new? *Eur. Endod. J.* **2019**, *4*, 111–117.
6. Ruiz-Sanchez, C.; Faus-Llacer, V.; Faus-Matoses, I.; Zubizarreta-Macho, A.; Sauro, S.; Faus-Matoses, V. The influence of niti alloy on the cyclic fatigue resistance of endodontic files. *J. Clin. Med.* **2020**, *9*, 3755. [CrossRef] [PubMed]
7. Machado, R.; Junior, C.S.; Colombelli, M.F.; Picolli, A.P.; Junior, J.S.; Cosme-Silva, L.; Garcia, L.; Alberton, L.R. Incidence of protaper universal system instrument fractures—A retrospective clinical study. *Eur. Endod. J.* **2018**, *3*, 77–81. [CrossRef] [PubMed]
8. Patnana, A.K.; Chugh, A.; Chugh, V.K.; Kumar, P. The incidence of nickel-titanium endodontic hand file fractures: A 7-year retrospective study in a tertiary care hospital. *J. Conserv. Dent.* **2020**, *23*, 21–25. [CrossRef]
9. Han, Q.G.; Zhang, H.Z.; Fang, L.; Wang, Y.Q. Tribological properties of nanocrystalline cu coatings produced by electroless plating under various lubrication conditions. *Rare Met. Mater. Eng.* **2011**, *40*, 356–359.
10. Gorka, J.; Czuprynski, A.; Zuk, M.; Adamiak, M.; Kopysc, A. Properties and structure of deposited nanocrystalline coatings in relation to selected construction materials resistant to abrasive wear. *Materials* **2018**, *11*, 1184. [CrossRef]
11. Matsui, I.; Mori, H.; Kawakatsu, T.; Takigawa, Y.; Uesugi, T.; Higashi, K. Mechanical behavior of electrodeposited bulk nanocrystalline fe-ni alloys. *Mater. Res. Ibero Am. J.* **2015**, *18*, 95–100. [CrossRef]
12. Latha, N.; Raj, V.; Selvam, M. Effect of plating time on growth of nanocrystalline ni-p from sulphate/glycine bath by electroless deposition method. *Bull. Mater. Sci.* **2013**, *36*, 719–727. [CrossRef]
13. Byun, M.H.; Cho, J.W.; Han, B.S.; Kim, Y.K.; Song, Y.S. Material characterization of electroplated-nanocrystalline nickel-iron alloys for micro electronic mechanical-system. *Jpn. J. Appl. Phys.* **2006**, *45*, 7084–7090. [CrossRef]
14. See, S.H.; Seet, H.L.; Li, X.P.; Lee, J.Y.; Lee, K.Y.T.; Teoh, S.H.; Lim, C.T. Effect of nanocrystalline electroplating of nife on the material permeability. *Mater. Sci. Forum* **2003**, *437–438*, 53–56. [CrossRef]
15. Hu, J.J.; Chai, L.J.; Xu, H.B.; Ma, C.P.; Deng, S.B. Microstructural modification of brush-plated nanocrystalline cr by high current pulsed electron beam irradiation. *J. Nano Res. Sw.* **2016**, *41*, 87–95. [CrossRef]
16. Kosta, I.; Vicenzo, A.; Muller, C.; Sarret, M. Mixed amorphous-nanocrystalline cobalt phosphorous by pulse plating. *Surf. Coat. Tech.* **2012**, *207*, 443–449. [CrossRef]
17. Protsenko, V.S.; Danilov, F.I.; Gordiienko, V.O.; Baskevich, A.S.; Artemchuk, V.V. Improving hardness and tribological characteristics of nanocrystalline cr-c films obtained from cr(iii) plating bath using pulsed electrodeposition. *Int. J. Refract. Met. Hard Mater.* **2012**, *31*, 281–283. [CrossRef]
18. Zhang, H.Z.; Fang, L.; Zhou, Y.G.; Shi, G.L. Microstructures and corrosion properties of nanocrystalline cu coatings synthesized by electroless plating. *Rare Met. Mat. Eng.* **2012**, *41*, 239–242.
19. Dmitriev, A.I.; Nikonov, A.Y.; Osterle, W. Molecular dynamics sliding simulations of amorphous ni, ni-p and nanocrystalline ni films. *Comp. Mater. Sci.* **2017**, *129*, 231–238. [CrossRef]
20. Tsyntsaru, N.; Kaziukaitis, G.; Yang, C.; Cesiulis, H.; Philipsen, H.G.G.; Lelis, M.; Celis, J.P. Co-w nanocrystalline electrodeposits as barrier for interconnects. *J. Solid State Electr.* **2014**, *18*, 3057–3064. [CrossRef]
21. Umapathy, G.; Senguttuvan, G.; Berchmans, L.J.; Sivakumar, V.; Jegatheesan, P. Influence of cerium substitution on structural, magnetic and dielectric properties of nanocrystalline ni-zn ferrites synthesized by combustion method. *J. Mater. Sci. Mater. El.* **2017**, *28*, 17505–17515. [CrossRef]
22. Farshbaf, P.A.; Bostani, B.; Yaghoobi, M.; Farshbaf, P.A.; Bostani, B.; Yaghoobi, M.; Ahmadi, N.P. Evaluation of corrosion resistance of electrodeposited nanocrystalline ni-fe alloy coatings. *Trans. Inst. Met. Finish.* **2017**, *95*, 269–275. [CrossRef]
23. Bigos, A.; Beltowska-Lehman, E.; Kot, M. Studies on electrochemical deposition and physicochemical properties of nanocrystalline ni-mo alloys. *Surf. Coat. Tech.* **2017**, *317*, 103–109. [CrossRef]

24. Wang, C.S.; Su, H.J.; Guo, Y.A.; Guo, J.T.; Zhou, L.Z. Solidification characteristics and segregation behavior of a ni-fe-cr based alloy. *Rare Met. Mat. Eng.* **2018**, *47*, 3816–3823. [CrossRef]
25. Sharma, A.; Bhattacharya, S.; Das, S.; Das, K. A study on the effect of pulse electrodeposition parameters on the morphology of pure tin coatings. *Met. Mater. Trans. A* **2014**, *45*, 4610–4622. [CrossRef]
26. Kolonits, T.; Czigany, Z.; Peter, L.; Bakonyi, I.; Gubicza, J. Influence of bath additives on the thermal stability of the nanostructure and hardness of ni films processed by electrodeposition. *Coatings* **2019**, *9*, 644. [CrossRef]
27. Torabinejad, V.; Aliofkhazraei, M.; Assareh, S.; Allahyarzadeh, M.H.; Rouhaghdam, A.S. Electrodeposition of ni-fe alloys, composites, and nano coatings-a review. *J. Alloy. Compd.* **2017**, *691*, 841–859. [CrossRef]
28. Tripkovic, D.V.; Strmcnik, D.; van der Vliet, D.; Stamenkovic, V.; Markovic, N.M. The role of anions in surface electrochemistry. *Faraday Discuss.* **2008**, *140*, 25–40. [CrossRef]
29. Sun, J.; Du, D.X.; Lv, H.F.; Zhou, L.; Wang, Y.G.; Qi, C.G. Microstructure and corrosion resistance of pulse electrodeposited ni-cr coatings. *Surf. Eng.* **2015**, *31*, 406–411. [CrossRef]
30. Muller, B.; Ferkel, H. Properties of nanocrystalline ni/al2o3 composites. *Z. Metallkd.* **1999**, *90*, 868–871.
31. Oberle, R.R.; Scanlon, M.R.; Cammarata, R.C.; Searson, P.C. Processing and hardness of electrodeposited ni/al2o3 nanocomposites. *Appl. Phys. Lett.* **1995**, *66*, 19–21. [CrossRef]
32. Muller, B.; Ferkel, H. Al2o3-nanoparticle distribution in plated nickel composite films. *Nanostructured Mater.* **1998**, *10*, 1285–1288. [CrossRef]
33. Nicolenco, A.; Mulone, A.; Imaz, N.; Tsyntsaru, N.; Sort, J.; Pellicer, E.; Klement, U.; Cesiulis, H.; Garcia-Lecina, E. Nanocrystalline electrodeposited fe-w/al2o3 composites: Effect of alumina sub-microparticles on the mechanical, tribological, and corrosion properties. *Front. Chem.* **2019**, *7*, 241. [CrossRef] [PubMed]
34. Arzt, E. Overview no. 130—Size effects in materials due to microstructural and dimensional constraints: A comparative review. *Acta Mater.* **1998**, *46*, 5611–5626. [CrossRef]
35. Zimmerman, A.F.; Clark, D.G.; Aust, K.T.; Erb, U. Pulse electrodeposition of ni-sic nanocomposite. *Mater. Lett.* **2002**, *52*, 85–90. [CrossRef]
36. Hsu, H.J.; Abd Waris, R.; Ruslin, M.; Lin, Y.H.; Chen, C.S.; Ou, K.L. An innovative alpha-calcium sulfate hemihydrate bioceramic as a potential bone graft substitute. *J. Am. Ceram. Soc.* **2018**, *101*, 419–427. [CrossRef]
37. Gittens, R.A.; McLachlan, T.; Olivares-Navarrete, R.; Cai, Y.; Berner, S.; Tannenbaum, R.; Schwartz, Z.; Sandhage, K.H.; Boyan, B.D. The effects of combined micron-/submicron-scale surface roughness and nanoscale features on cell proliferation and differentiation. *Biomaterials* **2011**, *32*, 3395–3403. [CrossRef]
38. Jafari, S.; Mahyad, B.; Hashemzadeh, H.; Janfaza, S.; Gholikhani, T.; Tayebi, L. Biomedical applications of tio2 nanostructures: Recent advances. *Int. J. Nanomed.* **2020**, *15*, 3447–3470. [CrossRef]
39. Mohammadi, M.; Hesaraki, S.; Hafezi-Ardakani, M. Investigation of biocompatible nanosized materials for development of strong calcium phosphate bone cement: Comparison of nano-titania, nano-silicon carbide and amorphous nano-silica. *Ceram. Int.* **2014**, *40*, 8377–8387. [CrossRef]
40. Ibrahim, M.A.M. Black nickel electrodeposition from a modified watts bath. *J. Appl. Electrochem.* **2006**, *36*, 295–301. [CrossRef]

Review

Inorganic Finishing for Textile Fabrics: Recent Advances in Wear-Resistant, UV Protection and Antimicrobial Treatments

Silvia Sfameni [1,2,†], Mariam Hadhri [3,†], Giulia Rando [2,4], Dario Drommi [4], Giuseppe Rosace [3], Valentina Trovato [3,*] and Maria Rosaria Plutino [2,*]

[1] Department of Engineering, University of Messina, Contrada di Dio, S. Agata, 98166 Messina, Italy
[2] Institute for the Study of Nanostructured Materials, ISMN–CNR, Palermo, c/o Department of ChiBioFarAm, University of Messina, Viale F. Stagno d'Alcontres 31, Vill. S. Agata, 98166 Messina, Italy
[3] Department of Engineering and Applied Sciences, University of Bergamo, Viale Marconi 5, 24044 Dalmine, Italy
[4] Department of ChiBioFarAm, University of Messina, Viale F. Stagno d'Alcontres 31, Vill. S. Agata, 98166 Messina, Italy
* Correspondence: valentina.trovato@unibg.it (V.T.); mariarosaria.plutino@cnr.it (M.R.P.)
† These authors contributed equally to this work.

Abstract: The surface modification of textile fabrics and therefore, the development of advanced textile materials featuring specific implemented and new properties, such as improved durability and resistance, is increasingly in demand from modern society and end-users. In this regard, the sol–gel technique has shown to be an innovative and convenient synthetic route for developing functional sol–gel coatings useful for the protection of textile materials. Compared with the conventional textile finishing process, this technique is characterized by several advantages, such as the environmentally friendly approaches based on one-step applications and low concentration of non-hazardous chemicals. The sol–gel method, starting from inorganic metal alkoxides or metal salts, leads to inorganic sols containing particles that enable a chemical or physical modification of fiber surfaces, giving rise to final multifunctional properties of treated textile fabrics. This review considered the recent developments in the synthesis of inorganic nanoparticles and nanosols by sol–gel approach for improving wear and UV resistance, as well as antibacterial or antimicrobial effects for textile applications.

Keywords: inorganic coatings; functional coatings; stimuli-responsive polymers; sol–gel; antimicrobial; wear-resistance; photo-catalytic activity; UV protection

1. Introduction

During the last decades, chemical finishing treatments for natural and synthetic fibers or fabrics have been widely used to improve their specific properties and confer particular functionalities. In this regard, textile materials are treated with various functional chemical finishes with antimicrobial, durable press, repellent, flame retardant, soil release and antistatic properties as well as many others. Furthermore, the interest in developing functional fabrics mainly arises from the consumers' needs in terms of comfort, health, hygiene, protection against chemical, thermal or mechanical agents and easy care.

In this perspective, functional high-tech fabrics represent a challenge for the textile industry, with interesting implications in terms of future benefits and technological advancement, also related to the lengthening of the average life of textile products in compliance with the need for saving raw materials. Simultaneously, with the conferment of functional properties to the fabrics, the maintenance of textile properties, such as washing durability, feel and appearance should be guaranteed. Indeed, the employed finishing process (i.e., preparation method, if batch or continuous, dyes and dyeing methods, finishing agents and application techniques), mechanical and thermal combination treatments (sueding, sanding, calendering, brushing, sanforizing, etc.) have a relevant impact on the wearing and performance characteristics of textiles.

A quick look at recent literature [1] shows that various chemicals are frequently used as finishes to improve fabric performances. However, due to the intense pressure to ban harmful chemicals, such as halogen-containing flame retardants or chemical anti-microbial agents, different attempts have recently been made to reduce or replace questionable safety chemicals with new environmentally friendly solutions. Furthermore, the selection of the finishing process influences the industrial closing costs, making it another critical parameter to consider. To fulfil all these requirements, advanced and innovative technologies should be involved. In this regard, textile modifications through the creation of polymeric nanocomposites mainly based on metallic or inorganic nanostructured materials are constantly being developed due to their distinct and quite unique characteristics. Until now, surface modification of textiles has been accomplished using a variety of techniques, including (i) use of plasma pre-treatment [2,3]; (ii) ultraviolet irradiation of textiles [4]; (iii) sputtering of nano-particles during plasma polymerization [5]; (iv) in situ synthesis of metal nanoparticles [6] involving cellulose nanoporous structure; (v) use of ion-exchange functionalized surfactant during the polymerization process [7]; (vi) loading of nano-particles into liposomes [8]; and (vii) conventional pad-dry-cure systems.

Among these, most recently, the sol–gel process, which results in the formation of self-assembled (nano) layers on the fiber surface, has notably demonstrated its unique potential for the synthesis of new coatings with high molecular homogeneity and outstanding physical-chemical properties [9–11]. From 1984, more than 24,000 scientific papers reported sol–gel protective coatings and finishings. The sol–gel is a versatile synthetic route based on a two-step reaction (hydrolysis and condensation), generally starting from (semi) metal alkoxides (e.g., tetraethoxysilane (TEOS), tetramethoxysilane, titanium tetraisopropoxide), which conduct to the formation of completely inorganic or hybrid organic–inorganic coatings at, or near, room temperature [12,13]. These coatings can protect the polymer surface by acting as an insulator, thus improving the advanced performance of treated materials, such as flame retardancy [14–16], antimicrobial or UV radiation protection [16–19], anti-wrinkle finishing [20], washing fastness of dyes [21,22], bio-molecule immobilization [23] and super-hydrophobicity [24,25]. Moreover, the sol–gel technique has recently been investigated for novel applications, such as for imparting sensing [26–29] or self-cleaning [30] properties for fabrics, or for hydrogen production via water photo splitting [31]. This environmentally friendly method, in particular, has several advantages over other methods of film deposition, including the ability to develop a protective thin hybrid layer on the surface of a textile with well-defined physical characteristics, optical transparency and excellent chemical stability, as well as the ability to preserve the mechanical and chemical properties of the fibers, thereby introducing durability and highly performant functionalities [32–37]. Furthermore, the developed film, which is useful for the previously mentioned textile applications, does not have cytotoxic effects on human skin cells, providing additional benefits in the application of this technology [38]. An easy functionalization approach in order to improve the mechanical and thermal features, but also the chemical resistance and functional properties of the final coating, concerns the blend of the sols with proper hybrid modifiers. In this regard, a wide range of functional nanofillers can be employed as metals [39,40] and metal oxide nanoparticles [41], carbon-based nanomaterials [42,43] and silica-based nanofillers [44].

Although the majority of the literature data refers to sol–gel reactions for synthesizing hybrid organic–inorganic silane-based materials, on the other hand, this technology also contemplates the involvement of other inorganic precursors, such as Al, Ti, Zr, Sn, V salts or other alkoxides and organometallic substrates, in which the metal is linked to organic moieties able to hydrolyze and then condensed [45,46].

As already mentioned, the sol–gel process is based on the transition phase from a liquid (sol) into a dense system (gel), which is then dried and heated to form a hybrid organic–inorganic porous matrix. The involved organic and inorganic precursors can interact via weak bonds, such as ionic, hydrogen, or van der Waals (Class I) or covalent, coordination, or iono-covalent bonds (Class II) [33,47,48]. Therefore, the covalent interaction of organic and inorganic moieties results in final supramolecular Hybrid Organic-Inorganic Materials (HOIM), based on organofunctional alkoxysilane precursors (Class II) of the general chemical structure R'_n-$Si(OR)_{4-n}$ (where R' is an alkyl or organo-functional group and R is mainly methyl, propyl, or butyl group) useful in the chemical modification of textile fabrics. In particular, these three hydrolyzable and polymerizable R groups, together with a different functional group, such as an epoxy, vinyl, or methacryloxy group [33,47,49], are involved in forming well-oriented three-dimensional (3D) network structures.

As shown in Figure 1a–d, silica alkoxide precursors are hydrolyzed in water by acids (Figure 1a,b) or bases (Figure 1c,d), followed by parallel condensation reactions that result in the formation of a final 3D network bearing Si–O–Si bonds (Figure 2).

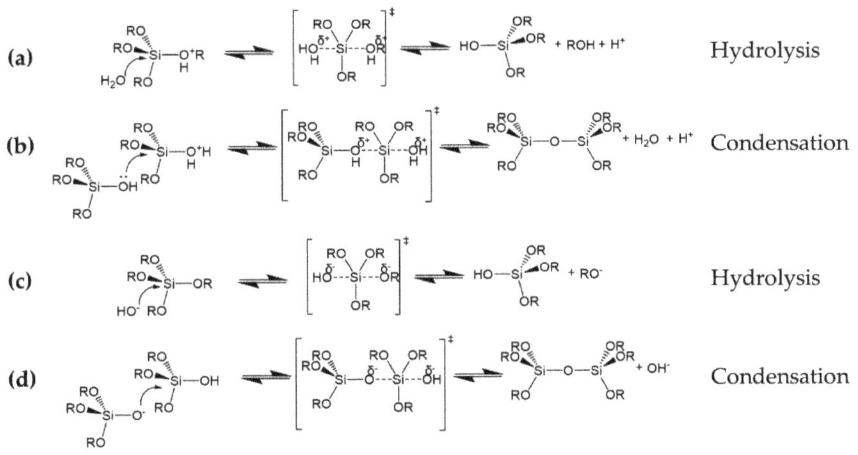

Figure 1. Hydrolysis and condensation reactions of silica alkoxide precursors in acid (**a,b**) and alkaline (**c,d**) conditions.

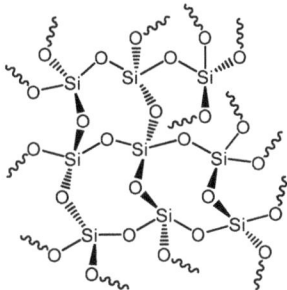

Figure 2. The obtained 3D network after the hydrolysis and condensation reactions of silica precursors.

When compared with other techniques, the sol–gel approach has several advantages, including low-temperature conditions during the process, easily removable solvents during the handling period and a high purity of the obtained products. Furthermore, it is possible to control the physical properties of the final inorganic matrix to obtain several porous materials, such as glass, dry gel, polycrystalline powder or coating films by tailoring the synthesis conditions, such as concentrations, temperature, time, pH, reactant ratios, solvents and catalysts [50]. Accordingly, the addition of acidic or basic catalysts is normally used to achieve complete and rapid hydrolysis, and the polymerization process is also strongly pH-dependent. Moreover, the type of hydrolysis catalyst influences both the pH of the sols and the nature of the sol–gel polymer aggregates towards the fabrication of films, powders, or monoliths [50–52]. Acidic hydrolyzed nanosols may form small-particle aggregates with larger pores, whereas base-catalyzed sols form weakly cross-linked condensation products with a denser layer structure [53,54]. It should be noted that the hydrolysis and drying conditions govern the density, porosity and mechanical properties of the layers when coating textiles (and other material surfaces) and critical thickness for cracking. In particular, thermal post-coating treatment is required to achieve strongly adhesive and stable coatings on textiles, increasing the washing speed and mechanical properties of the obtained films.

Nonetheless, for most textiles, the temperature of thermal treatments must not exceed 180 °C to avoid the thermal destruction and degradation of polymeric fibers. The resulting xerogel layers are very interesting for functionalizing textiles because the coatings can be easily modified chemically or physically [55,56], allowing a broad range of changes in the textile properties by reacting with different opportune nanosol precursors. Moreover, because of the low temperature of sol–gel material processing and the porosity of the resulting 3D network, the inorganic matrix can be doped with various organic/inorganic functional molecules, resulting in advanced functional hybrid materials [57].

Furthermore, chemical modification is possible through co-condensation with additives that can form covalent bonds with the metal oxide matrix as co-reactants (Figure 3a) [46,58] or between different metal alkoxides (e.g., the reaction of tetraethoxysilane $Si(OC_2H_5)_4$ (TEOS) with other metal alkoxides with generic formula $M(OR)_n$ (M = Al, Ti, Zr, Zn, etc.) [59] (see Figure 3b). When the hydrolysis rates of the precursors of mixed nanosols differ significantly, selective complexation of the more reactive component is necessary to coordinate the rate. Furthermore, trialkoxysilanes $R-Si(OR)_3$ containing an organic substituent R that is chemically bonded to the silica matrix (according to hydrolysis and condensation) can also be used for functional chemical modifications (Figure 3c).

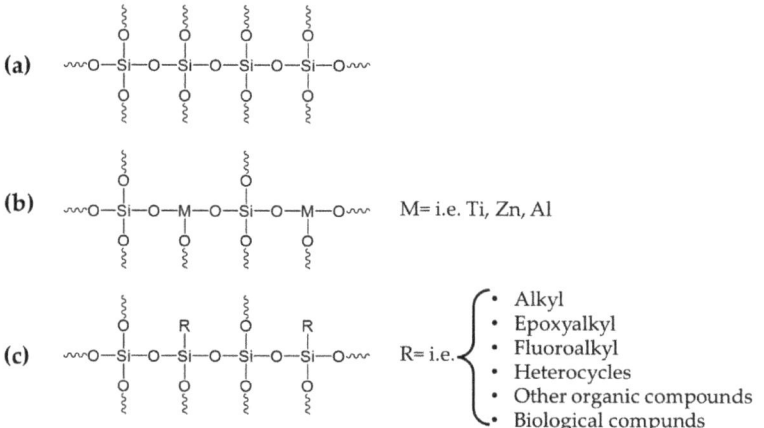

Figure 3. Examples of the chemical functionalization of nanosols including the co-condensation with additives that can form covalent bonds with the metal oxide matrix as co-reactant (**a**), between different metal alkoxides (**b**) and other trialkoxysilanes containing an organic substituent (**c**).

The nanosol coatings can be modified in various ways using the incorporated substituent R. For example, alkyl or perfluoroalkyl containing trialkoxysilanes can be used to achieve hydrophobic or oleophobic properties [60–63]. On the other hand, modified nanosols with epoxy alkyl or acrylate R groups favor the thermal or photochemical coating crosslinking and linking to the textile surface, thus significantly improving the mechanical properties of the coating as well as its adhesion to the textile fiber [58]. Physical modification of the oxide matrix is a universal method for preparing coatings with immobilized additives (Ad), such as inorganic colloidal metals [64,65], dyes [66–68], oxides [69] and pigments [70], biomolecules [71–73] or organic polymers [74,75] and living cells [71,72,76]. The immobilization can be performed by adding the ingredients either before (Figure 4a) or after (Figure 4b) hydrolysis of the precursors. Since encapsulation occurs during the condensation step, both variants "a" and "b" have nearly identical composite structures and immobilization behavior. Therefore, it is very efficient to immobilize the additives within the inorganic matrix, and it can be controlled by several parameters, among which are the composition and structure of the oxide matrix, the Ad:oxide ratio and the addition of pore-forming agents [77,78].

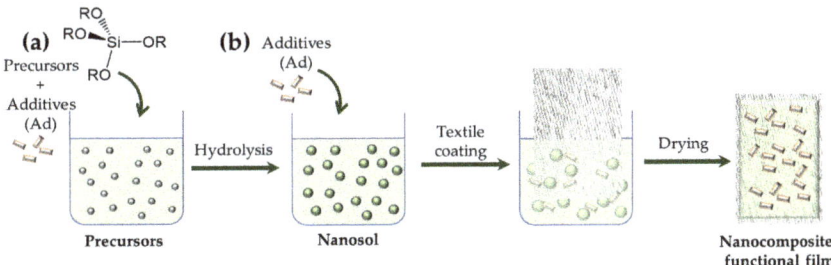

Figure 4. Physical modification of nanosol coatings by doping the precursor solution (**a**) or the nanosol after the hydrolysis reaction (**b**).

As a result, combining chemical and physical modifications offers limitless potential for developing and applying inorganic nanosol coatings for textile functionalization (Figure 5) [79].

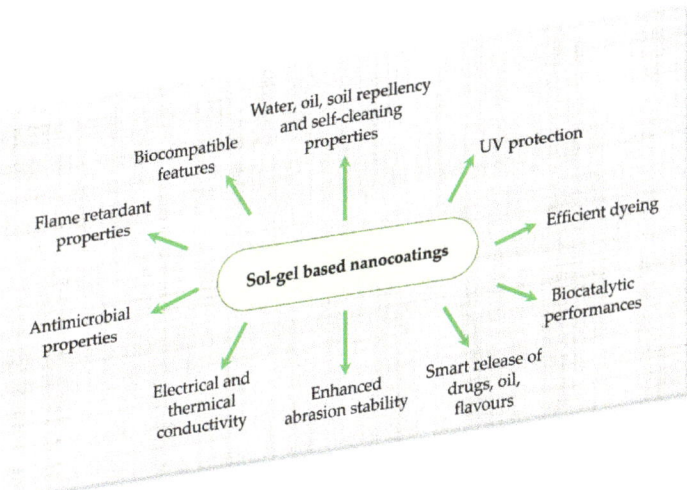

Figure 5. Some possibilities of textile functionalizations by proper sol–gel-based nanocoatings.

In this wide panorama of sols, sols-modification techniques and functional applications for textile materials, the following review aimed to provide an overview of inorganic sols based on nanoparticular-modified silica and other metal oxides for textile finishing.

The relevance of inorganic sol coatings in the functionalization of textiles can be ascribed to the following [80]:

1. The formation of well-adhering transparent oxide layers on textiles due to particle-diameter dimensions (smaller than 50 nm) of the modified silica or other metal oxide-based nanosols;
2. High stability of these inorganic oxide layers against heat, light, microbial and chemical attacks;
3. The capability of these oxide coatings to improve the mechanical properties of the textiles (e.g., wear and abrasion resistance), thus further offering possible methods of varying the surface properties and high mechanical strength;
4. Layer porosity and the degree of immobilization of the embedded compounds (e.g., organic, or biological compounds, inorganic particles, and polymers) for whom oxide coatings can serve as carriers, can be easily controlled;
5. Common textile finishing processes (i.e., pad, exhaustion, dip-coating or spraying) can be used for coating application at room temperature and normal pressure.

The enormous interest in textile materials with even higher added values for application in several fields, such as industrial, workers' protection or healthcare and medical uses continuously prompts scientific research towards innovative and efficient solutions for introducing specific functionalities in textile fibers. Particularly, to the best of our knowledge, this review considered the most relevant applications, as well as the most recent developments [81–98], of nanostructured materials for textile and polymeric material modifications to improve the wear- or UV-resistance properties or to introduce antibacterial effects. With this aim, the inorganic materials involved, as well as the textile material and functionalities achieved from the application of the inorganic coatings were described, taking into consideration the unique advantages and high potentialities of the nanostructured materials for producing high-performance textiles.

2. Inorganic-Based Materials for Improving Wear Resistance of Textile Fabrics

As commonly mentioned, textile fibers are very sensitive to wear generated from external mechanical effects, such as cellulose which, contrary to synthetic fibers, is less sensitive to such wear. In this regard, applying protective films on the fibers' surfaces increases their durability against mechanical actions. For this purpose, sol–gel technology is very promising since it leads to the formation of surface ceramic-like inorganic nanocoatings characterized by high hardness and resistance to abrasion [99,100]. On the other hand, however, for producing inorganic protective coatings for fabrics, which allow them to improve their abrasion resistance, some fundamental aspects must be taken into account, meaning this application is not always easy to carry out. In particular, applying such coatings, also in several layers, must not significantly affect the natural soft, pleasant handle or other aspects relating to the "fibrous" nature of the fabric, and also must not increase its stiffness or negatively affect its aesthetic or hygienic properties.

This can be accomplished by applying inorganic xerogel coatings as thin as possible, as well as being elastic, transparent and having durable characteristics ensured by establishing chemical bonds between the coating and the fibers. Conversely, thick coating on textiles can deteriorate the abrasion resistance instead of improving it by accelerating the wear compared with untreated fabrics. Accordingly, inorganic sol–gel coatings with thicknesses in the range of 150–500 nm are already reported in the literature [13,53,101].

Moreover, these inorganic colloidal suspensions (i.e., in water, alcohol, water/alcohol medium) should have good stability, and, as a consequence, they should contain particles smaller than 50 nm. This condition is applied not only to inorganic sols but also to doping nanoparticles incorporated in the crosslinked xerogel for increasing the coating abrasion resistance (e.g., Al_2O_3 nanoparticles) [13,53,102].

When silica sol is applied to fabric, it forms an inorganic glass-like network that contains siloxane and silanol groups and is highly interconnected with fibers via stable covalent bonds as well as weak interactions, including dipolar–dipolar and hydrogen-bond interactions. In this regard, the greater the dipolar–dipolar and hydrogen-bond interaction, the stronger the coating adhesion must be. Additionally, the polar chain structure, combined with the former cotton-fiber chemical structure, is one of the primary factors for dipolar–dipolar and hydrogen bond interaction, so a high add-on percentage is expected. During the treatment, the diffusion of silica nanoparticles or nanosols from the surface into the fabric texture, particularly in the first layer, controls the contribution to adhesion [103]. Table 1 reports some systems and precursors involving the production of functional nanosols with the objective of improving the wear resistance of different fabrics.

Table 1. Preparation and properties comparison of some inorganic nanosols employed as wear-resistant finishings for different textile substrates.

Precursors and Additives	Synthetic Approach	Textile Substrate	Deposition Approach	Wear-Resistance Tests Performed	Ref.
Methyltrimethoxysilane	Sol–gel	Continuous filament polyester double-knit interlock fabric	Soaking	Custom aging-abrasion-impact, wicking and laundry testing protocols	[104]
GPTMS *, TEOS and 1,2,3,4-butanetetracarboxylic acid	Sol–gel	Cotton fabrics	Impregnation	Abrasion resistance (dry crease-recovery angle test)	[105]
Aluminum isopropoxide	Sol–gel	Nonwoven PET fabrics	Impregnation	Mechanical resistance (breaking strength and elongation at break)	[102]
TEOS and Dibutyltindiacetate	Sol–gel	Cotton fabrics	Pad–dry–cure process	Mechanical resistance (tensile strength and elongation)	[106]
TEOS, GPTMS, Aluminum (III) isopropoxide	Sol–gel	Commercial cotton woven fabric	Pad–dry–cure process	Abrasion and laundering tests	[100]

* GPTMS = (3-Glycidyloxypropyl)trimethoxysilane

Anti-abrasion-coated textiles have relevant applications in several industries, such as polyester sieves in paper production, home textiles, carpets or furniture. For example, alkyl-modified silica nanosols coated on polyester sieves used in paper production can improve their abrasion properties due to the excellent smooth surface structure of the treated fibers and the homogeneous films that completely cover the fibers [80]. SEM characterizations of mixed-cotton textiles treated with different sol–gel coatings yielded similar results [80]. For this purpose, polyethylene terephthalate (PET) fabrics were also treated with methyltrimethoxysilane in an alkaline alcoholic solution containing ammonium hydroxide (Figure 6a) as the catalyst to achieve excellent resistance against different kinds of wear damage by mimicking superhydrophobic biological systems (Figure 6b,c) [104].

Indeed, several fiber materials, such as cellulose, polyamide and glass fibers have reported improved abrasion resistance and tensile strength when coated with inorganic nanosol [105,107]. However, as far as cotton is concerned, experimental findings demonstrated that the adding epoxy silanes, such as (3-Glycidyloxypropyl)trimethoxysilane (GPTMS), increases the adhesion of the organic-inorganic nanosol coating on the fibers, thereby improving their mechanical properties and the elasticity of the obtained coating.

However, in high concentrations, organic moieties can lead to adverse effects on the mechanical resistance of the final coating [13,53]. Certainly, the stiffness of nanosol-treated textiles is also affected by bridging or gel combinations between adjacent fibers that stabilize their mutual position and do not allow them to displace. Such bridging increases when excessively high concentrations of (or partially gelled) nanosols are applied on textile fibers, thus lowering the textile elasticity and increasing the coating thickness [100].

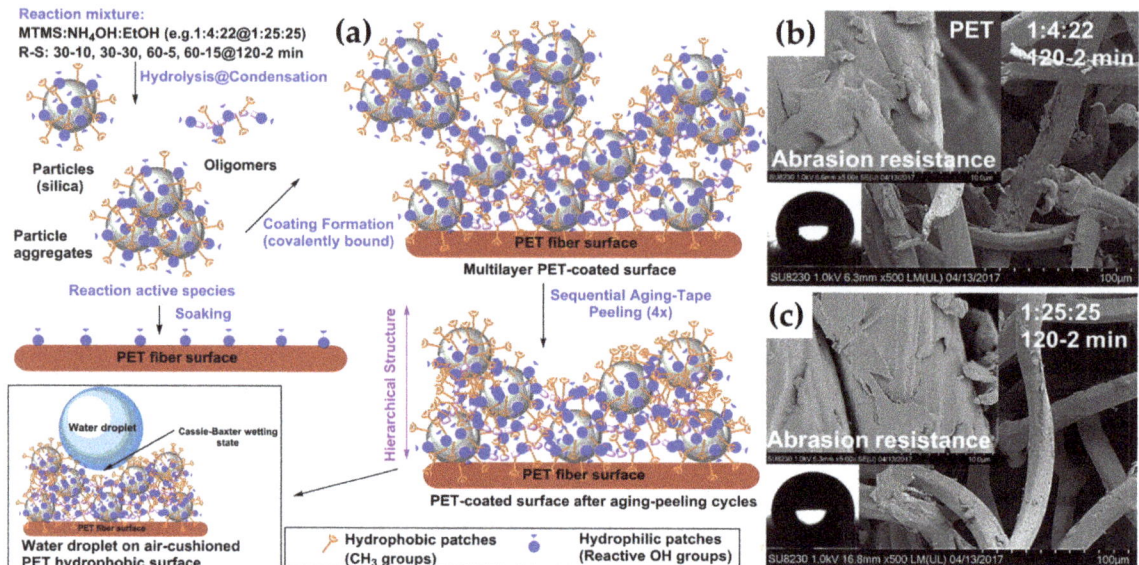

Figure 6. Schematic representation of methyltrimethoxysilane coating preparation for PET fabrics (**a**) and relative SEM images describing the appearance of the nanocoatings after standard abrasion-resistance testing: 1:4:22 (120-2 min) (**b**) and 1:25:25 (120-2 min) (**c**). Reproduced with permission from ref. [104] Copyright 2018 Elsevier.

An attempt to slightly ameliorate the stiffness properties of textile materials treated by inorganic TEOS-based coatings can consist of the addition of positively charged cationic softeners and uncharged nonionic softeners separately to silica-based sols [92]. In this regard, the mechanical properties of commercial denim fabrics were improved, showing an increase in abrasion resistance and tear strength when using nonionic and cationic softener-doped silica coatings.

Generally, using inorganic nanoparticles for surface modification of textiles makes coatings impermanent, and susceptible to washing. The majority of the stabilization methods of inorganic nanostructured materials on textile surfaces necessitate several preparatory steps (e.g., functionalization, drying, curing, final treatment and other ones) that are expensive and time-consuming for large-scale manufacturing, and/or environmentally harmful due to the use of numerous chemicals or organic solvents. Most of them result in a decrease in tensile properties, abrasion resistance, softness, appearance and other properties of textiles, as well as color change. For example, UV irradiation was used for generating radical groups on the textile surface, able to bond nanoparticles [4] or reduce metallic salts to nanoparticles in a polymeric matrix. Besides economic disadvantages, such as being costly and time-consuming, the presence of free radicals on the textile surfaces that can migrate during clothing usage results in some health risks for people. For example, ZnO is used to improve wear resistance and anti-sliding properties in composites [108] and reinforce polymeric nano-composites [109] due to its high elastic modulus and strength. However, the use of acrylic binders for the fixation of ZnO-soluble starch nano-composite particles on cotton fabrics [109] not only reduces the abrasion resistance and the cloth's comfort, but the possible decomposition of starch or acrylic binder during processing should also be of concern.

As an inorganic sol–gel precursor, TEOS, is widely used to prepare inorganic silica homogeneous films on textile materials due to its low temperature and cheap method of processing. Moreover, the TEOS gelation process through acid or alkaline catalysts [110,111] has been widely investigated by evaluating its viscoelastic behavior during the sol–gel transition.

In this regard, the influence of organic-tin compounds (i.e., DBTA-Dibutyltindiacetate) as polycondensation catalysts for TEOS was investigated in a multistep deposition process on cotton fabrics, thus leading to a coating well adhered to fabric surfaces [103]. The study was conducted by using two TEOS concentrations (0.03 M and 0.3 M) and evaluating the presence or not of the DBTA catalyst. In particular, the coating realized with the lowest TEOS concentration (0.03 M) showed the best mechanical properties when compared with untreated cotton or cotton treated with the highest TEOS concentration (0.3 M, Table 2). Although ensuring the highest thermal stability and washing fastness, this latter leads to surface coatings characterized by high rigidity that makes fiber movements difficult, thus lowering strength, yarn elongation and tensile strength (Table 1). On the other hand, DBTA was proven to increase the abrasion resistance, as well as the thermal and washing speed.

Table 2. Tensile strength and elongation for untreated (CO_UT) and treated cotton fabrics in both warp and weft directions (standard deviation lower than ±3%; nL = number of layers, C = catalyst). Adapted with permission from ref. [103] Copyright 2013 Elsevier and ref. [106] Copyright 2020 IOP Science.

	Tensile Strength (N)		Elongation (%)		Tensile Strength (N)		Elongation (%)		
	Warp	Weft	Warp	Weft	Warp	Weft	Warp	Weft	
CO_UT	240.3	265.5	17.0	16.0	240.3	265.5	17.0	16.0	CO_UT
CO_0.03M-1L	325.2	225.5	17.5 *	15.3	207.3	220.1	16.8	13.9	CO_0.3M-1L
CO_0.03M-3L	302.4	326.6	16.3	17.8	190.9	222.4	13.8	12.2	CO_0.3M-3L
CO_0.03M-6L	317.9	302.4	15.5	16.3	150.0	125.5	7.5	11.0	CO_0.3M-6L
CO_0.03M-1L-C	317.2	250.3	16.3	16.7	200.1	205.7	15.2	14.0	CO_0.3M-1L-C
CO_0.03M-3L-C	**357.1**	230.2	16.3	**17.9**	185.0	203.1	13.8	13.9	CO_0.3M-3L-C
CO_0.03M-6L-C	282.3	306.7	14.5	16.1	150.7	179.2	10.5	13.5	CO_0.3M-6L-C

* The best results are highlighted in bold.

The efficiency of inorganic sols was also proven for the protection of cellulose-based polymers in cultural heritage textiles, thus representing a valid alternative to existing materials [106]. However, the use of the DBTA catalyst should be further investigated in terms of its mechanism and replaced with more environmentally friendly alternatives.

Recently, the development of sustainable coatings of SiO$_2$ NPs-chitosan (with an average size of 150 nm) was proven to be efficient for multifunctional effects on polyester and viscose fabrics (both warp and braids structures) [91]. Concerning the mechanical properties, this nanocomposite led to an overall increase in both the elongation and tensile strength of the coated samples, thus measuring enhanced mechanical properties for increasing the SiO$_2$ NPs-chitosan mass loading. However, the further increase of the nanocomposite on textiles resulted in the failure of both tensile strength and elongation due to the inclusion of nanoparticles in the fibers rather than on the fabric surfaces.

3. Inorganic-Based Materials for UV Protection Textile Finishing

Some textile polymers are slightly resistant to ultraviolet radiations (UV) and thus decompose, as is the case of cellulose, polyester, phenylenebenzobisoxazole (PBO) and p-aramide fibers. However, the textile kind of use could necessitate increased UV resistance relative with the intrinsic resistance of the fabrics themselves. For example, the light stability of polyester textiles is enough only for clothing applications; the fibers of a textile roofing exposed for several years to sunlight can decompose, showing decreased mechanical strength over the years.

In this regard, several textile polymer fibers can be modified before the spinning process with UV-absorbing pigments (e.g., titania) or using organic molecules acting as radical scavengers to improve their UV resistance [13].

Moreover, ultraviolet (UV) protection finishes (or UV shielding) are some of the most important groups of chemical finishing agents used on textile materials to protect people and fabrics from the harmful effects of UV radiation. Indeed, the energy of UV radiation is significantly higher than that of visible light, having the potential to cause a variety of chemical reactions that are hazardous to human health, besides deteriorating textile fibers. Even though moderate sun exposure has health benefits, excessive exposure to UV radiation can cause serious harm since both UVA (320–400 nm) and UVB (280–320 nm) rays induce different cellular responses that can generate skin aging, sunburn, pigmentation, skin cancer and DNA damage [112]. In addition, long-term UV weathering of textiles is associated with the cleavage of different chemical bonds by absorbed UV radiation, which results in the photochemical degradation of textile fibers, color fading, increased crystallinity and other chemical and physical changes [113]. As a result, UV protection finishes are used to manufacture functional textiles for sportswear, high-altitude clothing, covering materials, wearable sensors and other high-value technical textiles.

UV protection of fabrics varies depending on their type, porosity, thickness and color, with white summer clothes, in particular, providing only low UV protection. It is worth noting that textiles of darker shades provide relatively higher UV protection since the used dyestuffs can more readily absorb the ultraviolet radiation from the sun [13]. Moreover, the kind of textile material can affect UV protection: cotton fabrics are more sensitive to UV light than polyester materials because of their aromatic structure, which is responsible for higher UV absorption [13]. Furthermore, parameters such as thickness or the setting of threads strongly influence the UV-protection properties [114]. However, laundry detergents doped with optical-brightening agents can improve a textile material's UV radiation blocking.

As a result, treating textiles to provide UV protection is a growing field of research. This can be accomplished through several approaches, such as the pigmentation (before the spinning) of the fibers with UV-absorbing inorganic pigments, such as titania nanoparticles (in rutile form) [13] or the covalent bonding of organic UV absorbers onto textile fibers. Recently, Attia et al. [91] synthesized a sustainable coating based on SiO_2 NPs-chitosan for polyester and viscose fabrics (both warp and braids structure), observing an improvement of 260% in the UPF (UV protection factor) of the coated fabrics with respect to the uncoated ones and demonstrated that the textile-type structure has a fundamental role in the performance of UV-shielding properties. Liang et al. [97] proposed an ultra-thin amorphous TiO_2 nano-film for silk fibers by the atomic layer deposition (ALD) method. The experimental findings revealed a decrease in the sample transmittance and an increase in the UPF values for increasing the film thickness, thus confirming the excellent UV-light protection of the TiO_2 nanofilm toward silk and the maintenance of the tensile strength of the original fiber after exposure to UV light.

Nanosols and the sol–gel technique are promising approaches for obtaining highly effective UV-protective coatings for textiles (Figure 7a–c). For this purpose, the nanosols should contain inorganic UV-absorbers, such as ZnO [107,115] or titanium dioxide [116–120], that readily absorb UV radiation. These nanosols reveal strongly increased absorption in the radiation range between 200 and 400 nm since ZnO and TiO_2 have extinction coefficients for ultraviolet radiation.

Moreover, at the transition from UV to visible light, the corresponding UV-Vis spectra reveal a steep drop in the absorption curves that ensure high protection against UV radiation and a colorless finish useful for maintaining the textile appearance [13]. Using TiO_2 or ZnO with particles sizes no greater than 50 nm not only ensures the coating to be colourless, but also ensures its transparency (only towards visible radiation) since no light scattering occurs within the filled coatings [13].

Zhai et al. [98] developed an aramid fiber chemically coated with uniform and a thickness-controlled amorphous TiO_2 coating by modified atomic layer deposition (mALD). The thickest coating (160 nm) achieved the highest mechanical performance and the best UV protection even after 10 accelerated washing cycles, thus confirming the UV-shielding effect and the good thermal stability of TiO_2-based layers.

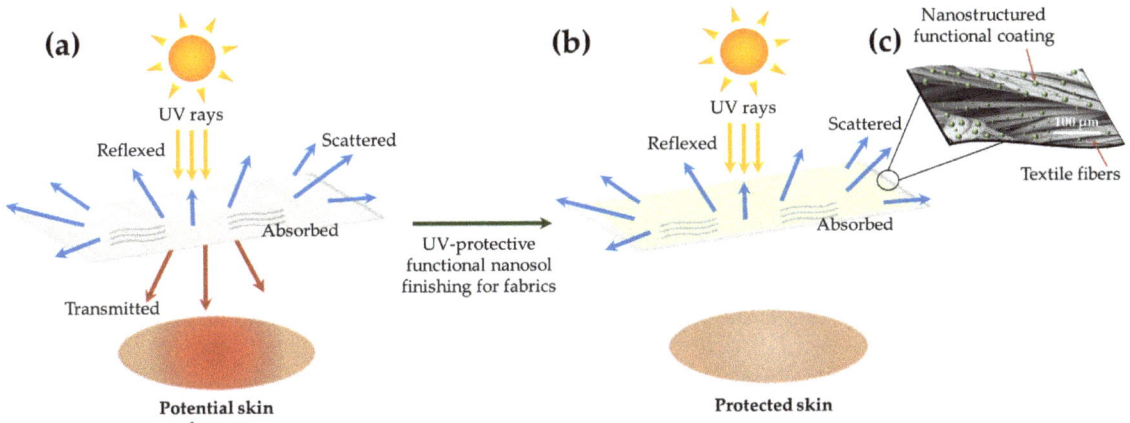

Figure 7. Schematic representation of the comparison of untreated (**a**), treated (**b**) fabrics with sol–gel based functional nanosol finishings featuring UV-protection properties and details of the possible appearance of the final nanostructured coating at SEM analysis (**c**).

Moreover, a dual-layer ultrathin Al_2O_3–TiO_2 coating with a thickness of 70–180 nm realized by a modified atomic-layer deposition (ALD) method on aramid fibers significantly improved their UV resistances, showing no yellowing effect after intense UV light for 90 min, and excellent washing durability with an ultraviolet protection factor higher than 70 after 50 commercial washing cycles [83].

Alternatively, the decomposition of gases [121], the effects of an electronic field on the optical properties of nano-TiO_2 [122], their structural and optical properties [123], the effects of UV irradiation on the hydrophilicity of TiO_2 treated fabric [124], cellulose decreasing with multi-functional organic TiO_2 catalytic effects on acid (carboxylic acids) under UV irradiation [122] and the UV-protective effect of TiO_2 on fabric [125] have been investigated.

ZnO has recently been discovered to be extremely appealing due to its significant potential applications in sensors, gas sensors, solar cells, varistors, displays, piezoelectric devices, electro-acoustic transducers, photo-diodes and UV light-emitting devices, sunscreens, anti-reflection coatings, catalysts, photo-catalysis and UV absorbers [126,127].

Behnajady et al. [128] demonstrated the ability of ZnO to remove dye from textile effluent under UVC light. Moreover, ZnO and TiO_2 are n-type semiconductors. Only these two metal oxides in the 3D transition metal oxide semi-conductor series have sufficient photo-excitation-state stability. TiO_2 and ZnO completely absorb light with energies greater than their gap energies (e.g., E_{gap}/TiO_2 3.0 eV; E_{gap}/ZnO 3.2 eV), which correspond to wavelengths of 413 nm and 387 nm, respectively [13].

The combination of nanotechnology and textiles will enable fabrics to become more multifunctional and valuable, thus resulting in a significant economic impact. Nanotechnology is currently being used to develop UV light resistance, antistatic properties, antibacterial properties and water and oil repellency [129]. Table 3 reports some systems and precursors involving the production of functional nanosols with the objective to improve the UV-protective features of different fabrics.

Table 3. Preparation and properties comparison of some inorganic nanosols employed as UV-protective finishings for different textile substrates.

Precursors and Additives	Synthetic Approach	Textile Substrate	Deposition Approach	UV-Response Features	Ref.
Titanium isopropoxide, TEOS and chitosan	Sol–gel	Acrylic acid binder-coated cotton fabrics	Spin-coating	Self-cleaning by photodegradation of methylene blue dye	[130]
Tetrabutyl titanate	Sol–gel	Cotton fabrics	Dip–pad–steam process	UV resistance and self-cleaning towards wine and coffee stains	[131]
Degussa P-25 TiO_2 powder and polyglycol	Aqueous dispersion	Cotton woven fabric and a polyester/cotton blend woven fabric	Pad–dry–cure process	Decomposition of gaseous ammonia	[121]
Zinc acetate dihydrate	Sol–gel	Cotton fabrics	Spin-coating	Self-cleaning by photodegradation of methyl orange dye	[132]
Copper sulphate, Zinc chloride and folic acid	In situ synthesis	Cotton fabrics	Impregnation	Self-cleaning by photodegradation of methylene blue dye	[133]
Rice husk silica nanoparticle-chitosan	In situ synthesis	Polyester-warp, polyester braids, viscose-warp, and viscose-braids textile fabrics	Impregnation	UPF improved by 260%	[91]
Titanium (IV) isopropoxide	-	Poly(m-phenyleneisophthalamide) fiber	Modified atomic layer deposition	54.08% retention of initial tenacity even exposure to UV irradiation	[98]
Zinc chloride and 4-aminobenzoic acid	In situ synthesis	Pre-oxidated cotton fabric	Sonochemical modification process	EUPF% * of 62.19% and 61.92% after the washing and abrasion processes	[96]

* EUPF% = UV protection factor percentage.

Due to its intriguing properties as a photocatalyst [134,135] for the degradation reaction of environmental organic pollutants [136,137], crystalline TiO_2 has received increasing attention in recent years, and many works have been devoted to the preparation and modification of this semiconductor. Additionally, TiO_2 nanoparticles have received special attention due to their photo-catalytic activities [80,138]. The photoactivity property is associated with structure, microstructure and powder-purification properties [139,140].

The use of semiconductor systems to initiate and control various photo-catalytic processes has sparked interest in understanding the photophysical properties of semiconductor colloids, as well as the dynamics of interfacial processes at semiconductor/electrolyte interfaces [141,142]. The photo-catalytic process activates TiO_2 by illuminating it with UV light with an energy greater than the band gap.

The photo-catalytic breakdown reaction of organic compounds proceeds in different steps, culminating in the mineralization of water, carbon dioxide and mineral acids. The formation of electron-hole pairs is the first step, followed by their separation. Electrons can be used in reduction processes, while holes can be used in oxidation processes.

Titania's photo-catalytic activity varies depending on its crystal phase, particle size and crystallinity. Anatase and rutile are the most active phases of titania, as opposed to brookite [143]. A review of recent literature reveals that there are many studies on the photocatalytic activity of anatase and rutile crystalline forms [144,145] and that only a few studies on amorphous TiO_2 thin films show rather low photo-catalytic activity, implying that crystallization treatment at high temperatures [146] or doping [147] is required to improve the photo-catalytic properties.

When exposed to UV light, TiO_2 excites electrons from the valance band to the conduction band whenever the light energy exceeds the band gap energy (e.g., 3.2 eV). The resulting electrons and holes can be separated and used in a redox reaction with chemicals from the outside world. When water is added, oxygen is reduced to a superoxide radical, which reacts with water molecules to produce hydroxide ions and peroxide radicals. These lasts combine with H$^+$ ions to form hydroxyl radicals. The combination of these radicals with highly oxidant species causes photocatalytic degradation of organic substrates as dyes (peroxide radicals). Because the resulting •OH radical is a strong oxidizing agent with a standard redox potential of +2.8 V, it can rapidly oxidize the majority of azo dyes to mineral end-products. In heterogeneous photocatalysis, reductive pathways have been observed to play a similar role in the degradation of several dyes, albeit with less importance than oxidation.

Furthermore, many factors influence titania crystallization, including annealing temperature, heating rate and atmosphere. These factors cannot be emphasized in the case of treatment to improve the amorphous-anatase transformation because they would destroy the textile polymer. From the standpoint of increasing the efficiency of the photo-catalytic process, it is clear that the tailoring and development of new and alternative photocatalysts are of great interest.

The photo-catalytic reactions activated by TiO_2 nanoparticles not only ensure UV-protection properties, but also self-cleaning, antibacterial effects, as well as ecological and economical wastewater treatment [148,149].

As an example of self-cleaning properties, TiO_2-SiO_2 nanocomposite powder was synthesized through a sol–gel process from titanium isopropoxide and TEOS as precursors with different molar ratios, diethanol amine (DEA) as the stabilizer, isopropanol as the solvent and chitosan in acetic acid as the dispersing agent [130]. Therefore, an acrylic acid binder-coated fabric was immersed in a TiO_2–SiO_2 suspension containing a surfactant and isopropanol and finally coated with the TiO_2–SiO_2 suspension by spin-coating with the aim of obtaining a homogeneous finishing. Rilda et al. [130] demonstrated that fabrics treated with TiO_2–SiO_2 powders (1:2 molar ratios) were responsible for the absorbance falling of methylene blue when UV-irradiated for 120 min, thus providing self-cleaning properties. Similarly, cotton fabric treated with TiO_2–SiO_2 colloidal solution showed increased discoloration of a wine stain under a solar light simulator for increased exposure times 0, 4, 8 and 24 h [150].

Cotton fabric, treated through a dip–pad–steam process with acidic TiO_2, prepared with titanium (IV) butoxide as the precursor, demonstrated excellent UV-protection properties, as well as a self-cleaning effect [131]. Moreover, Wang et al. [131] demonstrated that a higher photocatalytic activity of TiO_2 was obtained when prepared by a low-temperature steaming process instead of TiO_2 dried at 150 °C and that the higher the water amount in TiO_2 hydrosol, the higher the crystallinity and photocatalytic activity of TiO_2.

In another paper, a titania-based sol with the addition of PEG at acidic conditions was used for the TiO_2 finishing of fabrics with high photocatalytic activity, useful for self-cleaning or environmental applications [30]. Accordingly, different TiO_2 sols were synthesized from titanium tetraisopropoxide using three acidic solutions (nitric, hydrochloric and acetic acids) and two concentrations of PEG (0.025 M and 0.25 M). In particular, a titanium glycolate complex could be formed by adding PEG to the hydrolyzed precursor that is not present in the inorganic network of titania, as shown in Figure 8.

Indeed, as already demonstrated [151,152], PEGs are used as a template for obtaining high-porosity, crystalline TiO_2 (after calcination), thus realizing the TiO_2 cluster break from organic PEG structures instead of a complete inorganic matrix. This mechanism can explain the weak shift of the energy-gap value of amorphous titania. An optical analysis performed by the Tauc plot model showed an energy gap in a range of 3.3 eV to 3.5 eV, which is higher than the values for anatase (3.2 eV) [30]. It was demonstrated that the photocatalytic activity increased when the concentration of precursors increased, thus playing a fundamental role (Figure 9a–d).

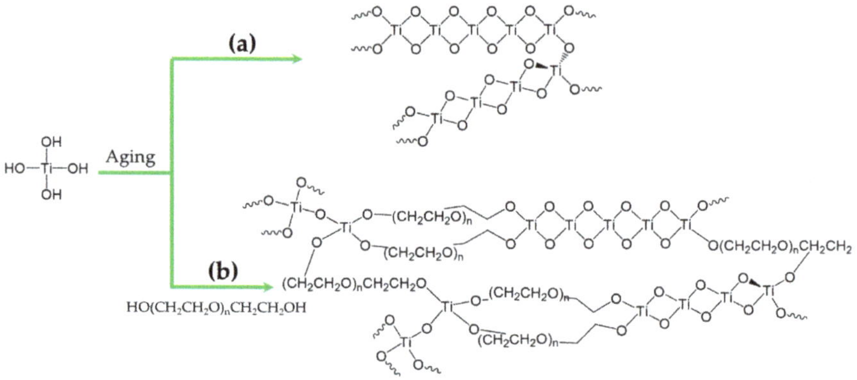

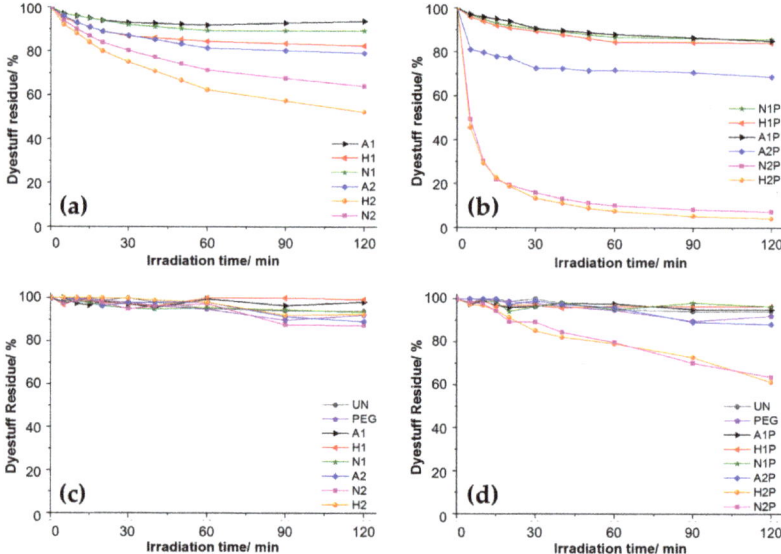

Figure 8. Illustrative scheme of the TiO$_2$ inorganic network (**a**) and possible titania glycolate cross-linking mechanism (**b**). Adapted with permission from ref. [30] Copyright 2012 Elsevier.

Figure 9. The photocatalytic activity of treated samples by 0.025 M TiO$_2$ (sample code 1) and 0.25 M TiO$_2$ sols (sample code 2) by UV (**a,b**) and visible (**c,d**) source. (A = acetic acid, H = hydrochloric acid, N = nitric acid, P = PEG). Adapted with permission from ref. [30] Copyright 2012 Elsevier.

Moreover, increased photocatalytic activity was proved by moving from acetic to nitric and finally to chloride acid, as well as when PEG was added in the presence of mineral acids and using a UV source rather than visible light. Indeed, the dye degradation rate under UV exposure dramatically increased when PEG was added to the mineral sol–gel solution. However, while the decomposition trend was maintained, its rate slowing down at longer wavelengths was evident. On the other hand, in the absence of PEG, the TiO$_2$ films revealed a rate of degradation linearly correlated with the pKa of the acid used. Shaban et al. [132] used a sol–gel spin-coating process to coat cotton fabrics with ZnO nanoparticles or ZnO solution containing zinc acetate dihydrate as a precursor, 2-methoxyethanol as a solvent and monoethanolamine as a stabilizer. Because of their high bandgap energy, ZnO nanoparticles had photocatalytic activity, especially under UV irradiation, and the methyl orange dye on cotton fabrics coated with ZnO nanoparticles or ZnO solution degraded under sunlight and 200 W lamp illumination [132]. In situ synthesis was used by Noorian et al. [133] to create precursor solutions based on copper sulfate and/or zinc chloride that contained folic acid, NaOH and water. Cotton fabric was immersed in the solutions, dried and finally washed. Authors stated that the combination of ZnO and Cu_2O, and the addition of folic acid, improved the UV protection and anti-crease properties of the fabric samples, with 87.31% enhanced UVP and 100.75° of the crease recovery angle [133]. In another study, Noorian et al. [96] imparted UV protection to cotton fabrics involving a sonochemical modification process, according to which the ZnO nanoparticles were synthesized in situ on pre-oxidated cotton fabric and further treated with 4-aminobenzoic acid (PABA) (UV-blocking agent). The latter generates more sites for the ZnO-NP growth and allows cross-linking to occur between the nanoparticles and oxidized cellulosic fibers, thus improving the coating's durability. In particular, the textile treated by the nanocomposite revealed an excellent level of UV protection factor percentage (EUPF %) of 62.19% and 61.92% after the washing and abrasion processes, with a decrease of 4.65% and 5.06%, respectively, with respect the unwashed samples. In a recent study, Sui et al. [84] proposed a nano-ZnO/fluorosiliconepolyacrylate emulsion finishing agent for the development of multifunctional linen fabric prepared by a semi-continuous seed emulsion polymerization method. The treated linen fabric showed excellent UV resistance, antibacterial properties and washing speed. In particular, by increasing the concentration of nano-ZnO, an increase in the UPF factor was also observed: the highest UV protection factor of 43.21 was measured for 1.0% of nano-ZnO, thus confirming the excellent UV protection performance. In another research, PET fabrics treated with ZnO-NPs revealed enhanced UPF as well as excellent protection for low concentrations of nanoparticles (0.5%) and nano-PU (50 g/L) [86]. In another experiment, Arik et al. [85] realized several sol–gel coatings with Zn salts (zinc acetate, zinc nitrate and zinc sulfate) and ZnO nanopowders dissolved in water and acetic acid/water [30:170], respectively. The sols were applied to linen fabrics using the pad–dry–cure method to improve the UV performance of textiles.

Otherwise, the effect of silver [153] and silver nano-particles (AgNPs) on the electrical conductivity of polymeric matrices [154,155], the effect of dyeing on the ultraviolet protection factor (UPF) [156] and the improvement of UV-protection properties [157] have also been reported. Babaahmadi et al. [93] developed a high-performance multi-layer nanocomposite coating for PET fabric by in situ synthesis of reduced graphene oxide-silver (rGO-Ag, average size of AgNPs lower than 100 nm) using ultrasonic and thermal treatment. The UV protection of this nanocoating on PET substrates was measured in terms of UPF by reaching 6145.66, up to 183-fold higher than uncoated PET fabric (UPF 33.63) thanks to the incorporation of the Ag-decorated rGO as a UV-blocking system. Porrawatkul et al. [94], for the first time, used *Averrhoa carambola* fruit extract as a reducing agent for the synthesis of Ag/ZnO composites using domestic microwave irradiation for imparting UV protection and antibacterial activity on cotton fabrics. According to the synthetic strategy, zinc acetate dihydrate ($Zn(CH_3CO_2)_2$) was used as a ZnO nanopowder precursor in the presence of the crude star fruit extract. Moreover, different concentrations of silver nitrate hexahydrate ($Ag(NO_3)_2 \cdot 6H_2O$) were added to the zinc acetate solution, and the resulting Ag/ZnO NPs were dissolved in a chitosan solution for padding cotton fabric. Ag/ZnO-coated fabric revealed a UPF value of 69.67 ± 1.53, which indicated an excellent ability to block UV radiation

The unique properties of carbon nanotubes (CNTs) have consistently attracted the interest of researchers [158,159]. In this regard, UV protection [160], increasing garment comfort [160], high stiffness and high strength properties by using low CNT content in polymeric nano-composites are described in literature [158,161]. In addition, $CNTs/TiO_2$ nanocomposite preparation has also been reported [162] with an expected synergistic photocatalytic effect. Additionally, the UV protection properties of a nonwoven polypropylene (PP) were improved using magnetron sputter deposition of Cu nanoparticles [163]. Good UV-protective properties were also reported for cotton fabrics treated by the pad–dry–cure method with a TEOS-based sol–gel coating containing CuO-NPs involving the chemical reaction between the –OH group of cotton and the Si–OMe group of TEOS [87].

Organic UV absorbers, such as phenyl-acrylates or benzotriazoles, can also be embedded in silica–sol coatings [164] and fixed on textiles, similarly to dyes. In fact, due to the typical absorption bands of molecular systems, organic UV absorbers absorb only UV light of specific wavelengths. As a result, in all of these cases, the UV protection of coatings containing only inorganic or organic UV absorbers is insufficient.

The sol–gel technique, which can embed both inorganic and organic UV absorbers in one and the same nanosol coating, can be used to create optimized UV-protective coatings with the full absorption of virtually all UV light, as demonstrated in principle for coatings on glass [123]. On textiles, analogous coatings can be prepared just as well.

Chemical bonding between the organic UV absorbers and the alkoxysilane compounds can be used to improve durability. The organic UV absorber can be directly covalently bonded to the inorganic matrix of the nanosol coatings using this method [13]. Moreover, nanosol coatings modified with both inorganic and organic UV absorbers improve UV absorption over a wide range [13]. Alternatively, a fluorescent whitening agent combined with a TiO_2 sol can be used to increase UV absorption, particularly in the UVA region. The addition of such a fluorescent whitening agents reduces the slight yellowing of the textile that occurs when titania-based nanosols are used for UV protection [13].

4. Inorganic-Based Materials for Antimicrobial Textile Finishing

Thanks to their large specific surfaces with good adhesion and water-storage properties, textiles provide ideal conditions for the settling of micro-organisms and offer ideal temperature conditions for this purpose, especially in the case of worn apparel. As a result, there is a high demand for antimicrobial finishes to prevent microbial degradation of textile fibers, limit the incidence of bacteria, reduce odor formation (due to microbial degradation of sweat) and protect users by preventing pathogen transfer and spread [13].

Antimicrobial coatings, also known as self-sterilizing [165] or hygienic coatings [166], are currently the most commercially important bioactive nanosol systems. In 2003, it was estimated that the potential European market for biocide textiles amounted to around 28,000 tons, or 180 million m^2 [80]. In the case of traditional clothing, sportswear, home textiles or outdoor textiles, such as tents or marquees, the antimicrobial property is critical for convenience and/or aesthetic reasons: a T-shirt should not have an odor, and a marquee should not be mildewed. On the other hand, the application and development of antimicrobial textiles that do not act selectively are especially important for medical applications.

Another important application of antimicrobial coatings is the use of antimicrobial wound dressing to prevent germ contamination of wounds [167]. In this regard, simple plasters or bandages could be modified with enough nanosols to aid the healing process. As an example, in innovative research, Maghimaa et al. [88] synthesized AgNPs from the aqueous extract of *Curcuma longa* leaf and the so-obtained AgNPs-coated cotton fabrics, revealing significant antimicrobial activity against *Staphylococcus aureus*, *Pseudomonas aeruginosa*, *Streptococcus pyogenes* and *Candida albicans* as well as potent wound-healing activity in the fibroblast (L929) cells.

Existing commercial products are made of special fibers with embedded organic biocides, such as bisphenols, biguanides or incorporated silver. Other products include blended fabrics with silver-coated nylon (e.g., X-StaticTM) [168] or chitosan fibers [169]. Despite the wide range of specific applications mentioned, a finishing material that can be used for any common fiber while providing a defined antimicrobial efficiency combined with a low biocide consumption is highly desirable.

Because of the particular interest in coatings to prevent biocontamination, the following sections describe various strategies for realizing an antimicrobial textile through nanosol treatment. As examples, nanosol coatings containing non-diffusible antimicrobial additives, as well as those employing controlled release of embedded biocides, or photoactive coatings that kill germs by oxidizing them during light exposure, should be mentioned.

Biocer (biological ceramics) refers to inorganic nanosol materials that contain embedded biological components, such as enzymes, proteins or even entire cells; by embedding biological components into inorganic xerogels, new composite materials are created [72,76,170]. Inorganic bioactive components should also be included in this definition, such as silver-containing nanosol coatings that show antimicrobial properties [18], titania coatings that interact with biotin and protein [171] and aluminosilicate coatings that improve biocompatibility [172].

Various biocides, such as silver, triclosan, chitosan, quaternary ammonium salts, N-halamines, biguanide derivatives, synthetic dyes and peroxyacids, have been applied to the surface of textiles to serve as antibacterial and antimicrobial finishes [173–175]. The combination of antibacterial biopolymers with titania and silica matrices is a novel method that has been shown to provide ecological benefits as well as antibacterial activity in the presence or absence of light. Arik and Seventekin [176] investigated the antibacterial activity of chitosan/titania and chitosan/silica hybrid coatings. The sol–gel method was used to prepare coating solutions, which were then applied to cotton fabric. According to these findings, hybrid coatings outperformed chitosan, titania and silica coatings in terms of antibacterial activity and washing resistance. Recently, a chitosan nanocomposite containing ZnO, Ag and Ag:ZnO was prepared by sol–gel method using GPTMS and TEOS for antimicrobial textile finishing [177]. The findings revealed chemical interactions between dispersed Ag and ZnO nanoparticles, confirming the attachment of chitosan with siloxane moieties resulting in the formation of a siloxane-polymer network. Good antimicrobial activity was observed for all samples, with a reduction of up to 50%–95% in the viability of bacteria.

Because of many chemical antimicrobial agents' potentially harmful or toxic effects, natural materials have recently been preferred for textile modification [178]. The use of inorganic nanoparticles and their nanocomposites would be a good substitute [179,180], potentially opening up a new alternative path for antimicrobial and multi-functional textile modification. The hydroxyapatite nano-ribbon spherites containing nano-silver were effectively synthesized by Liu et al. [181] due to the wide surface area and strong adsorption properties of hydroxyapatite nano-ribbon to adsorb bacteria and the high bioactivity of silver nano-particles. A composite coating of hydroxyapatite and silver was specifically developed to prevent bacterial infections following implant placement [182]. Magaña et al. created silver-modified montmorillonites through ion exchange with silver and tested the modified clay's antibacterial properties [183]. Copper nanoparticles were embedded in submicron Sepiolite ($Mg_8Si_{12}O_{30}(OH)_4(H_2O)_4 \cdot 8H_2O$) particles, and their antibacterial properties were compared to those of triclosan, thus demonstrating that both these composites and triclosan have strong bactericidal properties [184]. However, Le Pape et al. confirmed that copper nanoparticles have significantly lower antibacterial activity than silver nanoparticles [185]. The combination of a bio-pretreatment using laccase from *Cerrena unicolor* and a further modification with CuO–SiO_2 hybrid oxide microparticles by a dip-coating method of woven linen fabric results in strong antibacterial activity against *Staphylococcus aureus*, *Escherichia coli* and the fungus *Candida albicans*, besides good UV protection (UPF 40), thanks to the presence of CuO-SiO_2 particles on the flax fiber surface [90]. Hu et al. [186] investigated the antibacterial activity of ion-exchanged montmorillonite with Cu ions. On the other hand, the antibacterial properties of a nano-layer structured organo-clay-loaded fiber matrix have not been reported. Moreover, Kang et al. reported the first evidence of the antimicrobial activity of carbon nanotubes (CNTs) against *E. coli* using single-walled carbon nanotubes (SWNTs). The CNT antibacterial mechanism is defined as cell membrane damage caused by direct contact between bacteria cells and CNTs [187]. The highly stronger antibacterial efficiency of SWNTs over multi-walled carbon nanotubes (MWNTs) against *E. coli* is emphasized based on their subsequent findings, which involved the flattening of bacterial cells and the comprising of their integrity [188].

As an alternative method, sol–gel technology can provide benefits, such as ecological treatment, less chemical use, low-temperature processing, low toxicity to human health, protection of inherent textile material properties and excellent washing and usage durability in this finishing process [80,173,189]. Indeed, various types of sol–gel systems feature antibacterial or antimicrobial properties. Photoactive titania coatings with anatase modification and sol–gel coatings with embedded colloidal metal or metal compounds, such as silver, silver salts, copper compounds, zinc or biocidal quaternary ammonium salts are examples of these systems [190–192]. On the other hand, they have several drawbacks, such as high operating temperatures to produce highly photoactive thin films and the use of strong acids to keep aqueous sols in the peptized state, which causes textile destruction. Furthermore, titania coatings require UV radiation to be antimicrobial or antibacterial [116,193]. In addition to titania-based sol–gel coatings, silica-based coatings are being investigated for antibacterial or antimicrobial effectiveness. Polycationic components are incorporated within the silica layer matrix to enable the antibacterial or antimicrobial effect, and positively charged polycationic components interact with negatively charged microbial cell membranes, damaging their cellular functions [18,80,190]. Overall, nanostructured materials based on inorganic active agents with good antimicrobial activity on textile materials (Figure 10) can be divided into two categories: inorganic nano-structured materials and their nanocomposites and inorganic nano-structured-loaded organic carriers.

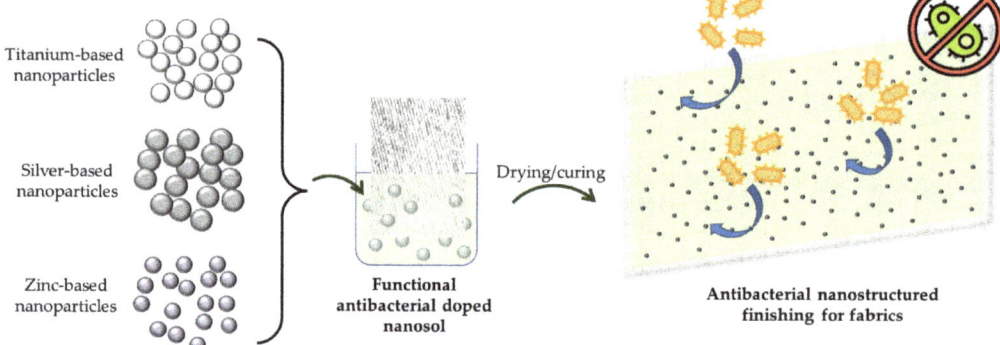

Figure 10. Schematization of the procedure based on inorganic active agents for the preparation of textile antibacterial nanosols and coatings.

Mahltig et al. reported that sol–gel processing exhibited antimicrobial activity based on controlled release, contact action, or photocatalytic activity [189]. Rivero and Goicoechea presented a detailed review of the sol–gel process for improving antibacterial properties of textiles [194]. Finally, Zhang et al. classified the application methods on cotton for colloidal suspensions of metal nanoparticles and precursor solutions of metal ions, respectively, as sol and a solution to gain antimicrobial properties [173].

In the next paragraphs, an overview of the most relevant and most recent inorganic antimicrobial coatings for textile fabrics is provided to the best of our knowledge, as a function of the active metal nanoparticles involved. A comparison of these examples is given in Table 4.

Table 4. Preparation and properties comparison of some inorganic nanosols employed as antibacterial finishings for different textile substrates.

Precursors and Additives	Synthetic Approach	Textile Substrate	Deposition Approach	Antibacterial Properties	Ref.
Chitosan/titanium-IV-isopropoxide and chitosan/TEOS	Sol–gel	Cotton	Pad–cure method	Staphylococcus aureus and Klebsiella pneumoniae	[176]
AgNO$_3$ and nano-SiO$_2$	Adsorption	Wool	Impregnation	Escherichia coli and Staphylococcus aureus	[195]
AgNO$_3$, TEOS and 3-glycidyloxypropyltriethoxysilane	Sol–gel	Polyamide fabrics	Pad–dry–cure process	Escherichia coli	[196]
ZnO and GPTMS	Sol–gel	Half-bleached pure cotton	Pad–cure method	Escherichia coli and Micrococcus luteus	[197]
Zinc acetate dihydrate and GPTMS	Sol–gel	Cotton	Pad–dry–cure method	Staphylococcus aureus and Klebsiella pneumoniae	[190]
Zinc chloride and 4-aminobenzoic acid	In situ synthesis by sonochemical process	Bleached cotton fabric	Immersion	Escherichia coli and Staphylococcus aureus	[96]
ZnO nanopowder, polyether polyurethane emulsion and folic acid	Water dispersion	Bleached PET fabrics	Coating	Escherichia coli and Staphylococcus aureus	[86]
Copper nitrate (V) trihydrate, sodium silicate and laccase	Precipitation method	Raw linen woven fabrics	Pad–dry–cure method	Staphylococcus aureus, Escherichia coli and the fungus Candida albicans	[90]

4.1. Titanium-Based Antimicrobial Textile Finishing

Nowadays, TiO$_2$ nanoparticles have been developed as a novel approach for exceptional applications as an appealing multifunctional material. Thus, TiO$_2$ nanoparticles have been used to obtain self-cleaning [2], antibacterial [116], UV protection [125], hydrophilic [124] or ultra-hydrophobic properties [198]. Moreover, they are useful in a variety of fields, such as environmental purification [199], dye degradation in fabric effluent [200], water and air purifiers [201,202], gas sensors and high-efficiency solar cells [139,140], as well as it being used as a nano-catalyst for crosslinking cellulose with polycarboxylic acids [122,203]. As already reported, the sol–gel process is the most common among several methods available for synthesizing TiO$_2$ nanoparticles [204]. As an example, textiles coated by TiO$_2$ nanoparticles in combination with ZnO nanoparticles synthesized from titanyl sulphate and zinc nitrate hexahydrate by sol–gel technology revealed a suppression level of *E. Coli* of more than 99% [89]. Moreover, the method was tested on a semi-industrial scale in roll-to-roll experiments, thus confirming the stable antibacterial properties of the so-coated textile and the suitability of the method for upscale and industrial use.

Actually, when exposed to UV light, crystalline TiO$_2$ coatings exhibit antimicrobial properties due to photo mineralisation processes taking place on the semiconductive TiO$_2$ surface [80]. Furthermore, many studies have confirmed that the addition of noble metals, such as silver and gold, increases the photo-catalytic activity of titanium dioxide by extending the light absorption range of TiO$_2$ from UV to visible light [6,205,206].

4.2. Silver-Based Antimicrobial Textile Finishing

Since olden times, silver has been the most extensively studied antimicrobial agent used to fight infections and prevent spoilage [207]. For this reason, many researchers are currently focusing on the antibacterial and multifunctional properties of silver nanoparticles [207,208].

For the manufacturing of nano-silver, several methods have been used, such as photo-catalytic reduction [206], matrix chemistry [209], chemical-reduction processes [210], photochemical or radiation-chemical reduction, metallic wire explosion, sono-chemical, polyols [211], photo-reduction [212], reverse micelle-based methods [213] and even biological synthesis [214,215].

With respect to other organic antimicrobial agents [216], which have been avoided since they are hazardous to the human body [178], silver is a safer antimicrobial agent. In this regard, it has long been known as 'oligo-dynamic' due to its ability to exert a bactericidal effect on silver-containing products, owing to its antimicrobial activities and low toxicity to human cells [217]. Its therapeutic property has been demonstrated against a wide range of microorganisms [218,219], including over 650 disease-causing organisms in the body at low concentrations [220]. Silver has also been shown to prevent the formation of biofilms [221]. Furthermore, silver ions and nanoparticles have been linked to a similar mechanism [218].

Silver nanoparticles are non-toxic and non-tolerant disinfectants [222,223], and their use may maximize antibacterial effects [224] by increasing the number of particles per unit area.

According to a general silver antimicrobial mechanism, metal ions destroy or pass through the cell membrane and bind to the -SH group of cellular enzymes [178]. The resulting critical decrease in enzymatic activity alters micro-organism metabolisms and inhibits cell growth, eventually leading to cell death. Metal ions also catalyze the formation of oxygen radicals, which oxidize bacteria's molecular structure. A chemical reaction results in the formation of active oxygen (Equation (1)).

$$H_2O + \tfrac{1}{2} O_2 \xrightarrow{\text{Metal ion}} H_2O_2 \rightarrow H_2O + (O) \tag{1}$$

Since the active oxygen created by the antimicrobial agent diffuses from the fiber to the surrounding environment, a direct interaction between the antimicrobial agent and bacteria is not necessary for this method. As a result, metal ions prevent micro-organisms from multiplying. Since bacteria are not always exposed to oxygen radicals, the ionic addition does not appear to aid in the selection of resistant strains of bacteria [216,225].

Latterly, the antimicrobial mechanism of AgNPs, according to recent research by Kim et al. [225], is linked to the production of free radicals and subsequent free radical-induced membrane damage. They validated that the antimicrobial properties of silver nanoparticles and silver nitrate was influenced by NAC (N-acetylcysteine). They have also mentioned the possibility of free radicals that may have been produced from the AgNPs surface, which was responsible for antibacterial activity via ESR (electron spin resonance) [225]. According to research on the bio-innoxiousness of silver, smaller silver particles at the same concentration are less harmful to the skin than larger ones. The anti-fungal efficiency of polyamide fabrics treated with AgNPs, and retreated polyester, was described by Ilic et al. [226].

Scientists are still interested in silver and silver-based antibacterial compounds. Silk fibers were sequentially dipped in $AgNO_3$ and NaCl to produce silver chloride nanocrystals. AgCl crystals have been pointed to exhibit antibacterial properties [227]. Wang et al. studied the antibacterial effectiveness of Ag/SiO_2 grafted on wool [195], and Simoncic et al. prepared silver nanoparticles through in situ methods and colloidal solutions to attain antibacterial activity [228].

Mohamed et al. [229] prepared colloidal solutions of silver nanoparticles (10–25 nm) synthesized by a chemical reduction process using dextran as stabilizing and reducing agent and modified TEOS using ascorbic acid as a scavenging agent and (3-aminopropyl)trimethoxysilane (APTES) by using 1,2,3,4-butanetetracarboxylic acid (BTCA) or vinyltriethoxysilane (VTEOS) and applied it to cotton fabrics by the sol–gel process by dip coating. A photoinitiator was used to cure silver nanoparticles modified with TEOS and VTEOS. They demonstrated that the modified fabrics had antibacterial activity against *S. aureus* and *E. coli*, with nearly 90% bacterial reduction even after 20 washing cycles, as well as thermoregulating properties [229]. The novel Ag decorated conjugated PET-rGO hybrid materials developed by Babaahmadi et al. [93] were proven to show antibacterial activity toward Gram-positive and negative bacteria, thus revealing a good clear inhibition zone 1.33 and 2.16 mm against

S. aureus and *E. coli*, respectively. However, the nanocomposite showed better antibacterial effects on *E. coli* maybe due to the thinner cell wall since AgNPs can penetrate inside the cell membrane by disrupting its structure and inactivating the bacteria. The innovative Ag/ZnO NPs modified cotton through the use of *Averrhoa carambola* fruit extract as a reducing agent for the synthesis of Ag/ZnO composites using domestic microwave showed better inhibition of Gram-positive bacteria (*S. aureus*) compared with Gram-negative bacteria (*E. coli*), thus revealing zone of inhibition with diameters of 55.48 ± 2.52 nm and 36.24 ± 2.08, respectively [94]. In another research, a highly durable antimicrobial coating for cellulose fibers through the in situ synthesis of Ag-NPs was developed using an extract of sumac leaves as the reducing agent, by observing a 99%–100% reduction of both *E. Coli* and *S. aureus* bacteria [95].

4.3. Zinc-Based Antibacterial Textile Finishing

ZnO was proven to have strong antibacterial effects on a broad bacterial spectrum [230,231], as well as on high-temperature- and high-pressure-resistant spores. Moreover, it was demonstrated that a decrease in the particle size (i.e., increasing specific surface area) with a rise in the powder concentration [232] leads to enhanced ZnO powder antibacterial activity. Indeed, as reported by Tam et al. [230], better antibacterial activity was provided by smaller ZnO particles, which were not affected by surface modifications (e.g., silane surfactant, gold nanoparticle attachments). Karunakaran et al. [233] demonstrated that sol–gel synthesized nanocrystalline ZnO has more extensive bactericidal activities against *Escherichia coli* than commercial ZnO nanoparticles. Furthermore, ZnO nanoparticles seem to be the most promising unconventional antibacterial agent that can fight methicillin-resistant *Staphylococcus aureus* (MRSA) and other drug-resistant bacteria [234]. However, thanks to its capability to degrade dirt, ZnO could also be used as a cheap self-cleaning substance [235]. Other advantages of this inorganic material are its no drug resistance generation, heat resistance and long-lasting, which make it of great interest [178]. Li and colleagues investigated the antibacterial activity and durability of nano-ZnO functionalized cotton fabric to sweat. They treated cotton fabrics at a concentration of 11 g/L ZnO and padded them to ensure 100% wet pick-up. Actually, the antibacterial activity of the finished fabric has been tested in alkaline, acidic and inorganic salt artificial sweat solutions. The treated fabrics demonstrated better salt and alkaline resistance than acid resistance [197]. ZnO nanoparticles have a negative surface charge and illumination can improve antibacterial performance compared with normal conditions [197]. Farouk et al. [236] demonstrated that the enhanced bioactivity of ZnO is attributed to the size of particles: the smaller their dimension, the higher the surface area to volume ratio, resulting in up to 98.8% of *E. coli* and 97.3% of *M. luteus* reduction within 5 h.

Recently, alcohol-based solutions applied at low temperatures on textile fabrics were investigated as alternative methods to overcome the harmful effects of strong acids and high temperatures on fabrics [189,190,196]. Accordingly, Poli et al. reported a less harmful method than conventional ones, as well as simple, reproducible and cheap, to prepare zinc-based coatings for cotton fabrics by sol–gel method in neutral hydro-alcoholic medium for obtaining antibacterial activity [190]. Additionally, also in this case, the one-step coating system used for preparing transparent antibacterial surfaces is simple and reproducible when compared with the conventional methods (by starting from zinc precursors in alkaline solution with reflux heating for obtaining ZnO powder). Moreover, this method highlights that the precipitation of ZnO is not necessary and other advantages are the low process temperature, the neutral media and the formation of the unique morphology on the treated textile fabrics. In particular, the acetate group of the Zn precursor plays a significant role in this innovative sol–gel finishing, in the absence of other nucleophilic species competing for the Zn^{2+} Lewis acid center (i.e., HO- or bidentate ligand). Furthermore, as reported in a previous study [237], colloids or precipitates are formed as a result of the progressive condensation of hydrolyzed moieties, thus resulting in this case in stable acetate-capped colloidal nano-sized to sub-micrometer-sized particles in dilute solutions (Figure 11).

In this study, the nano-Zn acetate-based sol–gel finishing for cotton, in the presence or not of GPTMS, showed significant bactericidal and bacteriostatic activities against *E. coli*, *S. aureus* and *K. pneumoniae* bacteria, even after five cycles of washing in case of highest concentration (Figure 12).

In another study, the multifactional cotton treated with ZnO-NPs and PABA developed from Noorian et al. [96] reveals antibacterial activity against *E. coli* and *S. aureus* with a reduction bacteria percentage (R%) of more than 99.4% and 99.9%, respectively. Moreover, excellent antibacterial activities after the abrasion process were observed against *E. coli* and *S. aureus* bacteria with R % of 93.4% and 93.7%, respectively.

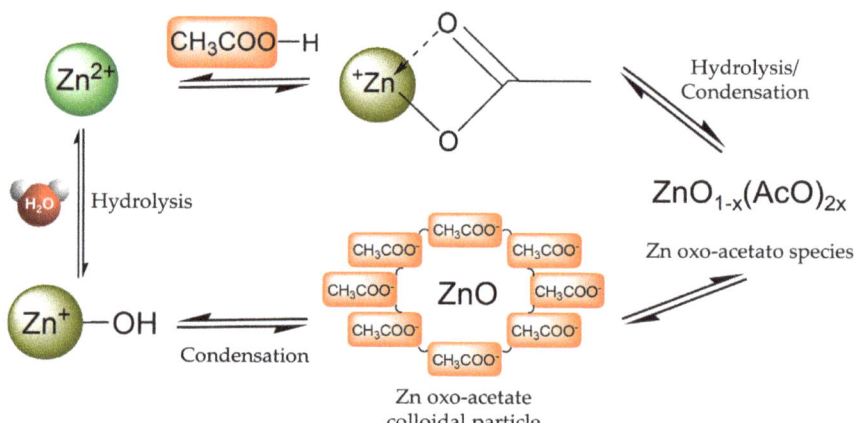

Figure 11. Schematic representation of the chemical equilibria of Zn-acetate precursor in neutral hydro-alcoholic media during sol–gel reactions [190].

Literature data report inorganic antimicrobial finishing for substrates different from cellulose, such as linen and PET. Indeed, the previously described linen fabric treated with nano-ZnO/fluorosiliconepolyacrylate emulsion showed also antibacterial activity due to the inhibition zone against *E. coli* of about 7.4 mm [84]. PET fabrics treated with ZnO NPs and nano-PU revealed efficient antimicrobial reductions of Gram-positive bacteria, Gram-negative bacteria and yeast even after 25 washing cycles [86].

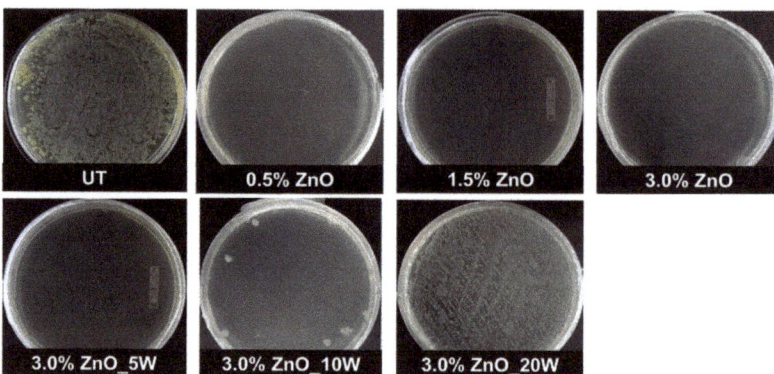

Figure 12. Comparison of the bactericidal effects of untreated cotton (UT) and cotton treated with Zn-acetate, at different concentrations and after washing cycles (5, 10, 20 washing), on the number of colony-forming units (CFU) of *Escherichia coli* after 24 h. Adapted with permission from ref. [190] Copyright 2014 Springer Nature.

5. Conclusions and Future Perspectives

In recent decades, the growing demand from modern society for more sustainable production and long-lasting, high-performance textile products has consistently supported the research for innovative functional textile finishes. In this regard, different examples and methodologies based on sol–gel inorganic coatings for the development of advanced and functional textile finishings were evaluated in this review. In particular, different sol–gel-based formulations were explored, employing a wide range of inorganic precursors and functional additives to fabricate textiles featuring different implemented properties. As a result, the as-obtained functional nanosols intended for treating natural or synthetic fabrics represent a more eco-friendly and safe approach than the most commonly employed formulations containing harmful substances. Furthermore, different deposition approaches lead to the production of thin protective films on textile substrates, enhancing their durability and general wear and washing resistance.

Moreover, by a proper rational design and using different inorganic precursors and additives, it is possible to obtain UV-protective and photo-catalytic coatings, addressing the challenge of covering a wide spectral range of UV light and leading to efficient protection of the skin. Finally, the preparation and application of nanosols based on different inorganic nanoparticles were described, focusing on those containing titanium, silver and zinc for the development of antibacterial textile fabrics, able to prevent the microbial degradation of fibers, odor formation and protect users by avoiding pathogen transfer and spread. Therefore, it is possible to suggest that sol–gel synthesis is a key technique for developing simple, scalable and sustainable approaches for functional textile finishings.

The application fields involving the use of these functional and technical textiles concern not only the individuals but embrace a wide range of sectors from the industrial to the public. In particular, the innovative coatings described in this review aim to offer a contribution to more well-performing textiles intended for automotive, naval, aero spatial, furnishing, common clothes and workwear (i.e., protective suits) applications. The main advantage comes from the ease of nanosol functionalization, as performed to obtain hybrid and nanocomposite textile coatings with improved properties in terms of wear resistance, UV-protectivity and antimicrobial traits, aiming at high-performance textiles for better comfort in everyday life.

The scientific developments in the field of inorganic coatings for functional textile materials presented in this review highlight the great potential of inorganic coatings and nanotechnologies in developing increasingly high-added-value and competitive fabrics. Indeed, the current high demand for features such as comfort, safety, aesthetics and functionality as well as greener and more efficient products and processes push scientific research towards increasingly innovative, cutting-edge and sustainable solutions. However, the development of increasingly advanced materials capable of conferring long-lasting or permanent multifunctional properties to textiles still represents a challenge. Other future challenges concern the implementation of these inorganic finishings on an industrial scale since they have not yet reached the market for some application fields, and the reduction of the environmental impact of the finishing processes. Moreover, future developments can be focused on innovative strategies, already partially underway, for the design of functional textiles through inorganic materials, taking into account consumer and environmental needs.

Author Contributions: Conceptualization, V.T., M.H., S.S., G.R. (Giulia Rando), D.D., G.R. (Giuseppe Rosace) and M.R.P.; resources, G.R. (Giuseppe Rosace) and M.R.P.; data curation, V.T., M.H., S.S., G.R. (Giulia Rando), D.D., G.R. (Giuseppe Rosace) and M.R.P.; writing original draft preparation, V.T., M.H., S.S., G.R. (Giulia Rando), G.R. (Giuseppe Rosace) and M.R.P.; writing—review and editing, V.T., M.H., S.S., G.R. (Giulia Rando), D.D., G.R. (Giuseppe Rosace) and M.R.P.; supervision, G.R. (Giuseppe Rosace) and M.R.P. All authors have read and agreed to the published version of the manuscript.

Funding: This research received no external funding.

Institutional Review Board Statement: Not applicable.

Informed Consent Statement: Not applicable.

Data Availability Statement: Not applicable.

Acknowledgments: This work was also conducted within the framework of the doctoral program of Giulia Rando, as financed by PON-MUR "Ricerca e Innovazione 2014–2020" RESTART project; MUR and CNR are gratefully acknowledged. All authors wish to thank F. Giordano, S. Romeo and G. Napoli for their technical and informatic assistance.

Conflicts of Interest: The authors declare no conflict of interest.

References

1. Schindler, W.D.; Hauser, P.J. *Chemical Finishing of Textiles*; Woodhead Publishing Ltd.: Cambridge, UK; CRC Press LLC: Cambridge, UK, 2004.
2. Bozzi, A.; Yuranova, T.; Kiwi, J. Self-cleaning of wool-polyamide and polyester textiles by TiO_2-rutile modification under daylight irradiation at ambient temperature. *J. Photochem. Photobiol. A Chem.* **2005**, *172*, 27–34. [CrossRef]
3. Gray, J.E.; Norton, P.R.; Alnouno, R.; Marold, C.L.; Valvano, M.A.; Griffiths, K. Biological efficacy of electroless-deposited silver on plasma activated polyurethane. *Biomaterials* **2003**, *24*, 2759–2765. [CrossRef] [PubMed]
4. Xu, B.; Niu, M.; Wei, L.; Hou, W.; Liu, X. The structural analysis of biomacromolecule wool fiber with Ag-loading SiO_2 nano-antibacterial agent by UV radiation. *J. Photochem. Photobiol. A Chem.* **2007**, *188*, 98–105. [CrossRef]
5. Hegemann, D.; Hossain, M.M.; Balazs, D.J. Nanostructured plasma coatings to obtain multifunctional textile surfaces. *Prog. Org. Coatings* **2007**, *58*, 237–240. [CrossRef]
6. He, J.; Kunitake, T.; Nakao, A. Facile In Situ Synthesis of Noble Metal Nanoparticles in Porous Cellulose Fibers. *Chem. Mater.* **2003**, *15*, 4401–4406. [CrossRef]
7. Singh, M.; Sharma, R.; Banerjee, U. Biotechnological applications of cyclodextrins. *Biotechnol. Adv.* **2002**, *20*, 341–359. [CrossRef]
8. Park, S.-H.; Oh, S.-G.; Mun, J.-Y.; Han, S.-S. Loading of gold nanoparticles inside the DPPC bilayers of liposome and their effects on membrane fluidities. *Colloids Surf. B Biointerfaces* **2006**, *48*, 112–118. [CrossRef]
9. Sfameni, S.; Rando, G.; Marchetta, A.; Scolaro, C.; Cappello, S.; Urzì, C.; Visco, A.; Plutino, M.R. Development of Eco-Friendly Hydrophobic and Fouling-Release Coatings for Blue-Growth Environmental Applications: Synthesis, Mechanical Characterization and Biological Activity. *Gels* **2022**, *8*, 528. [CrossRef]
10. Sfameni, S.; Rando, G.; Galletta, M.; Ielo, I.; Brucale, M.; De Leo, F.; Cardiano, P.; Cappello, S.; Visco, A.; Trovato, V.; et al. Design and Development of Fluorinated and Biocide-Free Sol–Gel Based Hybrid Functional Coatings for Anti-Biofouling/Foul-Release Activity. *Gels* **2022**, *8*, 538.
11. Trovato, V.; Colleoni, C.; Castellano, A.; Plutino, M.R. The key role of 3-glycidoxypropyltrimethoxysilane sol–gel precursor in the development of wearable sensors for health monitoring. *J. Sol-Gel Sci. Technol.* **2018**, *87*, 27–40. [CrossRef]
12. Sakka, S. *Sol-Gel Science and Technology: Topics and Fundamental Research and Applications*; Sakka, S., Ed.; Kluwer Academic Publishers: Norwell, MA, USA, 2003; ISBN 978-1402072918.
13. Mahltig, B.; Textor, T. *Nanosols and Textiles*; World Scientific: Singapore, 2008; ISBN 978-981-283-350-1.
14. Alongi, J.; Ciobanu, M.; Malucelli, G. Sol–gel treatments for enhancing flame retardancy and thermal stability of cotton fabrics: Optimisation of the process and evaluation of the durability. *Cellulose* **2011**, *18*, 167–177. [CrossRef]
15. Brancatelli, G.; Colleoni, C.; Massafra, M.R.; Rosace, G. Effect of hybrid phosphorus-doped silica thin films produced by sol-gel method on the thermal behavior of cotton fabrics. *Polym. Degrad. Stab.* **2011**, *96*, 483–490. [CrossRef]
16. Malucelli, G. Sol–Gel Flame Retardant and/or Antimicrobial Finishings for Cellulosic Textiles. In *Handbook of Renewable Materials for Coloration and Finishing*; Yusuf, M., Ed.; Scrivener Publishing LLC: Beverly, MA, USA, 2018; pp. 501–520.
17. Xing, Y.; Yang, X.; Dai, J. Antimicrobial finishing of cotton textile based on water glass by sol–gel method. *J. Sol-Gel Sci. Technol.* **2007**, *43*, 187–192. [CrossRef]
18. Mahltig, B.; Fiedler, D.; Böttcher, H. Antimicrobial Sol?Gel Coatings. *J. Sol-Gel Sci. Technol.* **2004**, *32*, 219–222. [CrossRef]
19. Mavrić, Z.; Tomšič, B.; Simončič, B. Recent advances in the ultraviolet protection finishing of textiles. *Tekstilec* **2018**, *61*, 201–220. [CrossRef]
20. Huang, K.S.; Nien, Y.H.; Hsiao, K.C.; Chang, Y.S. Application of $DMEU/SiO_2$ gel solution in the antiwrinkle finishing of cotton fabrics. *J. Appl. Polym. Sci.* **2006**, *102*, 4136–4143. [CrossRef]
21. Ak it, A.C.; Onar, N. Leaching and fastness behavior of cotton fabrics dyed with different type of dyes using sol-gel process. *J. Appl. Polym. Sci.* **2008**, *109*, 97–105. [CrossRef]
22. Mahltig, B.; Textor, T. Combination of silica sol and dyes on textiles. *J. Sol-Gel Sci. Technol.* **2006**, *39*, 111–118. [CrossRef]
23. Li, F.-Y.; Xing, Y.-J.; Ding, X. Immobilization of papain on cotton fabric by sol–gel method. *Enzyme Microb. Technol.* **2007**, *40*, 1692–1697. [CrossRef]
24. Xue, C.-H.; Jia, S.-T.; Chen, H.-Z.; Wang, M. Superhydrophobic cotton fabrics prepared by sol–gel coating of TiO_2 and surface hydrophobization. *Sci. Technol. Adv. Mater.* **2008**, *9*, 035001. [CrossRef]
25. Xing, L.; Zhou, Q.; Chen, G.; Sun, G.; Xing, T. Recent developments in preparation, properties, and applications of superhydrophobic textiles. *Text. Res. J.* **2022**, *92*, 3857–3874. [CrossRef]

26. Trovato, V.; Mezzi, A.; Brucale, M.; Rosace, G.; Rosaria Plutino, M. Alizarin-functionalized organic-inorganic silane coatings for the development of wearable textile sensors. *J. Colloid Interface Sci.* **2022**, *617*, 463–477. [CrossRef] [PubMed]
27. Trovato, V.; Teblum, E.; Kostikov, Y.; Pedrana, A.; Re, V.; Nessim, G.D.; Rosace, G. Sol-gel approach to incorporate millimeter-long carbon nanotubes into fabrics for the development of electrical-conductive textiles. *Mater. Chem. Phys.* **2020**, *240*, 122218. [CrossRef]
28. Trovato, V.; Teblum, E.; Kostikov, Y.; Pedrana, A.; Re, V.; Nessim, G.D.; Rosace, G. Electrically conductive cotton fabric coatings developed by silica sol-gel precursors doped with surfactant-aided dispersion of vertically aligned carbon nanotubes fillers in organic solvent-free aqueous solution. *J. Colloid Interface Sci.* **2021**, *586*, 120–134. [CrossRef] [PubMed]
29. Libertino, S.; Plutino, M.R.; Rosace, G. Design and development of wearable sensing nanomaterials for smart textiles. *AIP Conf. Proc.* **2018**, *1990*, 020016.
30. Colleoni, C.; Massafra, M.R.; Rosace, G. Photocatalytic properties and optical characterization of cotton fabric coated via sol–gel with non-crystalline TiO$_2$ modified with poly(ethylene glycol). *Surf. Coatings Technol.* **2012**, *207*, 79–88. [CrossRef]
31. Bhosale, R.R.; Shende, R.V.; Puszynski, J.A. Thermochemical water-splitting for H$_2$ generation using sol-gel derived Mn-ferrite in a packed bed reactor. *Int. J. Hydrog. Energy* **2012**, *37*, 2924–2934. [CrossRef]
32. Yin, Y.; Wang, C. Organic–inorganic hybrid silica film coated for improving resistance to capsicum oil on natural substances through sol–gel route. *J. Sol-Gel Sci. Technol.* **2012**, *64*, 743–749. [CrossRef]
33. Vasiljević, J.; Tomšič, B.; Jerman, I.; Simončič, B. Organofunkcionalni trialkoksisilanski prekurzorji sol-gel za kemijsko modifikacijo tekstilnih vlaken. *Tekstilec* **2017**, *60*, 198–213. [CrossRef]
34. Sfameni, S.; Del Tedesco, A.; Rando, G.; Truant, F.; Visco, A.; Plutino, M.R. Waterborne Eco-Sustainable Sol–Gel Coatings Based on Phytic Acid Intercalated Graphene Oxide for Corrosion Protection of Metallic Surfaces. *Int. J. Mol. Sci.* **2022**, *23*, 12021.
35. Abu Bakar, N.H.; Yusup, H.M.; Ismail, W.N.W.; Zulkifli, N.F. Sol-Gel Finishing for Protective Fabrics. *Biointerface Res. Appl. Chem.* **2022**, *13*, 283. [CrossRef]
36. Periyasamy, A.P.; Venkataraman, M.; Kremenakova, D.; Militky, J.; Zhou, Y. Progress in Sol-Gel Technology for the Coatings of Fabrics. *Materials* **2020**, *13*, 1838. [CrossRef]
37. Krzak, J.; Szczurek, A.; Babiarczuk, B.; Gąsiorek, J.; Borak, B. Sol–gel surface functionalization regardless of form and type of substrate. In *Handbook of Nanomaterials for Manufacturing Applications*; Elsevier: Amsterdam, The Netherlands, 2020; pp. 111–147.
38. Plutino, M.R.; Colleoni, C.; Donelli, I.; Freddi, G.; Guido, E.; Maschi, O.; Mezzi, A.; Rosace, G. Sol-gel 3-glycidoxypropyltriethoxysilane finishing on different fabrics: The role of precursor concentration and catalyst on the textile performances and cytotoxic activity. *J. Colloid Interface Sci.* **2017**, *506*, 504–517. [CrossRef]
39. Mohammed, M.K.A. Sol-gel synthesis of Au-doped TiO$_2$ supported SWCNT nanohybrid with visible-light-driven photocatalytic for high degradation performance toward methylene blue dye. *Optik* **2020**, *223*, 165607. [CrossRef]
40. Ramesan, M.T.; Varghese, M.P.J.; Periyat, P. Silver-Doped Zinc Oxide as a Nanofiller for Development of Poly(vinyl alcohol)/Poly(vinyl pyrrolidone) Blend Nanocomposites. *Adv. Polym. Technol.* **2018**, *37*, 137–143. [CrossRef]
41. Sivasamy, R.; Venugopal, P.; Mosquera, E. Synthesis of Gd$_2$O$_3$/CdO composite by sol-gel method: Structural, morphological, optical, electrochemical and magnetic studies. *Vacuum* **2020**, *175*, 109255. [CrossRef]
42. Rando, G.; Sfameni, S.; Plutino, M.R. Development of Functional Hybrid Polymers and Gel Materials for Sustainable Membrane-Based Water Treatment Technology: How to Combine Greener and Cleaner Approaches. *Gels* **2023**, *9*, 9. [CrossRef]
43. Anand, S.; Pauline, S.; Prabagar, C.J. Zr doped Barium hexaferrite nanoplatelets and RGO fillers embedded Polyvinylidenefluoride composite films for electromagnetic interference shielding applications. *Polym. Test.* **2020**, *86*, 106504. [CrossRef]
44. Akbarzadeh, S.; Sopchenski Santos, L.; Vitry, V.; Paint, Y.; Olivier, M.-G. Improvement of the corrosion performance of AA2024 alloy by a duplex PEO/clay modified sol-gel nanocomposite coating. *Surf. Coatings Technol.* **2022**, *434*, 128168. [CrossRef]
45. Serra, A.; Ramis, X.; Fernández-Francos, X. Epoxy Sol-Gel Hybrid Thermosets. *Coatings* **2016**, *6*, 8. [CrossRef]
46. Schubert, U.; Huesing, N.; Lorenz, A. Hybrid Inorganic-Organic Materials by Sol-Gel Processing of Organofunctional Metal Alkoxides. *Chem. Mater.* **1995**, *7*, 2010–2027. [CrossRef]
47. Sanchez, C.; Julián, B.; Belleville, P.; Popall, M. Applications of hybrid organic–inorganic nanocomposites. *J. Mater. Chem.* **2005**, *15*, e3592. [CrossRef]
48. Boury, B.; Corriu, R.J.P. Auto-organisation of hybrid organic–inorganic materials prepared by sol–gel chemistry. *Chem. Commun.* **2002**, *8*, 795–802. [CrossRef] [PubMed]
49. Schmidt, H. New type of non-crystalline solids between inorganic and organic materials. *J. Non. Cryst. Solids* **1985**, *73*, 681–691. [CrossRef]
50. Sanchez, C.; Rozes, L.; Ribot, F.; Laberty-Robert, C.; Grosso, D.; Sassoye, C.; Boissiere, C.; Nicole, L. "Chimie douce": A land of opportunities for the designed construction of functional inorganic and hybrid organic-inorganic nanomaterials. *Comptes Rendus Chim.* **2010**, *13*, 3–39. [CrossRef]
51. Ismail, W.N.W. Sol–gel technology for innovative fabric finishing—A Review. *J. Sol-Gel Sci. Technol.* **2016**, *78*, 698–707. [CrossRef]
52. Ielo, I.; Giacobello, F.; Sfameni, S.; Rando, G.; Galletta, M.; Trovato, V.; Rosace, G.; Plutino, M.R. Nanostructured Surface Finishing and Coatings: Functional Properties and Applications. *Materials* **2021**, *14*, 2733. [CrossRef]
53. Brinker, C.J.; Scherer, G.W. *Sol–Gel Science: The Physics and Chemistry of Sol–Gel-Processing*; A.P. Inc.: Boston, MA, USA, 1990.
54. Reisfeld, R.; Jorgenson, C.K. (Eds.) *Chemistry, Spectroscopy and Applications of Sol–Gel Glasses, Monograph Series Structure and Bonding*; Springer: Berlin, Germany, 1992.

55. Thim, G.P.; Oliveira, M.A.; Oliveira, E.D.; Melo, F.C. Sol–gel silica film preparation from aqueous solutions for corrosion protection. *J. Non. Cryst. Solids* **2000**, *273*, 124–128. [CrossRef]
56. Sanchez, C.; Soler-Illia, G.J.D.A.; Ribot, F.; Grosso, D. Design of functional nano-structured materials through the use of controlled hybrid organic–inorganic interfaces. *Comptes Rendus Chim.* **2003**, *6*, 1131–1151. [CrossRef]
57. Rando, G.; Sfameni, S.; Galletta, M.; Drommi, D.; Cappello, S.; Plutino, M.R. Functional Nanohybrids and Nanocomposites Development for the Removal of Environmental Pollutants and Bioremediation. *Molecules* **2022**, *27*, 4856. [CrossRef]
58. Schottner, G. Hybrid Sol−Gel-Derived Polymers: Applications of Multifunctional Materials. *Chem. Mater.* **2001**, *13*, 3422–3435. [CrossRef]
59. Francis, L.F. Sol-Gel Methods for Oxide Coatings. *Mater. Manuf. Process.* **1997**, *12*, 963–1015. [CrossRef]
60. Monde, T.; Fukube, H.; Nemoto, F.; Yoko, T.; Konakahara, T. Preparation and surface properties of silica-gel coating films containing branched-polyfluoroalkylsilane. *J. Non. Cryst. Solids* **1999**, *246*, 54–64. [CrossRef]
61. Haas, K.-H.; Amberg-Schwab, S.; Rose, K.; Schottner, G. Functionalized coatings based on inorganic–organic polymers (ORMOCER®s) and their combination with vapor deposited inorganic thin films. *Surf. Coatings Technol.* **1999**, *111*, 72–79. [CrossRef]
62. Jung, J.-I.; Bae, J.Y.; Bae, B.-S. Characterization and mesostructure control of mesoporous fluorinated organosilicate films. *J. Mater. Chem.* **2004**, *14*, 1988–1994. [CrossRef]
63. Sfameni, S.; Lawnick, T.; Rando, G.; Visco, A.; Textor, T.; Plutino, M.R. Functional Silane-Based Nanohybrid Materials for the Development of Hydrophobic and Water-Based Stain Resistant Cotton Fabrics Coatings. *Nanomaterials* **2022**, *12*, 3404. [CrossRef]
64. Mennig, M.; Schmitt, M.; Schmidt, H. Synthesis of Ag-Colloids in Sol-Gel Derived SiO_2-Coatings on Glass. *J. Sol-Gel Sci. Technol.* **1997**, *8*, 1035–1042. [CrossRef]
65. Prokopenko, V.B.; Gurin, V.S.; Alexeenko, A.A.; Kulikauskas, V.S.; Kovalenko, D.L. Surface segregation of transition metals in sol-gel silica films. *J. Phys. D Appl. Phys.* **2000**, *33*, 3152–3155. [CrossRef]
66. Dunn, B.; Zink, J.I. Optical properties of sol–gel glasses doped with organic molecules. *J. Mater. Chem.* **1991**, *1*, 903–913. [CrossRef]
67. Reisfeld, R. Prospects of sol–gel technology towards luminescent materials. *Opt. Mater.* **2001**, *16*, 1–7. [CrossRef]
68. Trovato, V.; Sfameni, S.; Rando, G.; Rosace, G.; Libertino, S.; Ferri, A.; Plutino, M.R. A Review of Stimuli-Responsive Smart Materials for Wearable Technology in Healthcare: Retrospective, Perspective, and Prospective. *Molecules* **2022**, *27*, 5709. [CrossRef] [PubMed]
69. Schmidt, H. Nanoparticles by chemical synthesis, processing to materials and innovative applications. *Appl. Organomet. Chem.* **2001**, *15*, 331–343. [CrossRef]
70. Böhmer, M.R.; Keursten, T.A.P.M. Incorporation of pigments in TEOS derived matrices. *J. Sol-Gel Sci. Technol.* **2000**, *19*, 361–364. [CrossRef]
71. Carturan, G.; Campostrini, R.; Diré, S.; Scardi, V.; De Alteriis, E. Inorganic gels for immobilization of biocatalysts: Inclusion of invertase-active whole cells of yeast (saccharomyces cerevisiae) into thin layers of SiO_2 gel deposited on glass sheets. *J. Mol. Catal.* **1989**, *57*, L13–L16. [CrossRef]
72. Livage, J.; Coradin, T.; Roux, C. Encapsulation of biomolecules in silica gels. *J. Phys. Condens. Matter* **2001**, *13*, R673–R691. [CrossRef]
73. Puoci, F.; Saturnino, C.; Trovato, V.; Iacopetta, D.; Piperopoulos, E.; Triolo, C.; Bonomo, M.G.; Drommi, D.; Parisi, O.I.; Milone, C.; et al. Sol−Gel Treatment of Textiles for the Entrapping of an Antioxidant/Anti-Inflammatory Molecule: Functional Coating Morphological Characterization and Drug Release Evaluation. *Appl. Sci.* **2020**, *10*, 2287. [CrossRef]
74. Novak, B.M. Hybrid Nanocomposite Materials?between inorganic glasses and organic polymers. *Adv. Mater.* **1993**, *5*, 422–433. [CrossRef]
75. Pomogailo, A.D. Hybrid polymer-inorganic nanocomposites. *Russ. Chem. Rev.* **2000**, *69*, 53–80. [CrossRef]
76. Böttcher, H.; Soltmann, U.; Mertig, M.; Pompe, W. Biocers: Ceramics with incorporated microorganisms for biocatalytic, biosorptive and functional materials development. *J. Mater. Chem.* **2004**, *14*, 2176–2188. [CrossRef]
77. Böttcher, H.; Kallies, K.-H.; Haufe, H. Model investigations of controlled release of bioactive compounds from thin metal oxide layers. *J. Sol-Gel Sci. Technol.* **1997**, *8*, 651–654. [CrossRef]
78. Wei, Y.; Xu, J.; Dong, H.; Dong, J.H.; Qiu, K.; Jansen-Varnum, S.A. Preparation and Physisorption Characterization of D-Glucose-Templated Mesoporous Silica Sol−Gel Materials. *Chem. Mater.* **1999**, *11*, 2023–2029. [CrossRef]
79. Ibrahim, N.A.; Eid, B.M.; Sharaf, S.M. Functional Finishes for Cotton-Based Textiles: Current Situation and Future Trends. In *Textiles and Clothing: Environmental Concerns and Solutions*; Shabbir, M., Ed.; Scrivener Publishing LLC: Beverly, MA, USA, 2019; pp. 131–190.
80. Mahltig, B.; Haufe, H.; Böttcher, H. Functionalisation of textiles by inorganic sol–gel coatings. *J. Mater. Chem.* **2005**, *15*, 4385–4398. [CrossRef]
81. Attia, N.F.; Mohamed, A.; Hussein, A.; El-Demerdash, A.-G.M.; Kandil, S.H. Bio-inspired one-dimensional based textile fabric coating for integrating high flame retardancy, antibacterial, toxic gases suppression, antiviral and reinforcement properties. *Polym. Degrad. Stab.* **2022**, *205*, 110152. [CrossRef]
82. Attia, N.F.; Osama, R.; Elashery, S.E.A.; Kalam, A.; Al-Sehemi, A.G.; Algarni, H. Recent Advances of Sustainable Textile Fabric Coatings for UV Protection Properties. *Coatings* **2022**, *12*, 1597. [CrossRef]

83. Chen, F.; Zhai, L.; Yang, H.; Zhao, S.; Wang, Z.; Gao, C.; Zhou, J.; Liu, X.; Yu, Z.; Qin, Y.; et al. Unparalleled Armour for Aramid Fiber with Excellent UV Resistance in Extreme Environment. *Adv. Sci.* **2021**, *8*, 2004171. [CrossRef]
84. Sui, Z.; Guo, Z.; Li, Y.; Zhang, Q.; Zu, B.; Zhao, X. Study on preparation and performance of multifunctional linen fabric finishing agent. *Text. Res. J.* **2022**, 93. [CrossRef]
85. Arik, B.; Karaman Atmaca, O.D. The effects of sol–gel coatings doped with zinc salts and zinc oxide nanopowders on multifunctional performance of linen fabric. *Cellulose* **2020**, *27*, 8385–8403. [CrossRef]
86. Abo El-Ola, S.M.; El-Bendary, M.A.; Mohamed, N.H.; Kotb, R.M. Substantial Functional Finishing and Transfer Printing of Polyester Fabric Using Zinc Oxide/Polyurethane Nanocomposite. *Fibers Polym.* **2022**, *23*, 2798–2808. [CrossRef]
87. Prilla, K.A.V.; Jacinto, J.M.; Ricardo, L.J.O.; Box, J.T.S.; Lim, A.B.C.; Francisco, E.F.; De Vera, G.I.N.; Yaya, J.A.T.; Natividad, V.V.M.; Awi, E.N.; et al. Flame retardant and uv-protective cotton fabrics functionalized with copper (II) oxide nanoparticles. *ANTORCHA* **2020**, *7*, 11–16.
88. Maghimaa, M.; Alharbi, S.A. Green synthesis of silver nanoparticles from *Curcuma longa* L. and coating on the cotton fabrics for antimicrobial applications and wound healing activity. *J. Photochem. Photobiol. B Biol.* **2020**, *204*, 111806. [CrossRef]
89. Abramova, A.V.; Abramov, V.O.; Bayazitov, V.M.; Voitov, Y.; Straumal, E.A.; Lermontov, S.A.; Cherdyntseva, T.A.; Braeutigam, P.; Weiße, M.; Günther, K. A sol-gel method for applying nanosized antibacterial particles to the surface of textile materials in an ultrasonic field. *Ultrason. Sonochem.* **2020**, *60*, 104788. [CrossRef] [PubMed]
90. Olczyk, J.; Sójka-Ledakowicz, J.; Walawska, A.; Antecka, A.; Siwińska-Ciesielczyk, K.; Zdarta, J.; Jesionowski, T. Antimicrobial Activity and Barrier Properties against UV Radiation of Alkaline and Enzymatically Treated Linen Woven Fabrics Coated with Inorganic Hybrid Material. *Molecules* **2020**, *25*, 5701. [CrossRef] [PubMed]
91. Attia, N.F.; Ebissy, A.A.E.; Morsy, M.S.; Sadak, R.A.; Gamal, H. Influence of Textile Fabrics Structures on Thermal, UV Shielding, and Mechanical Properties of Textile Fabrics Coated with Sustainable Coating. *J. Nat. Fibers* **2021**, *18*, 2189–2196. [CrossRef]
92. Sezgin Bozok, S.; Ogulata, R.T. Applying Softener Doped Silica Coating to Cotton Denim Fabrics. *J. Nat. Fibers* **2022**, *19*, 7566–7578. [CrossRef]
93. Babaahmadi, V.; Abuzade, R.A.; Montazer, M. Enhanced ultraviolet-protective textiles based on reduced graphene oxide-silver nanocomposites on polyethylene terephthalate using ultrasonic-assisted in-situ thermal synthesis. *J. Appl. Polym. Sci.* **2022**, *139*, 52196. [CrossRef]
94. Porrawatkul, P.; Pimsen, R.; Kuyyogsuy, A.; Teppaya, N.; Noypha, A.; Chanthai, S.; Nuengmatcha, P. Microwave-assisted synthesis of Ag/ZnO nanoparticles using *Averrhoa carambola* fruit extract as the reducing agent and their application in cotton fabrics with antibacterial and UV-protection properties. *RSC Adv.* **2022**, *12*, 15008–15019. [CrossRef]
95. Filipič, J.; Glažar, D.; Jerebic, Š.; Kenda, D.; Modic, A.; Roškar, B.; Vrhovski, I.; Štular, D.; Golja, B.; Smolej, S.; et al. Tailoring of Antibacterial and UV-protective Cotton Fabric by an in situ Synthesis of Silver Particles in the Presence of a Sol-gel Matrix and Sumac Leaf Extract. *Tekstilec* **2020**, *63*, 4–13. [CrossRef]
96. Noorian, S.A.; Hemmatinejad, N.; Navarro, J.A.R. Ligand modified cellulose fabrics as support of zinc oxide nanoparticles for UV protection and antimicrobial activities. *Int. J. Biol. Macromol.* **2020**, *154*, 1215–1226. [CrossRef]
97. Liang, Z.; Zhou, Z.; Li, J.; Zhang, S.; Dong, B.; Zhao, L.; Wu, C.; Yang, H.; Chen, F.; Wang, S. Multi-functional silk fibers/fabrics with a negligible impact on comfortable and wearability properties for fiber bulk. *Chem. Eng. J.* **2021**, *415*, 128980. [CrossRef]
98. Zhai, L.; Huang, Z.; Luo, Y.; Yang, H.; Xing, T.; He, A.; Yu, Z.; Liu, J.; Zhang, X.; Xu, W.; et al. Decorating aramid fibers with chemically-bonded amorphous TiO_2 for improving UV resistance in the simulated extreme environment. *Chem. Eng. J.* **2022**, *440*, 135724. [CrossRef]
99. Liu, H.-K. Investigation on the pressure infiltration of sol-gel processed textile ceramic matrix composites. *J. Mater. Sci.* **1996**, *31*, 5093–5099. [CrossRef]
100. Brzeziński, S.; Kowalczyk, D.; Borak, B.; Jasiorski, M.; Tracz, A. Applying the sol-gel method to the deposition of nanocoats on textiles to improve their abrasion resistance. *J. Appl. Polym. Sci.* **2012**, *125*, 3058–3067. [CrossRef]
101. Liu, J.; Berg, J.C. An aqueous sol–gel route to prepare organic–inorganic hybrid materials. *J. Mater. Chem.* **2007**, *17*, 4430–4435. [CrossRef]
102. Xiao, X.; Chen, F.; Wei, Q.; Wu, N. Surface modification of polyester nonwoven fabrics by Al_2O_3 sol-gel coating. *J. Coatings Technol. Res.* **2009**, *6*, 537–541. [CrossRef]
103. Colleoni, C.; Donelli, I.; Freddi, G.; Guido, E.; Migani, V.; Rosace, G. A novel sol-gel multi-layer approach for cotton fabric finishing by tetraethoxysilane precursor. *Surf. Coatings Technol.* **2013**, *235*, 192–203. [CrossRef]
104. Rosu, C.; Lin, H.; Jiang, L.; Breedveld, V.; Hess, D.W. Sustainable and long-time 'rejuvenation' of biomimetic water-repellent silica coating on polyester fabrics induced by rough mechanical abrasion. *J. Colloid Interface Sci.* **2018**, *516*, 202–214. [CrossRef]
105. Schramm, C.; Binder, W.H.; Tessadri, R. Durable Press Finishing of Cotton Fabric with 1,2,3,4-Butanetetracarboxylic Acid and TEOS/GPTMS. *J. Sol-Gel Sci. Technol.* **2004**, *29*, 155–165. [CrossRef]
106. Trovato, V.; Rosace, G.; Colleoni, C.; Sfameni, S.; Migani, V.; Plutino, M.R. Sol-gel based coatings for the protection of cultural heritage textiles. *IOP Conf. Ser. Mater. Sci. Eng.* **2020**, *777*, 012007. [CrossRef]
107. Textor, T.; Bahners, T.; Schollmyer, E. Modern Approaches for Intelligent Surface Modification. *J. Ind. Text.* **2003**, *32*, 279–289. [CrossRef]
108. Xu, T.; Xie, C.S. Tetrapod-like nano-particle ZnO/acrylic resin composite and its multi-function property. *Prog. Org. Coatings* **2003**, *46*, 297–301. [CrossRef]

109. Vigneshwaran, N.; Kumar, S.; Kathe, A.A.; Varadarajan, P.V.; Prasad, V. Functional finishing of cotton fabrics using zinc oxide–soluble starch nanocomposites. *Nanotechnology* **2006**, *17*, 5087–5095. [CrossRef]
110. Li, F.; Xing, Y.; Ding, X. Silica xerogel coating on the surface of natural and synthetic fabrics. *Surf. Coatings Technol.* **2008**, *202*, 4721–4727. [CrossRef]
111. Sequeira, S.; Evtuguin, D.V.; Portugal, I.; Esculcas, A.P. Synthesis and characterisation of cellulose/silica hybrids obtained by heteropoly acid catalysed sol–gel process. *Mater. Sci. Eng. C* **2007**, *27*, 172–179. [CrossRef]
112. GUGUMUS, F. *Developments in the UV-Stabilisation of Polymers*; Applied Science Publishers: London, UK, 1979.
113. Biswa, R. Das UV Radiation Protective Clothing. *Open Text. J.* **2010**, *3*, 14–21.
114. Osterwalder, U.; Schlenker, W.; Rohwer, H.; Martin, E.; Schuh, S. Facts and Ficton on Ultraviolet Protection by Clothing. *Radiat. Prot. Dosim.* **2000**, *91*, 255–259. [CrossRef]
115. Wang, R.H.; Xin, J.H.; Tao, X.M. UV-Blocking Property of Dumbbell-Shaped ZnO Crystallites on Cotton Fabrics. *Inorg. Chem.* **2005**, *44*, 3926–3930. [CrossRef]
116. Daoud, W.A.; Xin, J.H. Low Temperature Sol-Gel Processed Photocatalytic Titania Coating. *J. Sol-Gel Sci. Technol.* **2004**, *29*, 25–29. [CrossRef]
117. Xin, J.H.; Daoud, W.A.; Kong, Y.Y. A New Approach to UV-Blocking Treatment for Cotton Fabrics. *Text. Res. J.* **2004**, *74*, 97–100. [CrossRef]
118. Abidi, N.; Hequet, E.; Tarimala, S.; Dai, L.L. Cotton fabric surface modification for improved UV radiation protection using sol–gel process. *J. Appl. Polym. Sci.* **2007**, *104*, 111–117. [CrossRef]
119. Onar, N.; Ebeoglugil, M.F.; Kayatekin, I.; Celik, E. Low-temperature, sol–gel-synthesized, silver-doped titanium oxide coatings to improve ultraviolet-blocking properties for cotton fabrics. *J. Appl. Polym. Sci.* **2007**, *106*, 514–525. [CrossRef]
120. Xing, Y.; Ding, X. UV photo-stabilization of tetrabutyl titanate for aramid fibers via sol–gel surface modification. *J. Appl. Polym. Sci.* **2007**, *103*, 3113–3119. [CrossRef]
121. Dong, Y.; Bai, Z.; Zhang, L.; Liu, R.; Zhu, T. Finishing of cotton fabrics with aqueous nano-titanium dioxide dispersion and the decomposition of gaseous ammonia by ultraviolet irradiation. *J. Appl. Polym. Sci.* **2006**, *99*, 286–291. [CrossRef]
122. Wang, C.-C.; Chen, C.-C. Physical properties of the crosslinked cellulose catalyzed with nanotitanium dioxide under UV irradiation and electronic field. *Appl. Catal. A Gen.* **2005**, *293*, 171–179. [CrossRef]
123. Park, Y.R.; Kim, K.J. Structural and optical properties of rutile and anatase TiO_2 thin films: Effects of Co doping. *Thin Solid Films* **2005**, *484*, 34–38. [CrossRef]
124. Sawada, K.; Sugimoto, M.; Ueda, M. Chan Hun Park Hydrophilic Treatment of Polyester Surfaces Using TiO_2 Photocatalytic Reactions. *Text. Res. J.* **2003**, *73*, 819–822. [CrossRef]
125. Han, K.; Yu, M. Study of the preparation and properties of UV-blocking fabrics of a PET/TiO_2 nanocomposite prepared by in situ polycondensation. *J. Appl. Polym. Sci.* **2006**, *100*, 1588–1593. [CrossRef]
126. Pan, Z.W.; Dai, Z.R.; Wang, Z.L. Nanobelts of Semiconducting Oxides. *Science* **2001**, *291*, 1947–1949. [CrossRef]
127. Pinnavaia, T.J.; Beall, G.W. (Eds.) Polymer–Clay Nanocomposites. John Wiley: Hoboken, NJ, USA, 2000.
128. Behnajady, M.; Modirshahla, N.; Hamzavi, R. Kinetic study on photocatalytic degradation of C.I. Acid Yellow 23 by ZnO photocatalyst. *J. Hazard. Mater.* **2006**, *133*, 226–232. [CrossRef]
129. Ahmed, N.S.E.; El-Shishtawy, R.M. The use of new technologies in coloration of textile fibers. *J. Mater. Sci.* **2010**, *45*, 1143–1153. [CrossRef]
130. Rilda, Y.; Fadhli, F.; Syukri, S.; Alif, A.; Aziz, H.; Chandren, S.; Nur, H. Self-cleaning TiO_2-SiO_2 clusters on cotton textile prepared by dip-spin coating process. *J. Teknol.* **2016**, *78*. [CrossRef]
131. Wang, L.; Shen, Y.; Xu, L.; Cai, Z.; Zhang, H. Thermal crystallization of low-temperature prepared anatase nano-TiO_2 and multifunctional finishing of cotton fabrics. *J. Text. Inst.* **2016**, *107*, 651–662. [CrossRef]
132. Shaban, M.; Abdallah, S.; Khalek, A.A. Characterization and photocatalytic properties of cotton fibers modified with ZnO nanoparticles using sol–gel spin coating technique. *Beni-Suef Univ. J. Basic Appl. Sci.* **2016**, *5*, 277–283. [CrossRef]
133. Noorian, S.A.; Hemmatinejad, N.; Bashari, A. One-Pot Synthesis of Cu_2O/ZnO Nanoparticles at Present of Folic Acid to Improve UV-Protective Effect of Cotton Fabrics. *Photochem. Photobiol.* **2015**, *91*, 510–517. [CrossRef] [PubMed]
134. Mills, A.; Lee, S.-K. A web-based overview of semiconductor photochemistry-based current commercial applications. *J. Photochem. Photobiol. A Chem.* **2002**, *152*, 233–247. [CrossRef]
135. Langlet, M.; Kim, A.; Audier, M.; Herrmann, J.M. Sol-gel preparation of photocatalytic TiO_2 films on polymer substrates. *J. Sol-Gel Sci. Technol.* **2002**, *25*, 223–234. [CrossRef]
136. Kasprzyk-Hordern, B.; Ziółek, M.; Nawrocki, J. Catalytic ozonation and methods of enhancing molecular ozone reactions in water treatment. *Appl. Catal. B Environ.* **2003**, *46*, 639–669. [CrossRef]
137. Karkmaz, M.; Puzenat, E.; Guillard, C.; Herrmann, J.M. Photocatalytic degradation of the alimentary azo dye amaranth. *Appl. Catal. B Environ.* **2004**, *51*, 183–194. [CrossRef]
138. Lo, P.-H.; Kumar, S.A.; Chen, S.-M. Amperometric determination of H_2O_2 at nano-TiO_2/DNA/thionin nanocomposite modified electrode. *Colloids Surf. B Biointerfaces* **2008**, *66*, 266–273. [CrossRef]
139. Weibel, A.; Bouchet, R.; Knauth, P. Electrical properties and defect chemistry of anatase (TiO_2). *Solid State Ion.* **2006**, *177*, 229–236. [CrossRef]

140. Verran, J.; Sandoval, G.; Allen, N.S.; Edge, M.; Stratton, J. Variables affecting the antibacterial properties of nano and pigmentary titania particles in suspension. *Dye. Pigment.* **2007**, *73*, 298–304. [CrossRef]
141. Norris, J.R.; Meisel, D. *Photochemical Energy Conversion*; Elsevier: New York, NY, USA, 1989.
142. Gratzel, M. *Heterogenous Photochemical Electron Transfer*; CRC Press: Boca Raton, FL, USA, 2018; ISBN 9781351081658.
143. Kontos, A.I.; Arabatzis, I.M.; Tsoukleris, D.S.; Kontos, A.G.; Bernard, M.C.; Petrakis, D.E.; Falaras, P. Efficient photocatalysts by hydrothermal treatment of TiO$_2$. *Catal. Today* **2005**, *101*, 275–281. [CrossRef]
144. Šegota, S.; Ćurković, L.; Ljubas, D.; Svetličić, V.; Houra, I.F.; Tomašić, N. Synthesis, characterization and photocatalytic properties of sol-gel TiO$_2$ films. *Ceram. Int.* **2011**, *37*, 1153–1160. [CrossRef]
145. Kho, Y.K.; Iwase, A.; Teoh, W.Y.; Mädler, L.; Kudo, A.; Amal, R. Photocatalytic H$_2$ Evolution over TiO$_2$ Nanoparticles. The Synergistic Effect of Anatase and Rutile. *J. Phys. Chem. C* **2010**, *114*, 2821–2829. [CrossRef]
146. Randorn, C.; Irvine, J.T.S.; Robertson, P. Synthesis of visible-light-activated yellow amorphous TiO$_2$ photocatalyst. *Int. J. Photoenergy* **2008**, *2008*, 426872. [CrossRef]
147. Huang, J.; Li, Y.; Zhao, G.; Cai, X. Photocatalytic degradation characteristic of amorphous TiO$_2$-W thin films deposited by magnetron sputtering. *Trans. Nonferrous Met. Soc. China* **2006**, *16*, s280–s284. [CrossRef]
148. Onar, N.; Aksit, A.C.; Sen, Y.; Mutlu, M. Antimicrobial, UV-protective and self-cleaning properties of cotton fabrics coated by dip-coating and solvothermal coating methods. *Fibers Polym.* **2011**, *12*, 461–470. [CrossRef]
149. Fulekar, M.H. *Environmental Biotechnology*; CRC Press: Boca Raton, FL, USA, 2010; ISBN 9780429065040.
150. Yuranova, T.; Mosteo, R.; Bandara, J.; Laub, D.; Kiwi, J. Self-cleaning cotton textiles surfaces modified by photoactive SiO$_2$/TiO$_2$ coating. *J. Mol. Catal. A Chem.* **2006**, *244*, 160–167. [CrossRef]
151. Bu, S.; Jin, Z.; Liu, X.; Yang, L.; Cheng, Z. Fabrication of TiO$_2$ porous thin films using peg templates and chemistry of the process. *Mater. Chem. Phys.* **2004**, *88*, 273–279. [CrossRef]
152. Zhang, W.J.; Yang, B.; Bai, J.W. Photocatalytic Activity of TiO$_2$/Ti Film Prepared by Sol-Gel Method on Methyl Orange Degradation. *Adv. Mater. Res.* **2011**, *214*, 65–69. [CrossRef]
153. Boiteux, G.; Boullanger, C.; Cassagnau, P.; Fulchiron, R.; Seytre, G. Influence of Morphology on PTC in Conducting Polypropylene-Silver Composites. *Macromol. Symp.* **2006**, *233*, 246–253. [CrossRef]
154. Lee, H.-H.; Chou, K.-S.; Shih, Z.-W. Effect of nano-sized silver particles on the resistivity of polymeric conductive adhesives. *Int. J. Adhes. Adhes.* **2005**, *25*, 437–441. [CrossRef]
155. Xu, M.; Feng, J.Q.; Cao, X.L. Electrical properties of nano-silver/polyacrylamide/ethylene vinyl acetate composite. *J. Shanghai Univ.* **2008**, *12*, 85–90. [CrossRef]
156. Gorenšek, M.; Recelj, P. Nanosilver Functionalized Cotton Fabric. *Text. Res. J.* **2007**, *77*, 138–141. [CrossRef]
157. Jiang, S.; Newton, E.; Yuen, C.-W.M.; Kan, C.-W.; Jiang, S.-X.K. Application of Chemical Silver Plating on Polyester and Cotton Blended Fabric. *Text. Res. J.* **2007**, *77*, 85–91. [CrossRef]
158. Schartel, B.; Pötschke, P.; Knoll, U.; Abdel-Goad, M. Fire behaviour of polyamide 6/multiwall carbon nanotube nanocomposites. *Eur. Polym. J.* **2005**, *41*, 1061–1070. [CrossRef]
159. Xue, P.; Park, K.H.; Tao, X.M.; Chen, W.; Cheng, X.Y. Electrically conductive yarns based on PVA/carbon nanotubes. *Compos. Struct.* **2007**, *78*, 271–277. [CrossRef]
160. Mondal, S.; Hu, J.L. A novel approach to excellent UV protecting cotton fabric with functionalized MWNT containing water vapor permeable PU coating. *J. Appl. Polym. Sci.* **2007**, *103*, 3370–3376. [CrossRef]
161. Kim, H.-S.; Park, B.H.; Yoon, J.-S.; Jin, H.-J. Preparation and characterization of poly[(butylene succinate)-co-(butylene adipate)]/carbon nanotube-coated silk fiber composites. *Polym. Int.* **2007**, *56*, 1035–1039. [CrossRef]
162. Yen, C.-Y.; Lin, Y.-F.; Hung, C.-H.; Tseng, Y.-H.; Ma, C.-C.M.; Chang, M.-C.; Shao, H. The effects of synthesis procedures on the morphology and photocatalytic activity of multi-walled carbon nanotubes/TiO$_2$ nanocomposites. *Nanotechnology* **2008**, *19*, 045604. [CrossRef]
163. Wei, Q.; Yu, L.; Wu, N.; Hong, S. Preparation and Characterization of Copper Nanocomposite Textiles. *J. Ind. Text.* **2008**, *37*, 275–283. [CrossRef]
164. Dong, W.; Zhu, C.; Bongard, H.-J. Preparation and optical properties of UV dye DMT-doped silica films. *J. Phys. Chem. Solids* **2003**, *64*, 399–404. [CrossRef]
165. Tiller, J. Selbststerilisierende Oberflächen. *Nachr. Aus Der Chem.* **2007**, *55*, 499–502. [CrossRef]
166. Johns, K. Hygienic coatings: The next generation. *Surf. Coatings Int. Part B Coatings Trans.* **2003**, *86*, 101–110. [CrossRef]
167. Ovington, L.G. Battling Bacteria in Wound Care. *Home Healthc. Nurse J. Home Care Hosp. Prof.* **2001**, *19*, 622–630. [CrossRef]
168. MacKeen, P.C.; Person, S.; Warner, S.C.; Snipes, W.; Stevens, S.E. Silver-coated nylon fiber as an antibacterial agent. *Antimicrob. Agents Chemother.* **1987**, *31*, 93–99. [CrossRef]
169. Knittel, D.; Schollmeyer, E. Chitosan and its derivatives for textile finishing Part 4: Permanent finishing of cotton with ionic carbohydrates and analysis of thin layers obtained. *Melliand Textilb. Int.* **2002**, *83*, 58–61.
170. Gadre, S.Y.; Gouma, P.I. Biodoped Ceramics: Synthesis, Properties, and Applications. *J. Am. Ceram. Soc.* **2006**, *89*, 2987–3002. [CrossRef]
171. Huang, J.; Ichinose, I.; Kunitake, T. Biomolecular Modification of Hierarchical Cellulose Fibers through Titania Nanocoating. *Angew. Chemie Int. Ed.* **2006**, *45*, 2883–2886. [CrossRef]

172. Leivo, J.; Meretoja, V.; Vippola, M.; Levänen, E.; Vallittu, P.; Mäntylä, T.A. Sol–gel derived aluminosilicate coatings on alumina as substrate for osteoblasts. *Acta Biomater.* **2006**, *2*, 659–668. [CrossRef]
173. Zhang, Y.; Xu, Q.; Fu, F.; Liu, X. Durable antimicrobial cotton textiles modified with inorganic nanoparticles. *Cellulose* **2016**, *23*, 2791–2808. [CrossRef]
174. Liu, Y.; Ren, X.; Liang, J. Antimicrobial modification review. *BioResources* **2015**, *10*, 1964–1985.
175. Ielo, I.; Giacobello, F.; Castellano, A.; Sfameni, S.; Rando, G.; Plutino, M.R. Development of Antibacterial and Antifouling Innovative and Eco-Sustainable Sol–Gel Based Materials: From Marine Areas Protection to Healthcare Applications. *Gels* **2022**, *8*, 26.
176. Buket, A.R.I.K.; Seventekin, N. Evaluation of Antibacterial and Structural Properties of Cotton Fabric Coated By Chitosan/Titania and Chitosan/Silica Hybrid Sol-Gel Coatings. *Text. Appar.* **2011**, *21*, 107–115.
177. Buşilă, M.; Muşat, V.; Textor, T.; Mahltig, B. Synthesis and characterization of antimicrobial textile finishing based on Ag:ZnO nanoparticles/chitosan biocomposites. *RSC Adv.* **2015**, *5*, 21562–21571. [CrossRef]
178. Dastjerdi, R.; Montazer, M. A review on the application of inorganic nano-structured materials in the modification of textiles: Focus on anti-microbial properties. *Colloids Surf. B Biointerfaces* **2010**, *79*, 5–18. [CrossRef]
179. Chen, Q.; Shen, X.; Gao, H. One-step synthesis of silver-poly(4-vinylpyridine) hybrid microgels by γ-irradiation and surfactant-free emulsion polymerisation. The photoluminescence characteristics. *Colloids Surf. A Physicochem. Eng. Asp.* **2006**, *275*, 45–49. [CrossRef]
180. Dimitrov, D.S. Interactions of antibody-conjugated nanoparticles with biological surfaces. *Colloids Surf. A Physicochem. Eng. Asp.* **2006**, *282–283*, 8–10. [CrossRef]
181. Liu, J.-K.; Yang, X.-H.; Tian, X.-G. Preparation of silver/hydroxyapatite nanocomposite spheres. *Powder Technol.* **2008**, *184*, 21–24. [CrossRef]
182. Chen, W.; Liu, Y.; Courtney, H.; Bettenga, M.; Agrawal, C.M.; Bumgardner, J.D.; Ong, J.L. In vitro anti-bacterial and biological properties of magnetron co-sputtered silver-containing hydroxyapatite coating. *Biomaterials* **2006**, *27*, 5512–5517. [CrossRef]
183. Magaña, S.M.; Quintana, P.; Aguilar, D.H.; Toledo, J.A.; Ángeles-Chávez, C.; Cortés, M.A.; León, L.; Freile-Pelegrín, Y.; López, T.; Sánchez, R.M.T. Antibacterial activity of montmorillonites modified with silver. *J. Mol. Catal. A Chem.* **2008**, *281*, 192–199. [CrossRef]
184. Esteban-Cubillo, A.; Pecharromán, C.; Aguilar, E.; Santarén, J.; Moya, J.S. Antibacterial activity of copper monodispersed nanoparticles into sepiolite. *J. Mater. Sci.* **2006**, *41*, 5208–5212. [CrossRef]
185. Le Pape, H.; Solano-Serena, F.; Contini, P.; Devillers, C.; Maftah, A.; Leprat, P. Evaluation of the anti-microbial properties of an activated carbon fibre supporting silver using a dynamic method. *Carbon N. Y.* **2002**, *40*, 2947–2954. [CrossRef]
186. Hu, C.H.; Xu, Z.R.; Xia, M.S. Antibacterial effect of Cu2+-exchanged montmorillonite on Aeromonas hydrophila and discussion on its mechanism. *Vet. Microbiol.* **2005**, *109*, 83–88. [CrossRef]
187. Kang, S.; Pinault, M.; Pfefferle, L.D.; Elimelech, M. Single-Walled Carbon Nanotubes Exhibit Strong Antimicrobial Activity. *Langmuir* **2007**, *23*, 8670–8673. [CrossRef]
188. Kang, S.; Herzberg, M.; Rodrigues, D.F.; Elimelech, M. Antibacterial Effects of Carbon Nanotubes: Size Does Matter! *Langmuir* **2008**, *24*, 6409–6413. [CrossRef]
189. Mahltig, B.; Fiedler, D.; Fischer, A.; Simon, P. Antimicrobial coatings on textiles–modification of sol–gel layers with organic and inorganic biocides. *J. Sol-Gel Sci. Technol.* **2010**, *55*, 269–277. [CrossRef]
190. Poli, R.; Colleoni, C.; Calvimontes, A.; Polášková, H.; Dutschk, V.; Rosace, G. Innovative sol–gel route in neutral hydroalcoholic condition to obtain antibacterial cotton finishing by zinc precursor. *J. Sol-Gel Sci. Technol.* **2015**, *74*, 151–160. [CrossRef]
191. Tarimala, S.; Kothari, N.; Abidi, N.; Hequet, E.; Fralick, J.; Dai, L.L. New approach to antibacterial treatment of cotton fabric with silver nanoparticle–doped silica using sol–gel process. *J. Appl. Polym. Sci.* **2006**, *101*, 2938–2943. [CrossRef]
192. Mahltig, B.; Gutmann, E.; Meyer, D.C.; Reibold, M.; Dresler, B.; Günther, K.; Faßler, D.; Böttcher, H. Solvothermal preparation of metallized titania sols for photocatalytic and antimicrobial coatings. *J. Mater. Chem.* **2007**, *17*, 2367–2374. [CrossRef]
193. Mahltig, B.; Fischer, A. Inorganic/organic polymer coatings for textiles to realize water repellent and antimicrobial properties-A study with respect to textile comfort. *J. Polym. Sci. Part B Polym. Phys.* **2010**, *48*, 1562–1568. [CrossRef]
194. Rivero, P.J.; Goicoechea, J. *Sol-Gel Technology for Antimicrobial Textiles*; Elsevier Ltd.: Amsterdam, The Netherlands, 2016; ISBN 9780081005859.
195. Wang, S.; Hou, W.; Wei, L.; Jia, H.; Liu, X.; Xu, B. Antibacterial activity of nano-SiO$_2$ antibacterial agent grafted on wool surface. *Surf. Coatings Technol.* **2007**, *202*, 460–465. [CrossRef]
196. Mahltig, B.; Textor, T. Silver containing sol-gel coatings on polyamide fabrics as antimicrobial finish-description of a technical application process for wash permanent antimicrobial effect. *Fibers Polym.* **2010**, *11*, 1152–1158. [CrossRef]
197. Li, Q.; Chen, S.-L.; Jiang, W.-C. Durability of nano ZnO antibacterial cotton fabric to sweat. *J. Appl. Polym. Sci.* **2007**, *103*, 412–416. [CrossRef]
198. Rios, P.F.; Dodiuk, H.; Kenig, S.; McCarthy, S.; Dotan, A. Durable ultra-hydrophobic surfaces for self-cleaning applications. *Polym. Adv. Technol.* **2008**, *19*, 1684–1691. [CrossRef]
199. Ikezawa, S.; Homyara, H.; Kubota, T.; Suzuki, R.; Koh, S.; Mutuga, F.; Yoshioka, T.; Nishiwaki, A.; Ninomiya, Y.; Takahashi, M.; et al. Applications of TiO$_2$ film for environmental purification deposited by controlled electron beam-excited plasma. *Thin Solid Films* **2001**, *386*, 173–176. [CrossRef]

200. Mahmoodi, N.M.; Arami, M.; Limaee, N.Y.; Tabrizi, N.S. Kinetics of heterogeneous photocatalytic degradation of reactive dyes in an immobilized TiO$_2$ photocatalytic reactor. *J. Colloid Interface Sci.* **2006**, *295*, 159–164. [CrossRef]
201. Li, D.; Haneda, H.; Hishita, S.; Ohashi, N. Visible-Light-Driven N−F−Codoped TiO$_2$ Photocatalysts. 1. Synthesis by Spray Pyrolysis and Surface Characterization. *Chem. Mater.* **2005**, *17*, 2588–2595. [CrossRef]
202. Cermenati, L.; Pichat, P.; Guillard, C.; Albini, A. Probing the TiO$_2$ Photocatalytic Mechanisms in Water Purification by Use of Quinoline, Photo-Fenton Generated OH• Radicals and Superoxide Dismutase. *J. Phys. Chem. B* **1997**, *101*, 2650–2658. [CrossRef]
203. Nazari, A.; Montazer, M.; Rashidi, A.; Yazdanshenas, M.; Anary-Abbasinejad, M. Nano TiO$_2$ photo-catalyst and sodium hypophosphite for cross-linking cotton with poly carboxylic acids under UV and high temperature. *Appl. Catal. A Gen.* **2009**, *371*, 10–16. [CrossRef]
204. Keshmiri, M.; Mohseni, M.; Troczynski, T. Development of novel TiO$_2$ sol–gel-derived composite and its photocatalytic activities for trichloroethylene oxidation. *Appl. Catal. B Environ.* **2004**, *53*, 209–219. [CrossRef]
205. Fu, G.; Vary, P.S.; Lin, C.-T. Anatase TiO$_2$ Nanocomposites for Antimicrobial Coatings. *J. Phys. Chem. B* **2005**, *109*, 8889–8898. [CrossRef]
206. Chang, C.-C.; Lin, C.-K.; Chan, C.-C.; Hsu, C.-S.; Chen, C.-Y. Photocatalytic properties of nanocrystalline TiO$_2$ thin film with Ag additions. *Thin Solid Films* **2006**, *494*, 274–278. [CrossRef]
207. Rai, M.; Yadav, A.; Gade, A. Silver nanoparticles as a new generation of antimicrobials. *Biotechnol. Adv.* **2009**, *27*, 76–83. [CrossRef]
208. Yeo, S.Y.; Lee, H.J.; Jeong, S.H. Preparation of nanocomposite fibers for permanent antibacterial effect. *J. Mater. Sci.* **2003**, *38*, 2143–2147. [CrossRef]
209. Ayyad, O.; Muñoz-Rojas, D.; Oró-Solé, J.; Gómez-Romero, P. From silver nanoparticles to nanostructures through matrix chemistry. *J. Nanopart. Res.* **2010**, *12*, 337–345. [CrossRef]
210. Yu, D.-G. Formation of colloidal silver nanoparticles stabilized by Na+–poly(γ-glutamic acid)–silver nitrate complex via chemical reduction process. *Colloids Surf. B Biointerfaces* **2007**, *59*, 171–178. [CrossRef]
211. Nersisyan, H.H.; Lee, J.H.; Son, H.T.; Won, C.W.; Maeng, D.Y. A new and effective chemical reduction method for preparation of nanosized silver powder and colloid dispersion. *Mater. Res. Bull.* **2003**, *38*, 949–956. [CrossRef]
212. Courrol, L.C.; de Oliveira Silva, F.R.; Gomes, L. A simple method to synthesize silver nanoparticles by photo-reduction. *Colloids Surf. A Physicochem. Eng. Asp.* **2007**, *305*, 54–57. [CrossRef]
213. Xie, Y.; Ye, R.; Liu, H. Synthesis of silver nanoparticles in reverse micelles stabilized by natural biosurfactant. *Colloids Surf. A Physicochem. Eng. Asp.* **2006**, *279*, 175–178. [CrossRef]
214. Sathishkumar, M.; Sneha, K.; Won, S.W.; Cho, C.-W.; Kim, S.; Yun, Y.-S. Cinnamon zeylanicum bark extract and powder mediated green synthesis of nano-crystalline silver particles and its bactericidal activity. *Colloids Surf. B Biointerfaces* **2009**, *73*, 332–338. [CrossRef]
215. Durán, N.; Marcato, P.D.; De Souza, G.I.H.; Alves, O.L.; Esposito, E. Antibacterial effect of silver nanoparticles produced by fungal process on textile fabrics and their effluent treatment. *J. Biomed. Nanotechnol.* **2007**, *3*, 203–208. [CrossRef]
216. Dastjerdi, R.; Mojtahedi, M.R.M.; Shoshtari, A.M.; Khosroshahi, A. Investigating the production and properties of Ag/TiO$_2$/PP antibacterial nanocomposite filament yarns. *J. Text. Inst.* **2010**, *101*, 204–213. [CrossRef]
217. Dastjerdi, R.; Mojtahedi, M.R.M.; Shoshtari, A.M. Comparing the effect of three processing methods for modification of filament yarns with inorganic nanocomposite filler and their bioactivity against *Staphylococcus aureus*. *Macromol. Res.* **2009**, *17*, 378–387. [CrossRef]
218. Lok, C.-N.; Ho, C.-M.; Chen, R.; He, Q.-Y.; Yu, W.-Y.; Sun, H.; Tam, P.K.-H.; Chiu, J.-F.; Che, C.-M. Proteomic Analysis of the Mode of Antibacterial Action of Silver Nanoparticles. *J. Proteome Res.* **2006**, *5*, 916–924. [CrossRef] [PubMed]
219. Jeong, S.H.; Hwang, Y.H.; Yi, S.C. Antibacterial properties of padded PP/PE nonwovens incorporating nano-sized silver colloids. *J. Mater. Sci.* **2005**, *40*, 5413–5418. [CrossRef]
220. Jeong, S.H.; Yeo, S.Y.; Yi, S.C. The effect of filler particle size on the antibacterial properties of compounded polymer/silver fibers. *J. Mater. Sci.* **2005**, *40*, 5407–5411. [CrossRef]
221. Stobie, N.; Duffy, B.; McCormack, D.E.; Colreavy, J.; Hidalgo, M.; McHale, P.; Hinder, S.J. Prevention of Staphylococcus epidermidis biofilm formation using a low-temperature processed silver-doped phenyltriethoxysilane sol–gel coating. *Biomaterials* **2008**, *29*, 963–969. [CrossRef]
222. Wen, H.-C.; Lin, Y.-N.; Jian, S.-R.; Tseng, S.-C.; Weng, M.-X.; Liu, Y.-P.; Lee, P.-T.; Chen, P.-Y.; Hsu, R.-Q.; Wu, W.-F.; et al. Observation of Growth of Human Fibroblasts on Silver Nanoparticles. *J. Phys. Conf. Ser.* **2007**, *61*, 445–449. [CrossRef]
223. Jia, X.; Ma, X.; Wei, D.; Dong, J.; Qian, W. Direct formation of silver nanoparticles in cuttlebone-derived organic matrix for catalytic applications. *Colloids Surf. A Physicochem. Eng. Asp.* **2008**, *330*, 234–240. [CrossRef]
224. Yeo, S.Y.; Jeong, S.H. Preparation and characterization of polypropylene/silver nanocomposite fibers. *Polym. Int.* **2003**, *52*, 1053–1057. [CrossRef]
225. Kim, J.S.; Kuk, E.; Yu, K.N.; Kim, J.-H.; Park, S.J.; Lee, H.J.; Kim, S.H.; Park, Y.K.; Park, Y.H.; Hwang, C.-Y.; et al. Antimicrobial effects of silver nanoparticles. *Nanomed. Nanotechnol. Biol. Med.* **2007**, *3*, 95–101. [CrossRef]
226. Ilić, V.; Šaponjić, Z.; Vodnik, V.; Molina, R.; Dimitrijević, S.; Jovančić, P.; Nedeljković, J.; Radetić, M. Antifungal efficiency of corona pretreated polyester and polyamide fabrics loaded with Ag nanoparticles. *J. Mater. Sci.* **2009**, *44*, 3983–3990. [CrossRef]
227. Potiyaraj, P.; Kumlangdudsana, P.; Dubas, S.T. Synthesis of silver chloride nanocrystal on silk fibers. *Mater. Lett.* **2007**, *61*, 2464–2466. [CrossRef]

228. Simončič, B.; Klemenčič, D. Preparation and performance of silver as an antimicrobial agent for textiles: A review. *Text. Res. J.* **2016**, *86*, 210–223. [CrossRef]
229. Mohamed, A.L.; El-Naggar, M.E.; Shaheen, T.I.; Hassabo, A.G. Laminating of chemically modified silan based nanosols for advanced functionalization of cotton textiles. *Int. J. Biol. Macromol.* **2017**, *95*, 429–437. [CrossRef]
230. Tam, K.H.; Djurišić, A.B.; Chan, C.M.N.; Xi, Y.Y.; Tse, C.W.; Leung, Y.H.; Chan, W.K.; Leung, F.C.C.; Au, D.W.T. Antibacterial activity of ZnO nanorods prepared by a hydrothermal method. *Thin Solid Films* **2008**, *516*, 6167–6174. [CrossRef]
231. Zhang, L.; Ding, Y.; Povey, M.; York, D. ZnO nanofluids—A potential antibacterial agent. *Prog. Nat. Sci.* **2008**, *18*, 939–944. [CrossRef]
232. Yamamoto, O. Influence of particle size on the antibacterial activity of zinc oxide. *Int. J. Inorg. Mater.* **2001**, *3*, 643–646. [CrossRef]
233. Karunakaran, C.; Rajeswari, V.; Gomathisankar, P. Enhanced photocatalytic and antibacterial activities of sol–gel synthesized ZnO and Ag-ZnO. *Mater. Sci. Semicond. Process.* **2011**, *14*, 133–138. [CrossRef]
234. Ansari, M.A.; Khan, H.M.; Khan, A.A.; Sultan, A.; Azam, A. Characterization of clinical strains of MSSA, MRSA and MRSE isolated from skin and soft tissue infections and the antibacterial activity of ZnO nanoparticles. *World J. Microbiol. Biotechnol.* **2012**, *28*, 1605–1613. [CrossRef]
235. Wang, H.; Xie, C.; Zhang, W.; Cai, S.; Yang, Z.; Gui, Y. Comparison of dye degradation efficiency using ZnO powders with various size scales. *J. Hazard. Mater.* **2007**, *141*, 645–652. [CrossRef] [PubMed]
236. Farouk, A.; Moussa, S.; Ulbricht, M.; Schollmeyer, E.; Textor, T. ZnO-modified hybrid polymers as an antibacterial finish for textiles. *Text. Res. J.* **2014**, *84*, 40–51. [CrossRef]
237. Znaidi, L. Sol–gel-deposited ZnO thin films: A review. *Mater. Sci. Eng. B* **2010**, *174*, 18–30. [CrossRef]

Disclaimer/Publisher's Note: The statements, opinions and data contained in all publications are solely those of the individual author(s) and contributor(s) and not of MDPI and/or the editor(s). MDPI and/or the editor(s) disclaim responsibility for any injury to people or property resulting from any ideas, methods, instructions or products referred to in the content.

Review

Nanostructured Iridium Oxide: State of the Art

Francesca Scarpelli [1], Nicolas Godbert [1,2,*], Alessandra Crispini [1,2] and Iolinda Aiello [1,2,3]

[1] MAT-InLAB (Laboratorio di Materiali Molecolari Inorganici), LASCAMM-CR INSTM, Unità INSTM della Calabria, Dipartimento di Chimica e Tecnologie Chimiche, Università della Calabria, 87036 Rende, CS, Italy
[2] LPM-Laboratorio Preparazione Materiali, Star-Lab, Via Tito Flavio, 87036 Rende, CS, Italy
[3] CNR NANOTEC, Institute of Nanotechnology, U.O.S. Cosenza, 87036 Rende, CS, Italy
* Correspondence: nicolas.godbert@unical.it

Abstract: Iridium Oxide (IrO_2) is a metal oxide with a rutile crystalline structure, analogous to the TiO_2 rutile polymorph. Unlike other oxides of transition metals, IrO_2 shows a metallic type conductivity and displays a low surface work function. IrO_2 is also characterized by a high chemical stability. These highly desirable properties make IrO_2 a rightful candidate for specific applications. Furthermore, IrO_2 can be synthesized in the form of a wide variety of nanostructures ranging from nanopowder, nanosheets, nanotubes, nanorods, nanowires, and nanoporous thin films. IrO_2 nanostructuration, which allows its attractive intrinsic properties to be enhanced, can therefore be exploited according to the pursued application. Indeed, IrO_2 nanostructures have shown utility in fields that span from electrocatalysis, electrochromic devices, sensors, fuel cell and supercapacitors. After a brief description of the IrO_2 structure and properties, the present review will describe the main employed synthetic methodologies that are followed to prepare selectively the various types of nanostructures, highlighting in each case the advantages brought by the nanostructuration illustrating their performances and applications.

Keywords: iridium oxide; nanostructuration; IrO_2 applications; IrO_2 synthesis; OER

1. Introduction

Iridium Oxide (IrO_2) is a noble metal oxide with a rutile crystalline structure [1], with a $P4_2/mnm$ space group. The rutile structure is characterized by a tetragonal unit cell in which every Iridium ion is coordinated to six Oxygen ions, adopting an octahedral geometry (Figure 1).

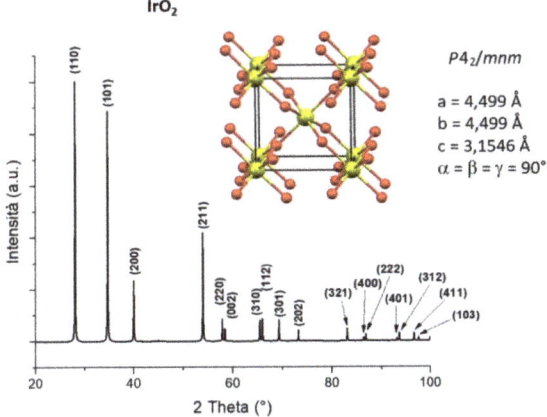

Figure 1. Unit cell, cell parameters and X-ray diffraction pattern with relative (hkl) indexation of IrO_2, information extracted from deposited crystal structure (yellow: Ir atoms, red: O atoms) [2].

Unlike other transition metal oxides, IrO_2 shows a metallic-type conductivity (~10^4 S·cm^{-1}) [3]. The reason for this behaviour was first investigated by Gillson et al. [1]: IrO_2, like other dioxides with a rutile-related structure, have incompletely filled d shells [IrO_2 ($5d^5$)]. Later, Verbist et al. [4], used X-ray Photoelectron Spectroscopy, to confirm the presence of a partially filled electron band in IrO_2, having a d character, just at the Fermi level, which could be responsible of the high conductivity of the metal oxide. Besides its high electrical conductivity, IrO_2 is also characterized by high chemical stability, low surface work function (4.23 eV) [5] and good stability under the influence of a high electric fields. Note that the work function (energy required for moving an electron from the Fermi level to the local vacuum level) is a surface sensitive property and typically depends on the electron chemical potential and the polarisation of the surface. Consequently, the work function can be modulated by several factors in particular the surface roughness and the orientation of the crystal lattice (i.e., which crystal face is mostly exposed). However, the more significant parameter to tune the work function of metal oxides would probably be the superficial oxygen vacancies [6]. Consequently, nanostructuration of metal oxides will have a great influence on the overall work function of the produced material. Table 1 reports the average work function of the most frequently studied metal oxides in comparison to IrO_2, considering thin films of polycrystalline metal oxide (i.e., without any preferred surface direction growth).

Table 1. Work functions and conductive characteristics of some metal-oxides.

Metal Oxide	Work Function (eV) *	Conductivity
IrO_2	4.23 [5]	Metallic (ca. 10^4 S·cm^{-1}) [7]
TiO	4.7 [8]	Metallic (5882 S·cm^{-1}) [9]
ReO_3	6.75 [10]	Metallic (ca. 10^5 S·cm^{-1}) [11]
MoO_2	5.9 [8]	Metallic (3355 S·cm^{-1}) [12]
RuO_2	5.2 [13]	Metallic (ca. $1,3.10^4$ S·cm^{-1}) [14]
NiO	6.3 [8]	p-type semiconductor, band gap 4 eV [15]
CuO	5.9 [8]	p-type semiconductor, band gap 1.5 eV [16]
Cu_2O	4.9 [8]	p-type semiconductor, band gap 2.40 eV [17]
Co_3O_4	6.3 [8]	p-type semiconductor, band gap 2.07 eV [18]
CoO	4.6 [8]	p-type semiconductor, band gap 2.6 eV [19]
Rh_2O_3	n.r. **	p-type semiconductor, band gap 1.4 eV for Rh_2O_3(I) and 1.2 for Rh_2O_3(III) [14]
TiO_2	5.4 [8]	n-type semiconductor, band gap anatase 3.2 eV, rutile 3.0 eV [20]
MoO_3	6.82 [8]	n-type semiconductor, α-MoO_3 band gap 3.2 eV [21]
WO_3	6.8 [8]	n-type semiconductor, band gap 2.75 eV [22]
SnO_2	4.75 [23]	n-type semiconductor, band gap 3.6 eV [24]
In_2O_3	5.0 [25]	n-type semiconductor, band gap 3.75 eV [26]
ZnO	4.71 [27]	n-type semiconductor, 3.3 eV [28]
ZrO_2	3.1 [29]	insulator
V_2O_5	7.0 [30]	insulator-metal transition (275 °C) [6]
V_2O_3	4.9 [8]	insulator-metal transition (−111 °C) [31]

* reported on polycrystalline thin film. ** not reported (to the best of our knowledge).

The appealing properties displayed by IrO_2 enable its use in many applications despite its relatively high cost. The main application fields of IrO_2 are summarized in Table 2 together with their challenging issues that are still nowadays to overcome.

In particular, IrO_2 plays a pivotal role in water electrolysis, the process in which water is split into hydrogen and oxygen gases, by means of the passing of an electric current. This process, if driven by renewable energy (solar or wind), can produce high-quality and clean hydrogen, as an energy source alternative to fossil fuels. Specifically, IrO_2 is considered one of the most active electrocatalysts for Oxygen Evolution Reaction (OER), the anodic reaction of water electrolysis, in which water is oxidized to molecular oxygen. OER represents the limiting process in water electrolysis, which determines the cell voltage and therefore the energy consumption of the process [32]. IrO_2 have been shown to possess very high electrocatalytic activity for OER, improving the process efficiency [33,34]. Several computational studies, electrochemical studies and characterization techniques, including, X-ray absorption near-edge structure, near-edge X-ray absorption fine structure, X-ray absorption, X-ray photoelectron, and Raman spectroscopies, have been devoted to understanding the OER mechanism over IrO_2 [35–37]. According to Nilsson et al. [37], IrO_2 surface in contact with water undergoes to a partial hydroxylation, showing hydroxide sites which coexist with the oxide sites. During OER, the hydroxide sites are converted into oxide sites, passing through an -OOH as intermediate. The simultaneous formation of some Ir(V) centres was ascertained, which could be responsible for the catalytic activity of IrO_2 toward OER, since its subsequent two-electrons reduction to Ir(III) can be sufficient to oxidize water in an oxygen molecule. However, several possible OER catalytic cycles over IrO_2, recently reviewed by Naito et al., have been proposed [35]. Specifically, the catalytic cycle can proceed via the oxidation-reduction reactions of either the Iridium centre or the absorbed O species or the Ir=O states. A catalytic cycle driven by the releasing and filling of an oxygen vacancy at the IrO_2 surface has also been suggested. Full understanding of the OER mechanism over Iridium is thus a contemporary and challenging issue that still requires attention for the design of future efficient IrO_2 based catalysts.

Moreover, IrO_2 is an electrochromic (EC) material that displays a reversible and persistent colour change under an external electric field, thus finding application in electrochromic devices. The change in its optical properties may be ascribed to the following reaction [38], as shown in Scheme 1:

$$Ir(OH)_3 \rightleftharpoons IrO_2 \cdot H_2O + H^+ + e^-$$

transparent   blue-black

Scheme 1. Redox reaction at the basis of IrO2 EC properties.

During colouring, electrons and protons are removed from the material by application of an anodic potential, whereas during bleaching electrons and protons are injected into the substrate [39]. In its lower oxidation states [Ir(III)], Iridium (hydr)oxide is transparent, while in its higher oxidation state [Ir(IV)], IrO_2 turns to a blue-black colour due to a strong absorption in the visible spectral region [40]. IrO_2 presents several ideal features for an EC material, such as fast colour change, good open-circuit memory, and long last durability (more than 10^7 cycle lives) [40,41], which promote its application in EC devices [41–44].

Notably, the transition between two oxidation states [Ir(III) and Ir(IV)] is also exploited for the fabrication of IrO_2 based pH sensors [45–48]. Owing to proton–electron double injection, IrO_2 is reduced to $Ir(OH)_3$ during pH detection [47]. IrO_2 provides a fast-potentiometric response to pH change, thanks to its high conductivity. IrO_2-based electrodes have useful properties, such as high stability in a wide range of temperature (from −20 °C to 250 °C) [49–51], linear response in a broad pH range (from pH = 0 to pH = 12) [50,52,53], great chemical stability, and low impedance [54]. Moreover, IrO_2-based electrodes could be used in many application fields, since their pH response is not affected by most anions present in environmental systems, such as Na^+, K^+, Li^+, Mg^{2+}, Ca^{2+}, Cl^-, Br^-, NO_3^-, nor by the main complexing agents present in biological systems, such as citrate, lactate and phosphate [53]. For these reasons, IrO_2 has been widely employed for the sensing of glucose, hydrogen peroxide, glutamate, metal ions, organophosphates, and pesticides [55]. Furthermore, the Food and Drug Administration approved IrO_2 as a high biocompatible material, facilitating its application in biosensors [56,57], probe for fluorescence imaging [58], photodynamic/photothermal therapy [59], and stimulating and recording electrodes [60,61]. To this end, due to its high charge capacity, for a given applied voltage pulse, IrO_2 is able to inject a very high charge density [62].

By virtue of its conductive nature, high chemical stability, low surface work function (4.23 eV) and stability under influence of high electric fields, IrO_2 has been used as field emitter cathode in vacuum microelectronics [63–65]. Indeed, IrO_2 doesn't suffer from the eventual presence of residual Oxygen in these devices, contrary to other metals, as Molybdenum, that reacts quickly with O_2 forming an insulating layer of oxide [66].

Further applications of IrO_2 include electrode material for direct methanol fuel cell (DMFC) [67], for supercapacitors [68], and for neural stimulation [69]. Specifically, the anodic reaction in a DMFC, in which methanol is oxidized to carbon dioxide, can be efficiently catalysed by IrO_2 [67]. Moreover, IrO_2, thanks to its ability to store electricity, can be an excellent negative electrode for electrochemical capacitors [68].

Table 2. Main applications of IrO_2-based materials.

Application	Main Features	Refs.	Current Challenge
Electrochromic devices	Fast colour change	[40,42,70]	Application in flexible devices (IrO_2 is a rigid material)
OER	High catalytic activity High stability in acidic media	[71–75]	Deep understanding of the OER mechanism over IrO_2
Sensing	Stability repeatability	[46,47,55] [76,77]	Standardization of electrode preparation methods (dependence of pH response of IrO_2) Improvement of stability over the pH range of 12–14 Improve sensing sensitivity Lowering the working temperature in gas sensing
Supercapacitor	High conductivity	[68,78]	Increase of the durability of the electrode (slight worsening of performance after 2000 charge/discharge cycles at 0.5 mA)
Field Emission Cathode	Low chemical reactivity Thermal stability Low work function	[65,79,80]	Achieve high aspect structures to enable operation at low applied fields Insure long-term device operation under adverse vacuum conditions

Noticeably, the above-described applications of IrO_2 can benefit from the use of nanostructures instead of bulk IrO_2. Indeed, all the intrinsic properties of IrO_2, as well as for other transition metal oxides, can be improved through its nanostructuration. Thus, the synthesis of IrO_2 in its nanostructured form (structures presenting at least one dimension on the nanoscale) has become an increasing field of interest in research. In particular, nanostructured materials that exhibit chemical, optical, mechanical and electrical properties modified with respect to the corresponding bulk material, due to size and quantum effects [81–86]. Several types of nanostructures, with different dimensionality, are known, including zero-dimensional (0D, quantum dots and nanopowder), one-dimensional (1D, nanotubes, nanowires and nanorods), and two-dimensional (2D, nanostructured films) nanomaterials [86]. Whatever the dimensionality of the nanomaterials, it is preferable that the nanostructures are well separated from each other, as an over-aggregation could determine the loss of the nanostructuration with the appearance of properties that are more reminiscent of those of a bulk material.

To this regard, IrO_2 can be synthesized in the form of various types of nanostructures, as nanoparticles with undefined shape, i.e., nanopowders, [87] and nanoparticles with a precise shape, including nanosheets [88], nanorods [5,89], nanotubes [76], nanowires [48,90], and nanoporous films [91,92]. Herein, we provide an overview on the nanostructuration of IrO_2, with special emphasis on the strategies pursued for the synthesis of IrO_2 nanostructures. Indeed, up to now many preparation procedures of IrO_2 nanostructures have been described in the literature, including soft- and hard-template routes, hydrothermal synthesis, colloidal methods, and many others. It is worth noting that, within the same preparation method, by finely tuning the experimental parameters, nanostructures with different shapes can also be synthetized. On this basis, we focused our attention on the shapes and morphologies that can be obtained through the different preparation methodologies, since the shape/morphology determine the physical and chemical properties of IrO_2 nanostructures. Furthermore, when available, the performances of these IrO_2 nanostructures in specific applications will be described.

2. IrO_2 Spherical Nanoparticles and Nanopowder

The synthesis methods, characterization and description of specific applications of IrO_2 spherical nanoparticles have been brilliantly reported recently by J. Quinson [93]. Interested readers are directed towards his report for a deeper overview, especially for what concerns the analysis about the difficulty to fully distinguish between Ir and IrO_x nanoparticles during their preparation [93]. Thus, in the following part of this review, we will only focus on the main synthesis pathways used to prepare IrO_2 as nanopowder. Several methods to synthesize IrO_2 nanopowder, meaning nanoparticles with undefined shape, have been experimented. In 2005, Marshall introduced a modification to the well-known polyol method, usually used for the preparation of metallic nanoparticles [94,95]. This procedure consists in dissolving or dispersing the metallic precursor, usually hexachloroiridic acid ($H_2IrCl_6 \cdot nH_2O$), in a polyol, such as ethylene glycol, which acts both as solvent and as reducing agent. Upon refluxing the reaction mixture, a metallic precipitate, composed of Ir nanoparticles with an average size of about 3 nm, is formed, which is then filtered and dried. The colloid is finally calcinated to ensure the full oxidation of the obtained product.

Another approach is represented by the Adams fusion method [96], in which the metallic precursor H_2IrCl_6 is melted together with $NaNO_3$. The possible reactions that may occur during the process are shown in Scheme 2. Basically, when H_2IrCl_6 is melted together with $NaNO_3$, $Ir(NO_3)_4$ is formed. This latter, at high temperature (ca. 460 °C) decomposes, thus generating IrO_2.

$$H_2IrCl_6 + 6NaNO_3 \longrightarrow 6NaCl + Ir(NO_3)_4 + 2HNO_3$$

$$Ir(NO_3)_4 \longrightarrow IrO_2 + 4NO_2 + O_2$$

Scheme 2. The hypothesized reactions taking place in the Adams fusion method [94].

After cooling, the mixture is thoroughly washed with water to remove salt residues, nitrites, and nitrates. Despite the simplicity of this method, a long purification step is required, and sodium traces may remain in the metal oxide powder.

Nanosized IrO_2 powder can also be synthesized through the colloidal method [97,98]. This process involves the addition of NaOH to a water solution of the Iridium precursor, $H_2IrCl_6 \cdot nH_2O$, to induce the formation of an Ir-hydroxide. The resulting colloidal solution is heated at 80–100 °C, then washed, dried, and calcinated at 400 °C to obtain colloidal IrO_2, composed of IrO_2 nanoparticles with an average size of diameter ca. 7 nm. However, although the simplicity of the method and the fact that no specific experimental set up is required, the final product remains of a colloid state, thus the nanoparticles are all aggregated to each other as shown in the reported Transmission Electron Microscopy (TEM) micrograph (Figure 2). Nevertheless, the as-synthesized nanopowder was tested as electrocatalyst for OER in a solid polymer electrolyte electrolyzer, demonstrating high stability and higher activity with respect to commercial IrO_2 powder, which does not feature any nanostructuration [97].

Figure 2. TEM micrograph of colloidal IrO_2 obtained through the colloidal method (Reprinted/adapted with permission from Ref. [97]).

The colloidal method, which involves an initial alkaline hydrolysis of the Iridium precursor (K_2IrCl_6) in water solution at high temperature, can also be carried out without the calcinations step, thus further simplifying the method, as reported by Khalil et al. [77]. By applying this procedure, IrO_2 nanoparticles with an average size of 1–2 nm have been obtained (Figure 3). However, also in this case, the particles are aggregated with each other, due to the absence of a capping agent. The same authors described the preparation of a pH electrode through the electrodeposition of the as-prepared IrO_2 nanopowder on an Au substrate and the evaluation of its potentiometric responses in pH buffer solutions between pH 1.68 to 12.36. The electrode demonstrated excellent pH sensitivity (-73.7 mV/pH unit), with a super-Nernstian response [77].

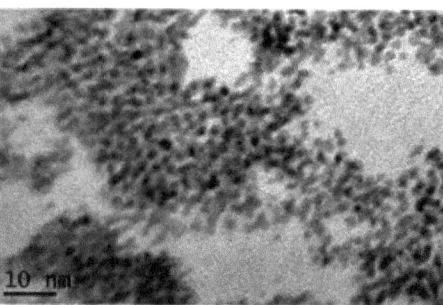

Figure 3. TEM image of colloidal iridium nanoparticles (Reprinted/adapted with permission from Ref. [77]).

Recently, the synthesis of colloidal IrO$_2$ has been also performed through a microwave-assisted route which combines the colloidal and polyol methods, and avoids the calcinations step [87]. Ethylene glycol was used as solvent, whereas NaOH was used as source of OH$^-$ anions, to generate the Ir-hydroxide intermediate. The reaction was carried out under microwave irradiation, until the colour solution turned to dark brown, indicating the formation of IrO$_2$ colloidal suspensions. Specifically, the metal oxide was formed when the solution reached its boiling point, without requiring a high temperature treatment in a muffle furnace. The obtained colloidal suspension was stable against flocculation and exhibited a low polydispersity. Moreover, the same method can be carried out in milder reaction conditions, by using alcohols with a low boiling point, such as methanol or ethanol, instead of ethylene glycol.

Finally, nanosized IrO$_2$ powder can be effectively synthesized through the soft-template route, by using an organic polymer as templating agent [99,100]. Specifically, Zhang et al. described a process in which nanostructuration of Iridium Oxide is achieved by the in-situ polymerization of pyrrole in a water solution containing the Iridium precursor (NH$_4$)$_2$IrCl$_6$, followed by thermal annealing of the nanocomposite (Figure 4) [99]. At a temperature value of 450 °C, the organic part of the composite is totally degraded, and Iridium Oxide nanoparticles, can be recovered. The high surface area displayed by these materials corresponds to a great number of accessible electrochemical sites. Indeed, the as-prepared nanosized Iridium Oxide exhibits a greater electrocatalytic efficiency towards OER, when compared to commercial IrO$_2$ [99]. In particular, the nanostructured iridium oxide exhibits low overpotential (291.3 ± 6 mV) to reach 10 mA/cm^2 current density towards OER and higher stability with respect to commercial IrO$_2$.

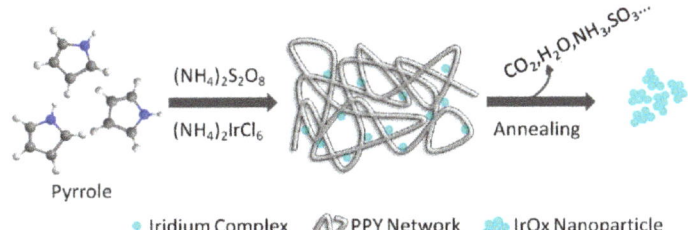

Figure 4. A schematic representation of the soft-template route for the synthesis of Iridium Oxide nanoparticles (Reprinted/adapted with permission from Ref. [99]).

3. IrO$_2$ 1D-Nanostructures

1D nanostructures present only one dimension greater than 100 nm, while the others are a few nanometers long. These nanostructures include nanotubes, nanorods and nanowires, and they usually display greater resistance to agglomeration with respect to nanopowder. These elongated structures, as well as nanopowder, exhibit superior properties with respect to bulk materials [86]. Moreover, theoretical studies recently have highlighted the superior electrocatalytic performances of IrO$_2$ 1D-nanostructures, specifically IrO$_2$ nanowires, with respect to IrO$_2$ nanospheres, thanks to their regular and periodic structure, without the tendency to agglomeration [71]. In the following part of the review, the IrO$_2$ 1D-nanostructures and the main methodologies for their synthesis will be described according to their specific shape.

3.1. IrO$_2$ Nanotubes

IrO$_2$ nanotubes have been built using a hard-templating route coupled with electrodeposition (Figure 5). Concretely, an anodic aluminium oxide (AAO) layer with large pores was first fabricated through the sputtering of an aluminium layer on a silicon substrate, followed by an anodization process. IrO$_2$ was electrodeposited on this template layer, generating nanotubes which grew along the walls of the AAO nanopores. At the end of the process, the AAO template was removed by dissolution in concentrated KOH solution [76].

As reported by Chiao et al., the shape and the length of the nanotubes prepared through this method depend on the morphology of the nanoporous AAO template, whereas the wall thickness of the nanotubes is finely controlled by the electrodeposition time [76]. The nanotubes obtained by Chiao et al., analysed by Scanning Electron Microscopy (SEM), show a diameter of 50 nm and length of around 750 nm. However, the not uncommon presence of defective nanopores in the AAO template, results in the formation of incomplete and collapsed nanotubes together with hollow nanotubes.

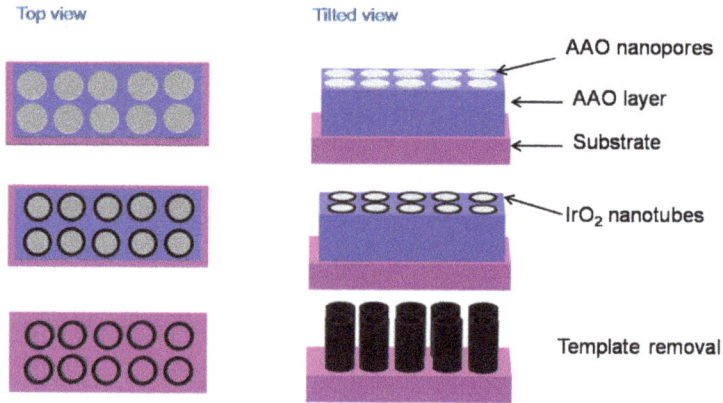

Figure 5. A schematic representation of the hard-template route for the synthesis of IrO_2 nanotubes.

A more uniformly distributed IrO_2 nanotubes array has been obtained by using the hard template route coupled to an acidic chemical bath, instead of electrodeposition, allowing to produce IrO_2 nanotubes [101]. Specifically, the AAO template has been soaked in a solution containing the Iridium precursor, Na_3IrCl_6, in addition to NaClO, HNO_3 and H_2O_2, generating, in 24 h and after acidic removal of the AAO template, a uniform film, with a thickness of 60 nm and a length of 400 nm, composed of IrO_2 nanotubes (Figure 6). The as-prepared nanotubes array exhibited a large charge storage capacity, measured through electrochemical analysis, proving its potential application as neural-electronic interface electrode [101].

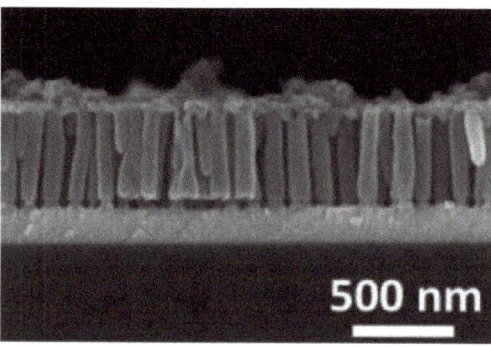

Figure 6. SEM image of IrO_2 nanotubes array (Reprinted/adapted with permission from Ref. [101]).

Recently, IrO_2 nanotubes have been synthesized without a hard template, by electrospinning and calcination techniques [102]. Specifically, a solution containing the Iridium precursor IrCl$_3$ was electrospun on an aluminium plate, then calcinated at 500 °C, under O$_2$ and He flow, recovering IrO$_2$ nanotubes after cooling. However, although the authors claim to prepare IrO$_2$ nanotubes without any template, the initial solution also contains poly(vinylpyrrolidone) (PVP), which, as stated by the authors, acts as a "framework", and is removed through the thermal annealing. The as-synthetized IrO$_2$ nanotubes were tested as electrode material for amperometric CO sensing, but they resulted in poorly electroactive towards electrochemical CO oxidation, in contrast to the metallic counterparts, i.e., Ir nanotubes that were produced by the reduction, under H$_2$ and Ar flow, of the as-prepared IrO$_2$ nanotubes.

Finally, IrO$_2$ nanotubes can be efficiently prepared through Metal-Organic Chemical Vapor Deposition (MOCVD) [103]. This method, however, requires a specifically designed reactor in which ultra-pure gases are injected to transport and react with the organometallic precursor. By applying this methodology, IrO$_2$ nanotubes have been successfully grown on a LiTaO$_3$ substrate using a low-melting Iridium source, (Methylcyclopentadienyl) (1,5-cyclooctadiene) Iridium(I), [(MeCp)Ir(COD)] (chemical structure reported in Figure 7), a high oxygen pressure (20–50 Torr), and a temperature of ca. 350 °C [103]. The as-obtained nanostructures, presenting an unusual square cross-section, present a tilt angle of 35° ca. with respect to the normal to the substrate surface and are perfectly aligned with each other (Figure 7a).

Figure 7. Chemical structure of [(MeCp)Ir(COD)] and SEM micrographs at different magnifications of IrO$_2$ nanotubes synthetized through MOCVD, (**a**) top view of IrO$_2$ nanotubes, (**b**) a wedge-shaped rod (Reprinted/adapted with permission from Ref. [103]).

Notably, by opportunely modifying the experimental parameters used during the MOCVD, the same authors were able to finely tune the shape of the obtained nanotubes, thus obtaining forms ranging from nearly-triangular nanorods, wedge-like nanorods (Figure 7b) and scrolled nanotubes of IrO$_2$. Indeed, the growth of these peculiar structures, and in particular their shape, is highly dependent on the substrate temperature and the degree of supersaturation, this latter being controlled by the temperature of the precursor reservoir [104]. Interestingly, these square hollowed IrO$_2$ nanotubes grown onto sapphire (100) substrate were successfully reduced to mixed Ir-IrO$_2$ nanotubes by high vacuum thermal annealing. Moreover, a Pt electrodeposition was carried-out on these nanotubes to generate a new nanostructured catalyst for methanol oxidation, with activity comparable to that of a commercial PtRu catalyst [105].

3.2. IrO$_2$ Nanorods

IrO$_2$ nanorods can be produced through several techniques, including the hard-template route [89], the "molten salt" method [106], and MOCVD [5]. In particular, similarly to IrO$_2$ nanotubes [76,101], arrays of IrO$_2$ nanorods have been synthetized using AAO membranes as hard template [89]. In these cases, the porous AAO membrane has been soaked in an alkaline chemical bath, containing IrCl$_4$, as Iridium precursor, and K$_2$CO$_3$ as source of OH$^−$ anions. The temperature was increased to 95 °C, allowing the reactions reported in Scheme 3 to take place. Basically, the OH$^−$ anions which derived from the reaction of K_2CO_3 with water, react with the Iridium precursor generating the Ir-hydroxyde intermediate which evolves towards the formation of IrO_2.

$$K_2CO_3 + H_2O \longrightarrow 2K^+ + 2OH^- + CO_2$$

$$IrCl_4 + OH^- \longrightarrow IrO_2 + H_2O + Cl^-$$

Scheme 3. The reactions taking place in the chemical bath method to synthetize IrO_2 nanorods on AAO membranes [89].

The IrO_2 nanoparticles are grown and packed inside the pores of the AAO membrane, forming well-aligned elongated nanostructures, i.e., nanorods, which have been characterized through electronic microscopy after the removal of the template with KOH (Figure 8). Specifically, the IrO_2 nanorods grew perpendicular to the substrate and present a diameter ranging from 80 to 100 nm, approximately corresponding to the AAO pores dimensions [107]. This IrO_2 nanorods array has been tested as neurotransmitter sensor, displaying a good response to dopamine, chosen as a neurotransmitter model [89].

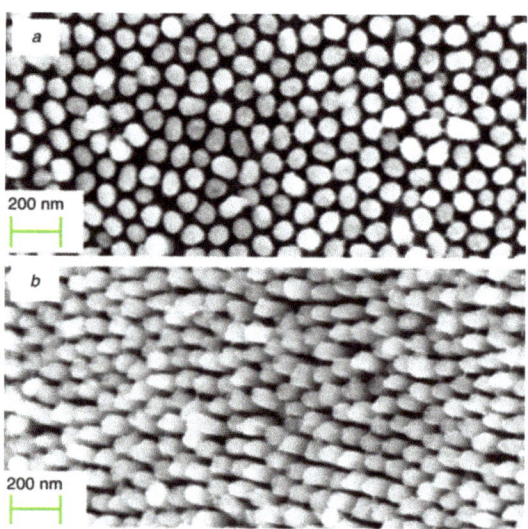

Figure 8. SEM micrographs of IrO_2 nanorods synthetized through the chemical bath route: (**a**) top view; (**b**) tilted view (Reprinted/adapted with permission from Ref. [107]).

IrO_2 nanorods can also be synthesized through the "molten salt" method, a synthetic route similar to the Adams fusions, consisting of the grinding of the Iridium precursor ($IrCl_4$) with NaCl and KCl, followed by the calcinations of the solid mixture at high temperature (600 °C) for 12 h [106]. Through this procedure, Mao et al. obtained nanostructures with an average diameter of 15 nm and length of ca. 200 nm (Figure 9) [106]. However, this method, similarly to the above-described Adams fusion synthesis of IrO_2 nanopowder, required long purification and drying processes. The same authors reported a high electrocatalytic activity of the produced IrO_2 nanorods towards OER, probably attributed to high specific area of the material. Indeed, IrO_2 NRs generate higher OER current density (70 mA/cm^2) than the commercial IrO_2 (58 mA/cm^2) at 0.6 V versus Ag/AgCl electrode in deaerated 0.5 M KOH electrolyte.

Interestingly, within the molten salt method, by carefully varying the $IrCl_4$: NaCl: KCl ratio, it is possible to tune the morphology of the IrO_2 nanostructures [108]. In particular, it has been reported that, using a IrCl$_4$: NaCl: KCl ratio of 1: 10: 10, nanocubes are obtained (Figure 10a); while changing the IrCl$_4$: NaCl: KCl ratio to 1: 30: 30, a mixture of IrO$_2$ nanocubes and nanorods is formed (Figure 10b); and ultimately, using a high salts percentage, specifically using a 1: 60: 60 ratio, a sample consisting predominantly of nanorods is obtained (Figure 10c). This experimental observation was explained by Mao et al. considering that the molten salts act both as solvent and as protective layers against aggregation of the formed nanoparticles [108]. Therefore, when the content of salts is low, the growth of nanostructures is allowed in all directions, thus nanocubes can be generated; conversely, when a higher salts percentage is used, the excess of salts block the growth of the nanostructures in all directions, except one, thus nanorods are preferentially formed. Also in this case, the obtained IrO$_2$ nanorods proved to be excellent electrocatalysts towards OER, displaying high current density, low overpotential, high stability and numerous accessible active sites [108]. The electrocatalytic performance of IrO$_2$ was again better than commercially available IrO$_2$.

Figure 9. TEM image of IrO$_2$ nanorods obtained through the molten salt method (Reprinted/adapted with permission from Ref. [106]).

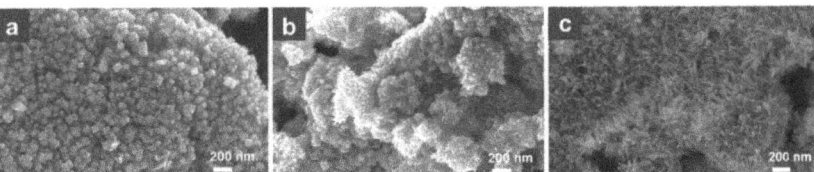

Figure 10. SEM image of the different IrO$_2$ nanostructures obtained through the molten salt method: (**a**) nanocubes, (**b**) mixture of nanocubes and nanorods, (**c**) nanorods (Reprinted/adapted with permission from Ref. [108]).

IrO$_2$ nanorods have been also produced through MOCVD technique. In this case, IrO$_2$ nanorods were grown on Si substrates using a low-melting Iridium source, [(MeCp)Ir(COD)], a high oxygen pressure (10–60 Torr) and a temperature of ca. 350 °C [5,109]. These as-produced IrO$_2$ nanorods present diameters between 75 and 150 nm, a wedge-shaped morphology and naturally formed sharp tips (Figure 11). However, these nanorods show a polycrystalline nature, characterized by many defects and dislocations.

Figure 11. Field-Emission Scanning Electron Microscopy images of the IrO_2 nanorods fabricated with this method (Reprinted/adapted with permission from Ref. [5]).

3.3. IrO_2 Nanowires/Nanofibres

The MOCVD technique can also be exploited to obtain another type of 1D structures, i.e., IrO_2 nanowires, when the experimental parameters are adequately tuned [104,110]. For this purpose, [(MeCp)Ir(COD)] as iridium source, oxygen as both carrier and reactant gas, high temperature (350–400 °C) and high pressure (33 Torr) were needed. By applying these conditions, Zhang et al. synthetized single crystal IrO_2 nanowires, having rutile structure, on plasma treated SiO_2 substrates covered with a thin metallic layer (Ti, Au, Ni, Co) [110]. The as-obtained nanowires have dimension ranging from 10 to 50 nm in diameter and 1–2 mm in length (Figure 12).

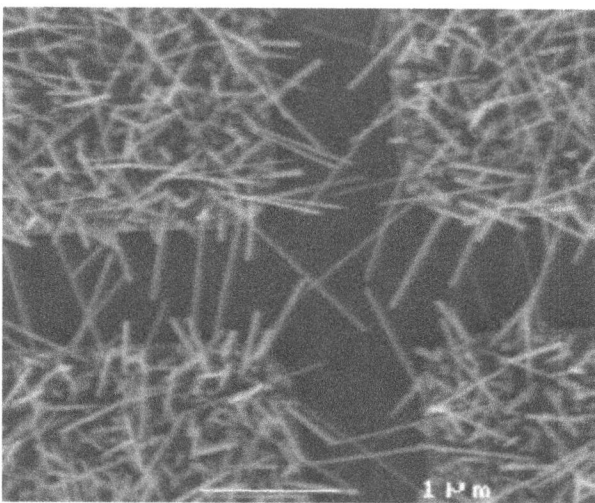

Figure 12. SEM image of IrO_2 nanowires obtained through MOCVD (Reprinted/adapted with permission from Ref. [110]).

IrO_2 nanowires were also grown on Au microwire and Si/SiO_2 substrates via Vapor Phase Transport [48]. IrO_2 powder was used as source material and placed in a quartz tube furnace under He and O_2 flow. Working at very high temperature (ca. 1000 °C) allows the precursor to sublimate and to be transported by the gas flow to the substrates, where recrystallization occurs in the form of nanowires. Kim et al. reported the preparation of single crystal IrO_2 nanowires, displaying lateral dimensions ranging from 20 to 100 nm near the nanowire tip, with the length extending up to tens of micrometers [48]. Through this method, the formation of nanowires is strongly affected by the O_2 flow. Indeed, without O_2 flow, IrO_2 nanowires are not formed, whereas with a high O_2 flow (50 sccm) polyhedral IrO_2 crystals are generated. IrO_2 nanowires, presenting random orientations have been obtained by using an O_2 flow rate within the range from 10 to 15 sccm. The electrochemical performance of the IrO_2 nanowires grown on the Au microwire was also tested as microelectrode, showing a linear pH response with a super-Nernstian behaviour [48].

IrO_2 nanofibres can be successfully formed through the electrospinning technique [111]. Iridium chloride can be dissolved in ethanol-water together with PVP. The electrospinning of the resulting solution allows the formation of nanofibres of diameter ranging between 50 and 150 nm of diameter (Figure 13a). These as-deposited nanofibres were directly employed for the electrochemical detection of ascorbic acid [112]. More, recently, based on the same synthetic protocol but carrying out a thermal annealing after deposition to ensure the complete removal of the polymer content, amorphous hollow nanofibres of IrO_2 were produced that were of an average diameter of ca. 60 nm (Figure 13b) and were successfully employed for the fabrication of a flexible solid-state gel symmetric supercapacitor [78].

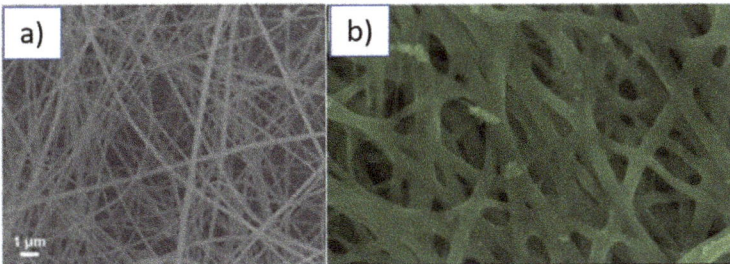

Figure 13. (**a**) nanofibres IrO_2/PVP synthesised through electrospinning (adapted from ref. [111]; (**b**) Hollow nanofibres of IrO_2 after thermal annealing at 300 °C (Reprinted/adapted with permission from Ref. [78]).

4. IrO_2 Nanostructures with Unusual Shapes

The preparation and characterization of IrO_2 nanostructures with unusual shapes have been recently reported. In particular, urchin-like IrO_2 nanostructures have been synthesized through a hydrothermal method, involving the pre-treatment of an aqueous solution of $IrCl_3$ with NaOH and H_2O_2 at 100 °C, followed by heat treatment of the solution at 160–200 °C in autoclave [113]. The urchin-like nanostructures are composed by more levels of structures, specifically short needles, and small cores (Figure 14). By monitoring the hydrothermal synthesis over time, Deng et al. demonstrated that first nanosized IrO_2 spheres with a rough surface are formed upon which thin needles grow in a fractal manner [113]. The urchin-like IrO_2 nanostructures thus obtained have been tested as a catalyst for OER in acidic medium, demonstrating excellent activity and stability in acidic medium, attributable to their hierarchical structure. Indeed, the authors demonstrated that these nanostructures possess improved electrochemical surface-active area with respect to plain spherical IrO_2 structures of a similar size.

Moreover, IrO_2 nanoneedles with an average diameter of 2 nm (Figure 15), have been produced through a modified Adams fusion route, implying the use of 2-mercaptoethylamine (chemical structure reported in Figure 15) in addition to $NaNO_3$ [75]. In this case, the presence of 2-mercaptoethylamine specifically determines the formation of nanoneedles rather than nanopowder. Indeed, without 2-mercaptoethylamine, unshaped and aggregate nanoparticles, like those obtained with the classical Adams fusion method, were obtained. Furthermore, by enhancing the amount of 2-mercaptoethylamine, an increase of the nanoneedles aspect ratio occurs. Also, for IrO_2 nanoneedles, the OER activity has been evaluated, verifying a better performance with respect to unshaped nanoparticles. However, the exact role of 2-mercaptoethylamine to direct the preferential growth into nanoneedles has not been reported.

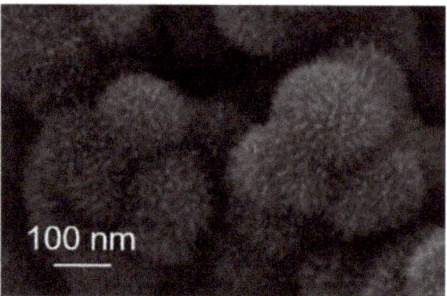

Figure 14. SEM image of urchin-like IrO$_2$ nanostructures (Reprinted/adapted with permission from Ref. [113]).

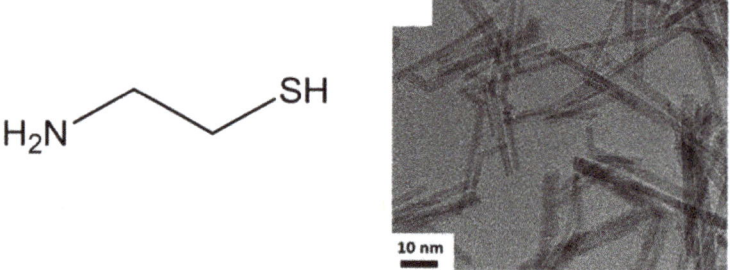

Figure 15. Chemical structure of 2-mercaptoethylamine and TEM image of the IrO$_2$ nanoneedles (Reprinted/adapted with permission from Ref. [75]).

5. IrO$_2$ Nanostructured Films

IrO$_2$ thin films can be prepared through several techniques, ranging from the hard template route [114], to spray pyrolysis [70], to reactive radio-frequency magnetron sputtering [43]. However, the easiest way to introduce a nanostructuration in a metal oxide film, also in an IrO$_2$ film, is the Evaporation Induced Self-Assembly method (EISA) (Figure 16).

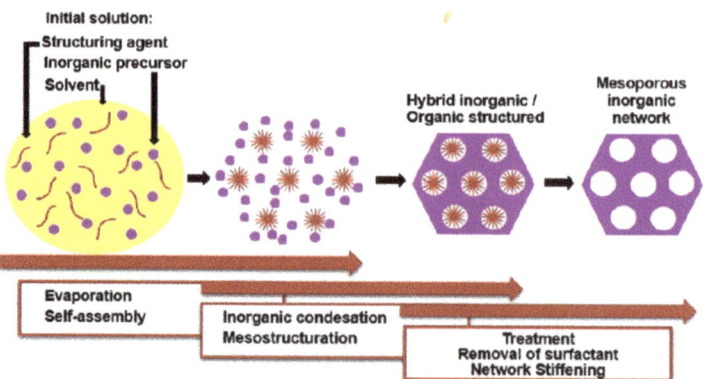

Figure 16. Schematic representation of EISA (Reprinted/adapted with permission from Ref. [115]).

The EISA route involves the use of a soft template, usually an ionic organic surfactant or non-ionic polymeric surfactant which self-assembles into a diversity of supramolecular structures. These latter can be formed of spherical micelles, hexagonal rods, lamellar liquid crystals or other assemblies in solution that self-organise through non-covalent weak interactions such as hydrogen bonding, van der Waals forces, electrostatic interactions and hydrophobic effect. Furthermore, these interactions are also driven by evaporation of the solvent that occurs in situ to the deposition. Hence, these assemblies are the structural directing agents for the formation of inorganic mesostructures. Indeed, the sol–gel precursor hydrolyzes and condenses around the mesostructured self-assembled phase. Subsequent thermal treatment induces the removal of the surfactant, the stiffening of the inorganic network and its crystallization. By varying the type of surfactant, its concentration in the starting solution, and the deposition conditions, it is possible to tune the pore structure and size of the porous materials. EISA is generally coupled with dip-coating or spin-coating deposition techniques, which allow the formation of a thin layer of precursor on different substrates. Through this procedure, Ortel et al. successfully synthesized IrO_2 thin films on different substrates by dip-coating, employing PEO_y-PB_x-PEO_y, (poly(ethylene oxide)-poly(butadiene)-poly(ethylene oxide, chemical structure reported in Figure 17), as templating agent [91,116]. These films presented nanocrystalline mesopores walls and some areas with locally ordered pores (Figure 17). Moreover, their electrocatalytic performance toward OER was tested and compared to untemplated IrO_2 films obtained with the same experimental procedure. The current response on templated IrO_2 films is about 2.1 times higher with respect to the untemplated IrO_2 films, demonstrating the nanostructuration advantages.

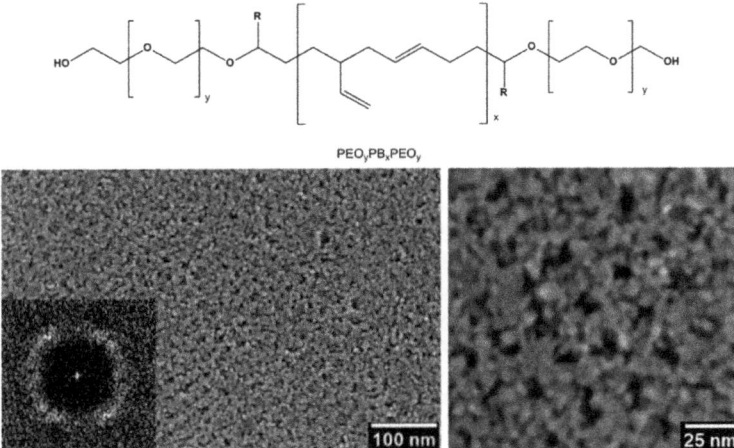

Figure 17. SEM images of IrO_2 mesoporous thin film template with PEO_y-PB_x-PEO_y and calcinated at 500 °C (Reprinted/adapted with permission from Ref. [91]).

Similarly, Chandra et al. developed mesoporous IrO_2 thin films choosing the triblock copolymer "Pluronic F127" ($PEO_{106}PPO_{70}PEO_{106}$, chemical structure reported in Figure 18), as structural directing agent and spin-coating as deposition technique [74,117]. They reported that samples calcinated at 400 °C present 2D hexagonal mesostructure (*p6mm* symmetry, Figure 18), but an increase in treatment temperature entails the transformation into a disordered mesostructure. The enhancement of the electrocatalytic performance toward OER with respect to the untemplated IrO_2 electrode was ca. 2 times higher for mesoporous structure and was ascribed to the larger accessible surface-to-volume ratio.

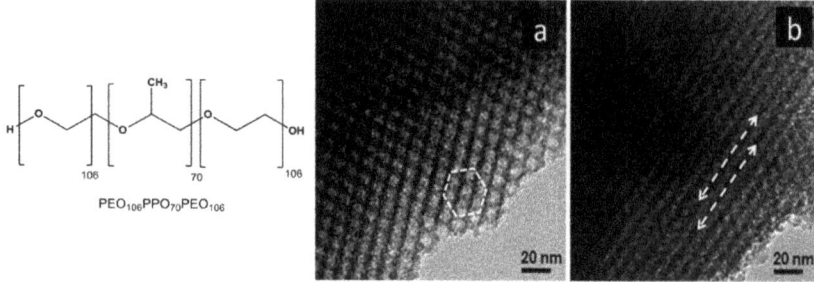

Figure 18. TEM images of mesoporous IrO$_2$ films template with PEO$_{106}$PPO$_{70}$PEO$_{106}$ recorded along the (**a**) [100] and (**b**) [110] axes of the 2D hexagonal structure (Reprinted/adapted with permission from Ref. [74]).

Although EISA appears as a versatile route to induce a nanostructuration into an inorganic material, allowing the modulation of size and shape of the nanostructures and not requiring sophisticated instrumental equipment [118], an important drawback of this technique is its high dependence on experimental parameters. Indeed, temperature, humidity, extraction time and velocity (for dip-coating), concentration of the precursor solution and speed (for spin-coating) would be required as given experimental data for the sake of reproducibility and understanding of the mechanism of nanostructuration. A small amount of variation of these conditions may drastically affect the final nanostructures [119].

The hard-templating route has been also adapted to the preparation of IrO$_2$ nanostructured films. In this case, colloidal SiO$_2$ microspheres were immersed in an ethanolic solution of (H$_2$IrCl$_6 \cdot n$H$_2$O) to allow the impregnation of the Iridium precursor within the template, then the suspension was dried and calcinated, and the template was removed by using a concentrated HF solution [98]. Chen et al. demonstrated that using SiO$_2$ microspheres with a mean diameter of 330 nm and very low polydispersity (ca. 0.5%), macroporous IrO$_2$, displaying an ordered honeycomb array of macropore, can be obtained (Figure 19) [98]. The macropores are typically 300 nm in diameter, which is slightly smaller than the size of SiO$_2$ microspheres, probably as a consequence of the contraction of the template during the heat treatment process. Although nicely achieved, the main drawback of this synthetical method is the drastic and highly toxic acidic condition (concentrated HF solution) required to remove the templating SiO$_2$ agent.

Nanostructured IrO$_2$ films have also been prepared employing an ordered supramolecular gel phase generated by an organometallic Ir(III) complex, used both as templating agent and metal source. Indeed, several Ir(III) complexes can self-assemble in highly organized supramolecular phases in water, including physical gels and lyotropic liquid-crystalline gels [120–122]. Moreover, many metal oxide (SiO$_2$, TiO$_2$, ZrO$_2$, ZnO and WO$_3$) nanostructures, such as nanotubes, nanoparticles, and nanowires, have been efficiently prepared taking advantage of a supramolecular gel as structural directing agent (SDA) [123]. In this context, the supramolecular gel phase of the Ir(III) compound [(ppy)$_2$Ir(bpy)]EtOCH$_2$CO$_2$ (chemical structure reported in Figure 20), where ppy is 2-phenylpiridine and bpy is 2,2'-bipyridine, which supramolecular architecture in water is built on a double 2D columnar system, was used as template and metal precursor for the preparation of IrO$_2$ films [122]. The highly ordered metallogel was deposited onto quartz substrates through spin-coating and was left to dry, obtaining the corresponding xerogel, that was calcinated at 600 °C for 4 h, obtaining a uniform IrO$_2$ thin film. As shown in Figure 19, the film prepared starting from the 5% w/w gel phase is composed of ordered vertical IrO$_2$ arrays that outline its nanostructure, whereas in the case of the 6% w/w gel phase, a self-assembled well-ordered multilayer thin film can be observed.

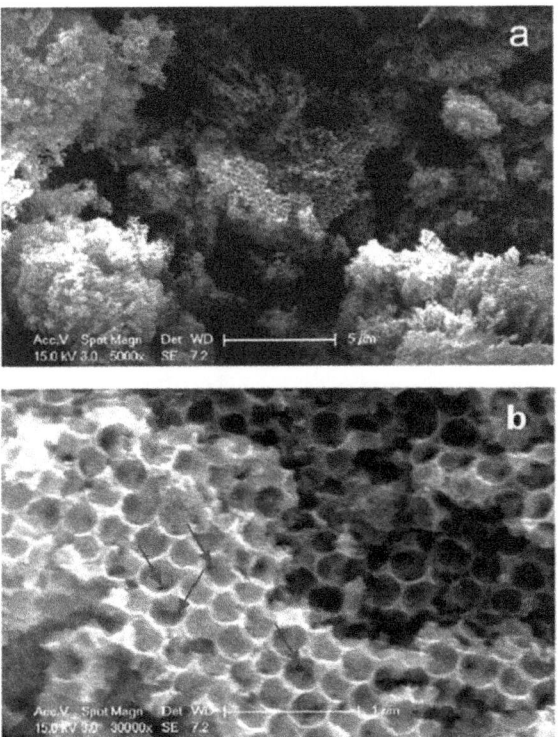

Figure 19. (**a**) Low and (**b**) high magnification SEM images of macroporous IrO_2 prepared by the hard-template method (Reprinted/adapted with permission from Ref. [98]).

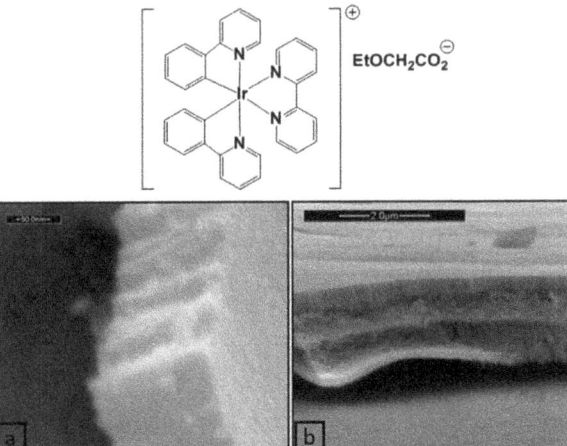

Figure 20. Chemical structure of $[(ppy)_2Ir(bpy)]EtOCH_2CO_2$ and SEM images of the IrO_2 film prepared starting from its 5% w/w (**a**) and 6% w/w (**b**) gel phase (Reprinted/adapted with permission from Ref. [122]).

Although this study was at its early stage, it clearly shows the possibility of using self-ordered lyotropic Ir(III) complexes for the production of ordered nanostructured thin films of IrO_2, opening a novel alternative route for their preparation.

6. Conclusions

Despite its intrinsic higher cost with respect to other semiconductive metal oxides, IrO_2 does present appealing characteristics that make it the ideal candidate for specific applications. IrO_2 is indeed the only active OER catalyst that is relatively stable in the acidic condition, which is a prerequisite for successful integration with photoanodes to reach optimal photoelectrochemical cells efficiency. As reviewed herein, IrO_2 can be obtained in various types of nanostructured forms allowing the boosting of its performances through mainly the increase of the active surface area owing to the nanoscale architecture. However, to reach such an increment in properties, severe experimental conditions are often required, specific templates or definite substrates must be employed, and a dedicated experimental set-up may also need to be designed. All these factors, of course, will further increase the effective cost of the active nanostructured metal-oxide. Efforts therefore must still be addressed in order to find more sustainable and environmentally friendly access to nanostructured IrO_2-based materials.

Author Contributions: Conceptualization, F.S. and N.G.; writing—original draft preparation, F.S.; reprint permission requests, I.A.; writ-ing—review and editing N.G., A.C. and I.A. All authors have read and agreed to the published version of the manuscript.

Funding: This research was supported by the project PON "Ricerca e Innovazione" 2014–2020–STAR 2–PIR01_00008 funded by MIUR (Ministero dell'Università e della Ricerca). FS is grateful to the project PON "Ricerca e Innovazione" 2014–2020, Asse IV "Istruzione e ricerca per il recupero," and Azione IV.6 "Contratti di ricerca su tematiche Green".

Conflicts of Interest: The authors declare no conflict of interest.

References

1. Rogers, D.B.; Shannon, R.D.; Sleight, A.W.; Gillson, J.L. Crystal Chemistry of Metal Dioxides with Rutile-Related Structures. *Inorg. Chem.* **1969**, *8*, 841–849. [CrossRef]
2. Cambridge Crystallographic Data Centre (CCDC), ICSD 640885, deposition number: 1759474. Available online: https://www.ccdc.cam.ac.uk/ (accessed on 2 July 2022).
3. Schultze, J.W.; Trasatti, S. (Eds.) Electrodes of Conductive Metallic Oxides, Part B. Elsevier Scientific Publishing Company, Amsterdam/New York 1981. 702 Seiten, Preis: US $ 83.00/Dfl 170.00. In *Berichte der Bunsengesellschaft für Physikalische Chemie*; John and Wiley and Sons: Hoboken, NJ, USA, 1981; Volume 85, p. 1085. [CrossRef]
4. Riga, J.; Tenret-Noël, C.; Pireaux, J.J.; Caudano, R.; Verbist, J.J.; Gobillon, Y. Electronic Structure of Rutile Oxides TiO_2, RuO_2 and IrO_2 Studied by X-ray Photoelectron Spectroscopy. *Phys. Scr.* **1977**, *16*, 351–354. [CrossRef]
5. Chen, R.S.; Huang, Y.S.; Liang, Y.M.; Tsai, D.S.; Tiong, K.K. Growth and Characterization of Iridium Dioxide Nanorods. *J. Alloys Compd.* **2004**, *383*, 273–276. [CrossRef]
6. Sheng, X.; Li, Z.; Cheng, Y. Electronic and Thermoelectric Properties of V_2O_5, MgV_2O_5, and CaV_2O_5. *Coatings* **2020**, *10*, 453. [CrossRef]
7. Liu, Y.; Masumoto, H.; Goto, T. Preparation of IrO_2 Thin Films by Oxydating Laser-ablated Ir. *Mater. Trans.* **2004**, *45*, 900. [CrossRef]
8. Greiner, M.T.; Lu, Z.-H. Thin-Film Metal Oxides in Organic Semiconductor Devices: Their Electronic Structures, Work Functions and Interfaces. *NPG Asia Mater.* **2013**, *5*, e55. [CrossRef]
9. Xu, B.; Sohn, H.Y.; Mohassab, Y.; Lan, Y. Structures, Preparation and Applications of Titanium Suboxides. *RSC Adv.* **2016**, *6*, 79706–79722. [CrossRef]
10. Yoo, S.-J.; Chang, J.-H.; Lee, J.-H.; Moon, C.-K.; Wu, C.-I.; Kim, J.-J. Formation of Perfect Ohmic Contact at Indium Tin Oxide/N,N'-Di(Naphthalene-1-Yl)-N,N'-Diphenyl-Benzidine Interface Using ReO_3. *Sci. Rep.* **2015**, *4*, 3902. [CrossRef]
11. Pearsall, T.P.; Lee, C.A. Electronic Transport in ReO_3: Dc Conductivity and Hall Effect. *Phys. Rev. B* **1974**, *10*, 2190–2194. [CrossRef]
12. Ben-Dor, L.; Shimony, Y. Crystal Structure, Magnetic Susceptibility and Electrical Conductivity of Pure and NiO-Doped MoO_2 and WO_2. *Mater. Res. Bull.* **1974**, *9*, 837–844. [CrossRef]
13. Lakshminarayana, G.; Kityk, I.V.; Nagao, T. Synthesis, Structural, and Electrical Characterization of RuO_2 Sol–Gel Spin-Coating Nano-Films. *J. Mater. Sci. Mater. Electron.* **2016**, *27*, 10791–10797. [CrossRef]

14. Murakami, Y.; Li, J.; Shimoda, T. Highly Conductive Ruthenium Oxide Thin Films by a Low-Temperature Solution Process and Green Laser Annealing. *Mater. Lett.* **2015**, *152*, 121–124. [CrossRef]
15. Wang, G.; Zheng, J.; Xu, B.; Zhang, C.; Zhu, Y.; Fang, Z.; Yang, Z.; Shang, M.-H.; Yang, W. Tailored Electronic Band Gap and Valance Band Edge of Nickel Oxide via P-Type Incorporation. *J. Phys. Chem. C* **2021**, *125*, 7495–7501. [CrossRef]
16. Jundale, D.M.; Joshi, P.B.; Sen, S.; Patil, V.B. Nanocrystalline CuO Thin Films: Synthesis, Microstructural and Optoelectronic Properties. *J. Mater. Sci. Mater. Electron.* **2012**, *23*, 1492–1499. [CrossRef]
17. Johan, M.R.; Suan, M.S.M.; Hawari, N.L.; Ching, H.A. Annealing Effects on the Properties of Copper Oxide Thin Films Prepared by Chemical Deposition. *Int. J. Electrochem. Sci.* **2011**, *6*, 6094–6104.
18. Chang, X.; Wang, T.; Zhang, P.; Zhang, J.; Li, A.; Gong, J. Enhanced Surface Reaction Kinetics and Charge Separation of p–n Heterojunction Co_3O_4/$BiVO_4$ Photoanodes. *J. Am. Chem. Soc.* **2015**, *137*, 8356–8359. [CrossRef]
19. Savio, A.K.P.D.; Fletcher, J.; Smith, K.; Iyer, R.; Bao, J.M.; Robles Hernández, F.C. Environmentally Effective Photocatalyst CoO–TiO_2 Synthesized by Thermal Precipitation of Co in Amorphous TiO_2. *Appl. Catal. B Environ.* **2016**, *182*, 449–455. [CrossRef]
20. Scarpelli, F.; Mastropietro, T.F.; Poerio, T.; Godbert, N. Mesoporous TiO_2 Thin Films: State of the Art. In *Titanium Dioxide—Material for a Sustainable Environment*; InTech: London, UK, 2018.
21. Carcia, P.F.; McCarron, E.M. Synthesis and Properties of Thin Film Polymorphs of Molybdenum Trioxide. *Thin Solid Films* **1987**, *155*, 53–63. [CrossRef]
22. González-Borrero, P.P.; Sato, F.; Medina, A.N.; Baesso, M.L.; Bento, A.C.; Baldissera, G.; Persson, C.; Niklasson, G.A.; Granqvist, C.G.; Ferreira da Silva, A. Optical Band-Gap Determination of Nanostructured WO_3 Film. *Appl. Phys. Lett.* **2010**, *96*, 061909. [CrossRef]
23. Feucht, D.L. Heterojunctions in Photovoltaic Devices. *J. Vac. Sci. Technol.* **1977**, *14*, 57–64. [CrossRef]
24. Batzill, M.; Diebold, U. The Surface and Materials Science of Tin Oxide. *Prog. Surf. Sci.* **2005**, *79*, 47–154. [CrossRef]
25. Pan, C.A.; Ma, T.P. Work Function of In_2O_3 Film as Determined from Internal Photoemission. *Appl. Phys. Lett.* **1980**, *37*, 714–716. [CrossRef]
26. Weiher, R.L.; Ley, R.P. Optical Properties of Indium Oxide. *J. Appl. Phys.* **1966**, *37*, 299–302. [CrossRef]
27. Wei, M.; Li, C.-F.; Deng, X.-R.; Deng, H. Surface Work Function of Transparent Conductive ZnO Films. *Energy Procedia* **2012**, *16*, 76–80. [CrossRef]
28. Srikant, V.; Clarke, D.R. On the Optical Band Gap of Zinc Oxide. *J. Appl. Phys.* **1998**, *83*, 5447–5451. [CrossRef]
29. Sotiropoulou, D.; Ladas, S. The Growth of Ultrathin Films of Copper on Polycrystalline ZrO_2. *Surf. Sci.* **2000**, *452*, 58–66. [CrossRef]
30. Meyer, J.; Zilberberg, K.; Riedl, T.; Kahn, A. Electronic Structure of Vanadium Pentoxide: An Efficient Hole Injector for Organic Electronic Materials. *J. Appl. Phys.* **2011**, *110*, 033710. [CrossRef]
31. Trastoy, J.; Kalcheim, Y.; del Valle, J.; Valmianski, I.; Schuller, I.K. Enhanced Metal–Insulator Transition in V_2O_3 by Thermal Quenching after Growth. *J. Mater. Sci.* **2018**, *53*, 9131–9137. [CrossRef]
32. Li, J. *Oxygen Evolution Reaction in Energy Conversion and Storage: Design Strategies under and Beyond the Energy Scaling Relationship*; Springer Nature Singapore: Singapore, 2022; Volume 14, ISBN 4082002200.
33. Suen, N.T.; Hung, S.F.; Quan, Q.; Zhang, N.; Xu, Y.J.; Chen, H.M. Electrocatalysis for the Oxygen Evolution Reaction: Recent Development and Future Perspectives. *Chem. Soc. Rev.* **2017**, *46*, 337–365. [CrossRef]
34. Ali, I.; AlGhamdi, K.; Al-Wadaani, F.T. Advances in Iridium Nano Catalyst Preparation, Characterization and Applications. *J. Mol. Liq.* **2019**, *280*, 274–284. [CrossRef]
35. Naito, T.; Shinagawa, T.; Nishimoto, T.; Takanabe, K. Recent Advances in Understanding Oxygen Evolution Reaction Mechanisms over Iridium Oxide. *Inorg. Chem. Front.* **2021**, *8*, 2900–2917. [CrossRef]
36. Nabor, G.S.; Hapiot, P.; Neta, P.; Harriman, A. Changes in the Redox State of Iridium Oxide Clusters and Their Relation to Catalytic Water Oxidation. Radiolytic and Electrochemical Studies. *J. Phys. Chem.* **1991**, *95*, 616–621. [CrossRef]
37. Sanchezcasalongue, H.G.; Ng, M.L.; Kaya, S.; Friebel, D.; Ogasawara, H.; Nilsson, A. InSitu Observation of Surface Species on Iridium Oxide Nanoparticles during the Oxygen Evolution Reaction. *Angew. Chem.-Int. Ed.* **2014**, *53*, 7169–7172. [CrossRef] [PubMed]
38. Abe, Y.; Ito, S.; Kim, K.H.; Kawamura, M.; Kiba, T. Electrochromic Properties of Sputtered Iridium Oxide Thin Films with Various Film Thicknesses. *J. Mater. Sci. Res.* **2016**, *6*, 44. [CrossRef]
39. Gottesfeld, S.; McIntyre, J.D.E. Electrochromism in Anodic Iridium Oxide Films: II. PH Effects on Corrosion Stability and the Mechanism of Coloration and Bleaching. *J. Electrochem. Soc.* **1979**, *126*, 742–750. [CrossRef]
40. Gottesfeld, S.; McIntyre, J.D.E.; Beni, G.; Shay, J.L. Electrochromism in Anodic Iridium Oxide Films. *Appl. Phys. Lett.* **1978**, *33*, 208–210. [CrossRef]
41. Dautremont-Smith, W.C. Transition Metal Oxide Electrochromic Materials and Displays: A Review. Part 2: Oxides with Anodic Coloration. *Displays* **1982**, *3*, 67–80. [CrossRef]
42. Shay, J.L.; Beni, G.; Schiavone, L.M. Electrochromism of Anodic Iridium Oxide Films on Transparent Substrates. *Appl. Phys. Lett.* **1978**, *33*, 942–944. [CrossRef]
43. Ito, S.; Abe, Y.; Kawamura, M.; Kim, K.H. Electrochromic Properties of Iridium Oxide Thin Films Prepared by Reactive Sputtering in O_2 or H_2O Atmosphere. *J. Vac. Sci. Technol. B Nanotechnol. Microelectron. Mater. Process. Meas. Phenom.* **2015**, *33*, 041204. [CrossRef]

44. Yamanaka, K. Anodically Electrodeposited Iridium Oxide Films (AEIROF) from Alkaline Solutions for Electrochromic Display Devices. *Jpn. J. Appl. Phys.* **1989**, *28*, 632–637. [CrossRef]
45. Nguyen, C.M.; Huang, W.D.; Rao, S.; Cao, H.; Tata, U.; Chiao, M.; Chiao, J.C. Sol-Gel Iridium Oxide-Based PH Sensor Array on Flexible Polyimide Substrate. *IEEE Sens. J.* **2013**, *13*, 3857–3864. [CrossRef]
46. Ges, I.A.; Ivanov, B.L.; Werdich, A.A.; Baudenbacher, F.J. Differential PH Measurements of Metabolic Cellular Activity in Nl Culture Volumes Using Microfabricated Iridium Oxide Electrodes. *Biosens. Bioelectron.* **2007**, *22*, 1303–1310. [CrossRef] [PubMed]
47. Kuo, L.M.; Chou, Y.C.; Chen, K.N.; Lu, C.C.; Chao, S. A Precise PH Microsensor Using RF-Sputtering IrO_2 and Ta_2O_5 Films on Pt-Electrode. *Sens. Actuators B Chem.* **2014**, *193*, 687–691. [CrossRef]
48. Lee, Y.; Kang, M.; Shim, J.H.; Lee, N.S.; Baik, J.M.; Lee, Y.; Lee, C.; Kim, M.H. Growth of Highly Single Crystalline IrO_2 Nanowires and Their Electrochemical Applications. *J. Phys. Chem. C* **2012**, *116*, 18550–18556. [CrossRef]
49. Hitchman, M.L.; Ramanathan, S. Thermally Grown Iridium Oxide Electrodes for PH Sensing in Aqueous Environments at 0 and 95 °C. *Anal. Chim. Acta* **1992**, *263*, 53–61. [CrossRef]
50. Dobson, J.V.; Snodin, P.R.; Thirsk, H.R. EMF Measurements of Cells Employing Metal—Metal Oxide Electrodes in Aqueous Chloride and Sulphate Electrolytes at Temperatures between 25–250 °C. *Electrochim. Acta* **1976**, *21*, 527–533. [CrossRef]
51. Bordi, S.; Carlá, M.; Papeschi, G.; Pinzauti, S. Iridium/Iridium Oxide Electrode for Potentiometric Determination of Proton Activity in Hydroorganic Solutions at Sub-Zero Temperatures. *Anal. Chem.* **1984**, *56*, 317–319. [CrossRef]
52. O'Hare, D.; Parker, K.H.; Winlove, C.P. Metal-Metal Oxide PH Sensors for Physiological Application. *Med. Eng. Phys.* **2006**, *28*, 982–988. [CrossRef]
53. Bezbaruah, A.N.; Zhang, T.C. Fabrication of Anodically Electrodeposited Iridium Oxide Film PH Microelectrodes for Microenvironmental Studies. *Anal. Chem.* **2002**, *74*, 5726–5733. [CrossRef]
54. Gláb, S.; Hulanicki, A.; Edwall, G.; Folke, F.; Ingman, I.; Koch, W.F. Metal-Metal Oxide and Metal Oxide Electrodes as PH Sensors. *Crit. Rev. Anal. Chem.* **1989**, *21*, 29–47. [CrossRef]
55. Dong, Q.; Sun, X.; He, S. Iridium Oxide Enabled Sensors Applications. *Catalysts* **2021**, *11*, 1164. [CrossRef]
56. Zhang, F.; Ulrich, B.; Reddy, R.K.; Venkatraman, V.L.; Prasad, S.; Vu, T.Q.; Hsu, S.T. Fabrication of Submicron IrO_2 Nanowire Array Biosensor Platform by Conventional Complementary Metal-Oxide-Semiconductor Process. *Jpn. J. Appl. Phys.* **2008**, *47*, 1147–1151. [CrossRef]
57. Quesada-González, D.; Sena-Torralba, A.; Wicaksono, W.P.; de la Escosura-Muñiz, A.; Ivandini, T.A.; Merkoçi, A. Iridium Oxide (IV) Nanoparticle-Based Lateral Flow Immunoassay. *Biosens. Bioelectron.* **2019**, *132*, 132–135. [CrossRef] [PubMed]
58. Zhang, H.; Zhang, L.X.; Zhong, H.; Niu, S.; Ding, C.; Lv, S. Iridium Oxide Nanoparticles-Based Theranostic Probe for in Vivo Tumor Imaging and Synergistic Chem/Photothermal Treatments of Cancer Cells. *Chem. Eng. J.* **2022**, *430*, 132675. [CrossRef]
59. Yuan, X.; Cen, J.; Chen, X.; Jia, Z.; Zhu, X.; Huang, Y.; Yuan, G. Liu, J. Iridium Oxide Nanoparticles Mediated Enhanced Photodynamic Therapy Combined with Photothermal Therapy in the Treatment of Breast Cancer. *J. Colloid Interface Sci.* **2022**, *605*, 851–862. [CrossRef]
60. Buchanan, R.A.; Lee, I.-S.; Williams, J.M. Surface Modification of Biomaterials through Noble Metal Ion Implantation. *J. Biomed. Mater. Res.* **1990**, *24*, 309–318. [CrossRef]
61. Cogan, S.F.; Ehrlich, J.; Plante, T.D.; Smirnov, A.; Shire, D.B.; Gingerich, M.; Rizzo, J.F. Sputtered Iridium Oxide Films for Neural Stimulation Electrodes. *J. Biomed. Mater. Res. Part B Appl. Biomater.* **2009**, *89B*, 353–361. [CrossRef]
62. Robblee, L.S.; Mangaudis, M.J.; Lasinsky, E.D.; Kimball, A.G.; Brummer, S.B. Charge Injection Properties of Thermally-Prepared Iridium Oxide Films. *MRS Proc.* **1985**, *55*, 303. [CrossRef]
63. Chalamala, B.R.; Wei, Y.; Reuss, R.H.; Aggarwal, S.; Gnade, B.E.; Ramesh, R.; Bernhard, J.M.; Sosa, E.D.; Golden, D.E. Effect of Growth Conditions on Surface Morphology and Photoelectric Work Function Characteristics of Iridium Oxide Thin Films. *Appl. Phys. Lett.* **1999**, *74*, 1394–1396. [CrossRef]
64. Chalamala, B.R.; Wei, Y.; Reuss, R.H.; Aggarwal, S.; Perusse, S.R.; Gnade, B.E.; Ramesh, R. Stability and Chemical Composition of Thermally Grown Iridium-Oxide Thin Films. *J. Vac. Sci. Technol. B Microelectron. Nanom. Struct.* **2000**, *18*, 1919. [CrossRef]
65. Chalamala, B.R.; Reuss, R.H.; Dean, K.A.; Sosa, E.; Golden, D.E. Field Emission Characteristics of Iridium Oxide Tips. *J. Appl. Phys.* **2002**, *91*, 6141–6146. [CrossRef]
66. Kubaschewski, O.; Hopkins, B.E. *Oxidation of Metals and Alloys*, 2nd ed.; Butterworths: London, UK, 1962.
67. Baglio, V.; Sebastián, D.; D'Urso, C.; Stassi, A.; Amin, R.S.; El-Khatib, K.M.; Aricò, A.S. Composite Anode Electrode Based on Iridium Oxide Promoter for Direct Methanol Fuel Cells. *Electrochim. Acta* **2014**, *128*, 304–310. [CrossRef]
68. Chen, Y.M.; Cai, J.H.; Huang, Y.S.; Lee, K.Y.; Tsai, D.S.; Tiong, K.K. A Nanostructured Electrode of IrO x Foil on the Carbon Nanotubes for Supercapacitors. *Nanotechnology* **2011**, *22*, 355708. [CrossRef] [PubMed]
69. Slavcheva, E.; Vitushinsky, R.; Mokwa, W.; Schnakenberg, U. Sputtered Iridium Oxide Films as Charge Injection Material for Functional Electrostimulation. *J. Electrochem. Soc.* **2004**, *151*, E226. [CrossRef]
70. Patil, P.S.; Kawar, R.K.; Sadale, S.B. Electrochromism in Spray Deposited Iridium Oxide Thin Films. *Electrochim. Acta* **2005**, *50*, 2527–2532. [CrossRef]
71. Cui, Z.; Qi, R. First-Principles Simulation of Oxygen Evolution Reaction (OER) Catalytic Performance of IrO_2 Bulk-like Structures: Nanosphere, Nanowire and Nanotube. *Appl. Surf. Sci.* **2021**, *554*, 149591. [CrossRef]
72. Chen, Z.; Duan, X.; Wei, W.; Wang, S.; Ni, B.J. Iridium-Based Nanomaterials for Electrochemical Water Splitting. *Nano Energy* **2020**, *78*, 105270. [CrossRef]

73. González, D.; Sodupe, M.; Rodríguez-Santiago, L.; Solans-Monfort, X. Surface Morphology Controls Water Dissociation on Hydrated IrO$_2$ nanoparticles. *Nanoscale* **2021**, *13*, 14480–14489. [CrossRef]
74. Chandra, D.; Abe, N.; Takama, D.; Saito, K.; Yui, T.; Yagi, M. Open Pore Architecture of an Ordered Mesoporous IrO$_2$ Thin Film for Highly Efficient Electrocatalytic Water Oxidation. *ChemSusChem* **2015**, *8*, 795–799. [CrossRef]
75. Lim, J.; Park, D.; Jeon, S.S.; Roh, C.W.; Choi, J.; Yoon, D.; Park, M.; Jung, H.; Lee, H. Ultrathin IrO$_2$ Nanoneedles for Electrochemical Water Oxidation. *Adv. Funct. Mater.* **2018**, *28*, 1704796. [CrossRef]
76. Nguyen, C.M.; Gurung, I.; Cao, H.; Rao, S.; Chiao, J.C. Fabrication of PH-Sensing Iridium Oxide Nanotubes on Patterned Electrodes Using Anodic Aluminum Oxide Nanotemplate. *Proc. IEEE Sensors 2013*, **2013**, 1–4. [CrossRef]
77. Khalil, M.; Liu, N.; Lee, R.L. Super-Nernstian Potentiometric PH Sensor Based on the Electrodeposition of Iridium Oxide Nanoparticles. *Int. J. Technol.* **2018**, *9*, 446–454. [CrossRef]
78. Beknalkar, S.A.; Teli, A.M.; Harale, N.S.; Patil, D.S.; Pawar, S.A.; Shin, J.C.; Patil, P.S. Fabrication of High Energy Density Supercapacitor Device Based on Hollow Iridium Oxide Nanofibers by Single Nozzle Electrospinning. *Appl. Surf. Sci.* **2021**, *546*, 149102. [CrossRef]
79. Park, T.J.; Jeong, D.S.; Hwang, C.S.; Park, M.S.; Kang, N.-S. Fabrication of Ultrathin IrO$_2$ Top Electrode for Improving Thermal Stability of Metal–Insulator–Metal Field Emission Cathodes. *Thin Solid Films* **2005**, *471*, 236–242. [CrossRef]
80. Chen, R.-S.; Huang, Y.-S.; Liang, Y.-M.; Hsieh, C.-S.; Tsai, D.-S.; Tiong, K.-K. Field Emission from Vertically Aligned Conductive IrO$_2$ Nanorods. *Appl. Phys. Lett.* **2004**, *84*, 1552–1554. [CrossRef]
81. Gleiter, H. Nanostructured Materials: Basic Concepts and Microstructure. *Acta Mater.* **2000**, *48*, 1–29. [CrossRef]
82. Argon, A.S.; Yip, S. The Strongest Size. *Philos. Mag. Lett.* **2006**, *86*, 713–720. [CrossRef]
83. Huang, C.; Chen, X.; Xue, Z.; Wang, T. Effect of Structure: A New Insight into Nanoparticle Assemblies from Inanimate to Animate. *Sci. Adv.* **2020**, *6*, eaba1321. [CrossRef]
84. Guo, D.; Xie, G.; Luo, J. Mechanical Properties of Nanoparticles: Basics and Applications. *J. Phys. D Appl. Phys.* **2014**, *47*, 013001. [CrossRef]
85. Khan, I.; Saeed, K.; Khan, I. Nanoparticles: Properties, Applications and Toxicities. *Arab. J. Chem.* **2019**, *12*, 908–931. [CrossRef]
86. Rafique, M.; Tahir, M.B.; Rafique, M.S.; Safdar, N.; Tahir, R. *Nanostructure Materials and Their Classification by Dimensionality*; Elsevier: Amsterdam, The Netherlands, 2020; ISBN 9780128211922.
87. Bizzotto, F.; Quinson, J.; Schröder, J.; Zana, A.; Arenz, M. Surfactant-Free Colloidal Strategies for Highly Dispersed and Active Supported IrO$_2$ Catalysts: Synthesis and Performance Evaluation for the Oxygen Evolution Reaction. *J. Catal.* **2021**, *401*, 54–62. [CrossRef]
88. Takimoto, D.; Fukuda, K.; Miyasaka, S.; Ishida, T.; Ayato, Y.; Mochizuki, D.; Shimizu, W.; Sugimoto, W. Synthesis and Oxygen Electrocatalysis of Iridium Oxide Nanosheets. *Electrocatalysis* **2017**, *8*, 144–150. [CrossRef]
89. Nguyen, C.M.; Rao, S.; Chiao, J.C.; Cao, H.; Li, A.; Peng, Y.B. Miniature Neurotransmitter Sensors Featured with Iridium Oxide Nanorods. *Sensors 2014*, **2014**, 1869–1872. [CrossRef]
90. Tao, Y.; Pan, Z.; Ruch, T.; Zhan, X.; Chen, Y.; Zhang, S.X.; Li, D. Remarkable Suppression of Lattice Thermal Conductivity by Electron-Phonon Scattering in Iridium Dioxide Nanowires. *Mater. Today Phys.* **2021**, *21*, 100517. [CrossRef]
91. Ortel, E.; Reier, T.; Strasser, P.; Kraehnert, R. Mesoporous IrO$_2$ Films Templated by PEO-PB-PEO Block-Copolymers: Self-Assembly, Crystallization Behavior, and Electrocatalytic Performance. *Chem. Mater.* **2011**, *23*, 3201–3209. [CrossRef]
92. Hu, J.; Abdelsalam, M.; Bartlett, P.; Cole, R.; Sugawara, Y.; Baumberg, J.; Mahajan, S.; Denuault, G. Electrodeposition of Highly Ordered Macroporous Iridium Oxide through Self-Assembled Colloidal Templates. *J. Mater. Chem.* **2009**, *19*, 3855–3858. [CrossRef]
93. Quinson, J. Iridium and IrOx Nanoparticles: An Overview and Review of Syntheses and Applications. *Adv. Colloid Interface Sci.* **2022**, *303*, 102643. [CrossRef]
94. Marshall, A.; Børresen, B.; Hagen, G.; Tsypkin, M.; Tunold, R. Preparation and Characterisation of Nanocrystalline Ir$_x$Sn$_{1-x}$O$_2$ Electrocatalytic Powders. *Mater. Chem. Phys.* **2005**, *94*, 226–232. [CrossRef]
95. Bonet, F.; Delmas, V.; Grugeon, S.; Herrera Urbina, R.; Silvert, P.Y.; Tekaia-Elhsissen, K. Synthesis of Monodisperse Au, Pt, Pd, Ru and Ir Nanoparticles in Ethylene Glycol. *Nanostructured Mater.* **1999**, *11*, 1277–1284. [CrossRef]
96. Adams, R.; Shriner, R.L. Platinum oxide as a catalyst in the reduction of organic compounds. iii. preparation and properties of the oxide of platinum obtained by the fusion of chloroplatinic acid with sodium nitrate. *J. Am. Chem. Soc.* **1923**, *45*, 2171–2179. [CrossRef]
97. Cruz, J.C.; Baglio, V.; Siracusano, S.; Ornelas, R.; Ortiz-Frade, L.; Arriaga, L.G.; Antonucci, V.; Aricò, A.S. Nanosized IrO$_2$ Electrocatalysts for Oxygen Evolution Reaction in an SPE Electrolyzer. *J. Nanoparticle Res.* **2011**, *13*, 1639–1646. [CrossRef]
98. Hu, W.; Wang, Y.; Hu, X.; Zhou, Y.; Chen, S. Three-Dimensional Ordered Macroporous IrO$_2$ as Electrocatalyst for Oxygen Evolution Reaction in Acidic Medium. *J. Mater. Chem.* **2012**, *22*, 6010–6016. [CrossRef]
99. Zhang, J.; Liu, S.; Wang, H.; Xia, Q.; Huang, X. Polypyrrole Assisted Synthesis of Nanosized Iridium Oxide for Oxygen Evolution Reaction in Acidic Medium. *Int. J. Hydrog. Energy* **2020**, *45*, 33491–33499. [CrossRef]
100. Diaz, C.; Valenzuela, M.L.; Cifuentes-Vaca, O.; Segovia, M.; Laguna-Bercero, M.A. Iridium Nanostructured Metal Oxide, Its Inclusion in Silica Matrix and Their Activity toward Photodegradation of Methylene Blue. *Mater. Chem. Phys.* **2020**, *252*, 123276. [CrossRef]
101. Chen, P.C.; Chen, Y.C.; Huang, C.N. Free-Standing Iridium Oxide Nanotube Array for Neural Interface Electrode Applications. *Mater. Lett.* **2018**, *221*, 293–295. [CrossRef]

102. Yu, A.; Kwon, T.; Lee, C.; Lee, Y. Highly Catalytic Electrochemical Oxidation of Carbon Monoxide on Iridium Nanotubes: Amperometric Sensing of Carbon Monoxide. *Nanomaterials* **2020**, *10*, 1140. [CrossRef]
103. Chen, R.S.; Huang, Y.S.; Tsai, D.S.; Chattopadhyay, S.; Wu, C.T.; Lan, Z.H.; Chen, K.H. Growth of Well Aligned IrO_2 Nanotubes on $LiTaO_3$(012) Substrate. *Chem. Mater.* **2004**, *16*, 2457–2462. [CrossRef]
104. Chen, R.S.; Korotcov, A.; Huang, Y.S.; Tsai, D.S. One-Dimensional Conductive IrO_2 nanocrystals. *Nanotechnology* **2006**, *17*, R67–R87. [CrossRef]
105. Shan, C.C.; Tsai, D.S.; Huang, Y.S.; Jian, S.H.; Cheng, C.L. Pt-Ir-IrO_2NT Thin-Wall Electrocatalysts Derived from IrO_2 Nanotubes and Their Catalytic Activities in Methanol Oxidation. *Chem. Mater.* **2007**, *19*, 424–431. [CrossRef]
106. Ahmed, J.; Mao, Y. Ultrafine Iridium Oxide Nanorods Synthesized by Molten Salt Method toward Electrocatalytic Oxygen and Hydrogen Evolution Reactions. *Electrochim. Acta* **2016**, *212*, 686–693. [CrossRef]
107. Nguyen, C.M.; Thumthan, O.; Huang, C.; Tata, U.; Hao, Y.; Chiao, J.C. Chemical Bath Method to Grow Precipitated Nanorods of Iridium Oxide on Alumina Membranes. *Micro Nano Lett.* **2012**, *7*, 1256–1259. [CrossRef]
108. Mohan, S.; Gupta, S.K.; Mao, Y. Morphology-Oxygen Evolution Activity Relationship of Iridium(Iv) Oxide Nanomaterials. *New J. Chem.* **2022**, *46*, 3716–3726. [CrossRef]
109. Chen, R.S.; Huang, Y.S.; Liang, Y.M.; Tsai, D.S.; Chi, Y.; Kai, J.J. Growth Control and Characterization of Vertically Aligned IrO_2 Nanorods. *J. Mater. Chem.* **2003**, *13*, 2525–2529. [CrossRef]
110. Zhang, F.; Barrowcliff, R.; Stecker, G.; Pan, W.; Wang, D.; Hsu, S.T. Synthesis of Metallic Iridium Oxide Nanowires via Metal Organic Chemical Vapor Deposition. *Jpn. J. Appl. Phys. Part 2 Lett.* **2005**, *44*, L398–L401. [CrossRef]
111. Lee, J.; Yang, H.-S.; Lee, N.-S.; Kwon, O.; Shin, H.-Y.; Yoon, S.; Baik, J.M.; Seo, Y.-S.; Kim, M.H. Hierarchically Assembled 1-Dimensional Hetero-Nanostructures: Single Crystalline RuO_2 Nanowires on Electrospun IrO_2 Nanofibres. *CrystEngComm* **2013**, *15*, 2367. [CrossRef]
112. Kim, S.; Kim, Y.L.; Yu, A.; Lee, J.; Lee, S.C.; Lee, C.; Kim, M.H.; Lee, Y. Electrospun Iridium Oxide Nanofibers for Direct Selective Electrochemical Detection of Ascorbic Acid. *Sens. Actuators B Chem.* **2014**, *196*, 480–488. [CrossRef]
113. Deng, Q.; Sun, Y.; Wang, J.; Chang, S.; Ji, M.; Qu, Y.; Zhang, K.; Li, B. Boosting OER Performance of IrO_2 in Acid via Urchin-like Hierarchical-Structure Design. *Dalt. Trans.* **2021**, *50*, 6083–6087. [CrossRef]
114. Comstock, D.J.; Christensen, S.T.; Elam, J.W.; Pellin, M.J.; Hersam, M.C. Synthesis of Nanoporous Activated Iridium Oxide Films by Anodized Aluminum Oxide Templated Atomic Layer Deposition. *Electrochem. Commun.* **2010**, *12*, 1543–1546. [CrossRef]
115. Machado, A.E.H.; Borges, K.A.; Silva, T.A.; Santos, L.M.; Borges, M.F.; Machado, W.A.; Caixeta, B.P.; Oliveira, S.M.; Trovó, A.G.; Patrocínio, A.O.T. Applications of Mesoporous Ordered Semiconductor Materials—Case Study of TiO_2. In *Solar Radiation Applications*; InTech: Rijeka, Croatia, 2015.TiO2. In *Solar Radiation Applications*; InTech: Rijeka, Croatia, 2015.
116. Bernicke, M.; Ortel, E.; Reier, T.; Bergmann, A.; Ferreira de Araujo, J.; Strasser, P.; Kraehnert, R. Iridium Oxide Coatings with Templated Porosity as Highly Active Oxygen Evolution Catalysts: Structure-Activity Relationships. *ChemSusChem* **2015**, *8*, 1908–1915. [CrossRef]
117. Chandra, D.; Sato, T.; Tanahashi, Y.; Takeuchi, R.; Yagi, M. Facile Fabrication and Nanostructure Control of Mesoporous Iridium Oxide Films for Efficient Electrocatalytic Water Oxidation. *Energy* **2019**, *173*, 278–289. [CrossRef]
118. Brinker, C.J.; Lu, Y.; Sellinger, A.; Fan, H. ChemInform Abstract: Evaporation-Induced Self-Assembly: Nanostructures Made Easy. *Adv. Mater.* **1999**, *11*, 579. [CrossRef]
119. Grosso, D.; Cagnol, F.; Soler-Illia, G.; Crepaldi, E.L.; Amenitsch, H.; Brunet-Bruneau, A.; Bourgeois, A.; Sanchez, C. Fundamentals of Mesostructuring through Evaporation-Induced Self-Assembly. *Adv. Funct. Mater.* **2004**, *14*, 309–322. [CrossRef]
120. Scarpelli, F.; Ricciardi, L.; La Deda, M.; Brunelli, E.; Crispini, A.; Ghedini, M.; Godbert, N.; Aiello, I. A Luminescent Lyotropic Liquid-Crystalline Gel of a Water-Soluble Ir(III) Complex. *J. Mol. Liq.* **2021**, *334*, 116187. [CrossRef]
121. Scarpelli, F.; Ionescu, A.; Ricciardi, L.; Plastina, P.; Aiello, I.; La Deda, M.; Crispini, A.; Ghedini, M.; Godbert, N. A Novel Route towards Water-Soluble Luminescent Iridium(III) Complexes: Via a Hydroxy-Bridged Dinuclear Precursor. *Dalt. Trans.* **2016**, *45*, 17264–17273. [CrossRef]
122. Scarpelli, F.; Ionescu, A.; Aiello, I.; La Deda, M.; Crispini, A.; Ghedini, M.; Brunelli, E.; Sesti, S.; Godbert, N. High Order in a Self-Assembled Iridium(III) Complex Gelator Towards Nanostructured IrO_2 Thin Films. *Chem.-Asian J.* **2017**, *12*, 2703–2710. [CrossRef] [PubMed]
123. Llusar, M.; Sanchez, C. Inorganic and Hybrid Nanofibrous Materials Templated with Organogelators. *Chem. Mater.* **2008**, *20*, 782–820. [CrossRef]

Review

A Comprehensive Study on the Applications of Clays into Advanced Technologies, with a Particular Attention on Biomedicine and Environmental Remediation

Roberto Nisticò

Department of Materials Science, University of Milano-Bicocca, Via R. Cozzi 55, 20125 Milano, Italy; roberto.nistico@unimib.it; Tel.: +39-02-6448-5111

Abstract: In recent years, a great interest has arisen around the integration of naturally occurring clays into a plethora of advanced technological applications, quite far from the typical fabrication of traditional ceramics. This "second (technological) life" of clays into fields of emerging interest is mainly due to clays' peculiar properties, in particular their ability to exchange (capture) ions, their layered structure, surface area and reactivity, and their biocompatibility. Since the maximization of clay performances/exploitations passes through the comprehension of the mechanisms involved, this review aims at providing a useful text that analyzes the main goals reached by clays in different fields coupled with the analysis of the structure-property correlations. After providing an introduction mainly focused on the economic analysis of clays global trading, clays are classified basing on their structural/chemical composition. The main relevant physicochemical properties are discussed (particular attention has been dedicated to the influence of interlayer composition on clay properties). Lastly, a deep analysis of the main relevant nonconventional applications of clays is presented. Several case studies describing the use of clays in biomedicine, environmental remediation, membrane technology, additive manufacturing, and sol-gel processes are presented, and results critically discussed.

Keywords: alumino-silicates; biomedicine; ceramics; clays; inorganic chemistry; environmental remediation; nanomaterials; porous materials

1. Introduction

Clays are naturally occurring 2D layered fine particles (less than 2 μm) extracted from earth and composed of phyllosilicates (clay minerals), showing plastic properties at appropriate water content and brittleness upon drying or firing (i.e., hardening processes) [1,2]. These very peculiar properties are strictly related to clays' chemical structure, and in particular to specific chemical reactions occurring at the surface/interface of these layered materials [3,4]. Such interfacial phenomena are the basis of the recent grown interest around clays (i.e., one of oldest class of materials handled and modeled by humans) as alternative (nanoscopic) materials. Scientific and engineering field of emerging interest where clays can be exploited such as biomedicine [5–9], environmental clean-up processes [10–18], membrane technologies [19–22], energetic and electronic applications [23–26], composite materials [27–33], sol-gel technology [34,35], 3D printing [36,37], and in developing traditional ceramics (i.e., earthenware, stoneware, porcelain, pottery, and so on) [38–40] are numerous.

The recent attention around the exploitation of 2D layered materials (e.g., graphene [41], and its derived graphene oxide (GO), carbon nitride (g-C_3N_4), molybdenum disulfide (MoS_2) [42], tungsten diselenide (WSe_2), tin sulfide (SnS_2), tin diselenide ($SnSe_2$), boron nitride (BN), and black phosphorous (P allotrope)) has also raised the interest concerning the exploitation of clays as naturally occurring layered materials. Furthermore, as pointed out in the literature [2,43,44], clays are largely diffused (almost worldwide in nature) in consistent amounts. The exploitation of clays guarantees a twofold advantage in terms of economic sustainability (i.e., low cost of raw materials) and environmental safety, coupled with their availability also in developing countries [45,46]. Therefore, the valorization of clays follows the principles at the basis of the circular economy [47–49], and analogously as in the cases of biomasses [50–54], biochars [55–59], and biofuels [60,61], clays can play a major role in the green chemistry-driven technological revolution.

From the economic viewpoint, actually clays represent raw materials of remarkable interest for the global market (i.e., clays accounted for 0.012% of total world trade in 2019). Countries leading in the extraction/exportation of clays are the US (USD 389 million, covering almost 18.2% of the global market), China (USD 260 million, 12.1%), Ukraine (USD 197 million, 9.2%), Germany (USD 166 million, 7.8%), India (USD 105 million, 4.9%), Turkey (USD 101 million, 4.7%), Spain (USD 93 million, 4.4%), and France (USD 84 million, 3.9%) [62]. Italy is the 7th largest European exporter with approximately USD 48 million (corresponding to approximately 2.2% of the global market). Interestingly, Italy is the largest global importer with a value of approximately 8.9% of the global market (ca. USD 191 million), followed by Germany (USD 180 million, 8.4%), the Netherlands (USD 142 million, 6.6%), Poland (USD 104 million, 4.9%), Japan (USD 95 million, 4.4%), and Canada (USD 93 million, 4.4%) [63]. Moreover, the growth of the previously cited nonconventional advanced applications is going to significantly increase the global trading of clays within the next few years.

The aim of this document is to provide a critical analysis of the potential nonconventional uses of clays into technological fields of emerging interest. Particular emphasis has been devoted to the chemical composition and structure of clays. In fact, it is with a correct understanding of the structure-mediated properties that it is possible to maximize the valorization of this class of naturally occurring materials and find novel, alternative (and advanced) applications.

2. Chemical Composition and Structure-Property Relationship of Clays

As stated in the previous paragraph, clays are fine-grained 2D earthen minerals. In general, the chemical composition of earthen crust is mainly silica (ca. 60%) and alumina (ca. 15%) with other mineral oxides constituting the residual fraction (approximately 25%) [2]. Hence, the chemical composition of clays reflects the Earth's crust, since clays are hydrous alumino-silicates (eventually containing some minor impurities, i.e., Na, K, Mg, Ca, Fe) [1,2]. Clays are made up of layered sheets organized as follows:

(i) 2D 1-1 layered clay minerals, formed by alternating layers made by one tetrahedral (Si-based) sheet and one octahedral (Al-based) sheet held together by hydrogen bonding [2,64].

(ii) 2D 2-1 layered clay minerals (organized into six subgroups), formed by alternating layers made by two tetrahedral (Si-based) sheets sandwiching one octahedral (Al-based) sheet in the middle. These sandwich structures are held together by an interlayer made by either water molecules or exchangeable cations (to maintain the electroneutrality of the system), or even both [2,64].

An extended classification of clays based on their chemical composition, structural stacking, and interlayer organization is summarized in Figure 1 [2]. In detail:

(i) Kaolinite-serpentine subgroup: 1-1 layered structures: This subgroup is characterized by electroneutral layered structures and the absence of cations at the interlayer (with very low CEC, 3–15 cmol/kg). In particular, surface charges of kaolinite derived from the presence of defects (such as isomorphic substitution and broken edges Al-O-Al and Si-O-Si). Tetrahedral and octahedral sheets are held together by either secondary forces (e.g., hydrogen bonding) or water molecules. The kaolinite-serpentine subgroup is characterized by the general chemical formulas $Al_2Si_2O_5(OH)_4$ (for the kaolinite subgroup) and $Mg_3Si_2O_5(OH)_4$ (for the serpentine subgroup) [65]. Examples of clays belonging to this subgroup are kaolinite, halloysite, dickite, nacrite, crysotile, antigorite, lizardite (the latter three belonging to the serpentine subgroup).

(i) Pyrophyllite-talc subgroup: 2-1 layered structures: This subgroup is characterized by electroneutral layered structures and the absence of cations at the interlayer (with very low CEC, below 1 cmol/kg). Weak secondary forces (e.g., van der Waals and/or dipolar interactions) that favor a loss of cohesion between the layers hold tetrahedral sheets together. The pyrophyllite-talc subgroup is characterized by the general chemical formulas $Al_2Si_4O_{10}(OH)_2$ (for the pyrophyllite subgroup) and $Mg_3Si_4O_{10}(OH)_2$ (for the talc subgroup) [66].

(ii) Smectite subgroup: 2-1 layered structures: This subgroup is characterized by having octahedral sheets partially substituted, thus generating weak negatively charged layers. In order to balance such negative charge and maintain the electroneutrality of the system, the smectites interlayer region contains miscellaneous cations, together with water molecules. The presence of such an interlayer (made by water molecules and cations) enhances the water affinity of smectites, thus favoring the hydraulic delamination and expansion. Furthermore, smectites show a high ion exchange capacity (i.e., CEC approximately 70–100 cmol/kg). The smectite subgroup is characterized by the chemical formula $(Na,Ca)_{0.33}(Al,Mg,Fe,Zn)_2Si_4O_{10}(OH)_2 \cdot nH_2O$ [67]. Examples of clays belonging to this subgroup are: montomorillonite, beidellite, laponite, saponite, and hectorite.

(iii) Vermiculite subgroup: 2-1 layered structures. This subgroup is characterized by having both tetrahedral and octahedral sheets partially substituted, thus generating a net negative charge in both layers. In order to balance such net negative charge and maintain the electroneutrality of the system, the vermiculites interlayer region contains two oriented water layers and magnesium cations, thus providing a limited expansion capacity and high ion exchange capacity (i.e., CEC approximately 100–150 cmol/kg). The vermiculite subgroup is characterized by the chemical formula $(Mg,Ca)_{0.3}(Mg,Fe)_3(Si,Al)_4O_{10}(OH)_2 \cdot 4H_2O)$ [68].

(iv) Mica subgroup: 2-1 layered structures: This subgroup is characterized by having the Si-based tetrahedral sheets partially substituted by aluminum atoms, thus generating a charge deficiency in the tetrahedral layers. In order to balance such strong negative charge and maintain the electroneutrality of the system, the micas interlayer region contains potassium cations occupying fixed positions at the tetrahedral sites surface. Such locked structure significantly limits the entry of water and micas ion exchange capacity (i.e., CEC approximately 10–40 cmol/kg). The mica subgroup is characterized by the chemical formula $(K,H)Al_2(Si,Al)_4O_{10}(OH)_2 \cdot nH_2O$ [69]. Examples of clays belonging to this subgroup are: muskovite, sericite, illite, biotite, and glauconite.

(v) Chlorites subgroup: 2-1 layered structures: This subgroup is characterized by having both tetrahedral and octahedral sheets partially substituted. In order to balance such negative charge and maintain the electroneutrality of the system, the chlorites interlayer region is made by hydroxide sheets, mainly constituted by brucite $Mg(OH)_2$, eventually partially substituted by iron atoms. The presence of hydroxyl functionalities at the interface between tetrahedral sheets and hydroxide-based interlayers induces the formation of hydrogen bonding that hold together the layered structures, generating a locked system characterized by having a poor ion exchange capacity (analogously as in the case of mica subgroup, namely: CEC approximately 10–40 cmol/kg). The chlorite subgroup is characterized by the chemical formula $(Mg,Fe)_3(Si,Al)_4O_{10}(OH)_2 \cdot ((Mg,Fe)_3(OH)_6)$ [70].

(vi) Inverted ribbons (palygorskite-sepiolite) subgroup: 2-1 layered structures: This subgroup is characterized by ribbons of 2-1 layered silicates presenting a periodic inversion of the apical oxygen atom in tetrahedral layers extending parallel to the layer directions, forming fibrous clays. These complex structures are characterized by the presence of nanometric channels (parallel-oriented respect to the direction of the layers) containing water molecules weakly bound to the magnesium ions forming the octahedral layers. The presence of these nanochannels guarantees high surface area (SSA higher than approximately 140–320 m^2/g) that allows their use as porous systems for advanced applications (e.g, controlled transport and/or release of chemicals, drug-delivery, separation science) [71]. The chemical formulas of palygorskite-sepiolite subgroup are the following, namely: $(Mg,Al)_2Si_4O_{10}OH \cdot 4H_2O$ (for the palygorskite subgroup), and $Mg_4Si_6O_{15}(OH)_2 \cdot 6H_2O$, (for the sepiolite subgroup) [72,73].

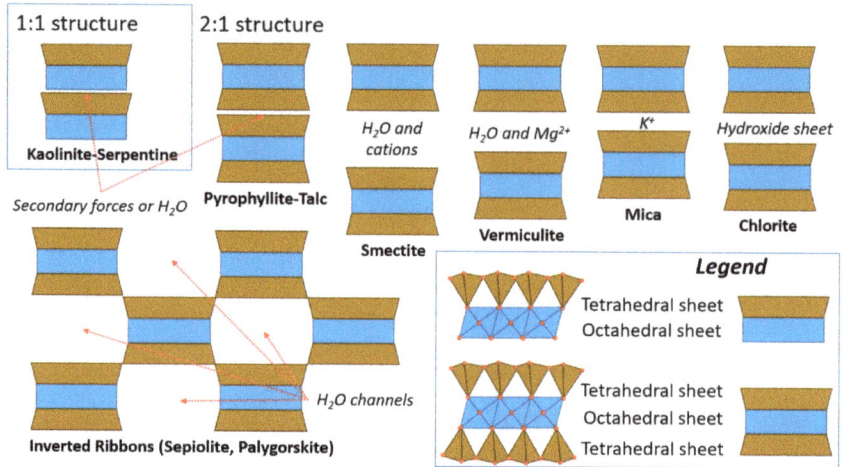

Figure 1. Schematic representation of 1:1 (inset) and 2:1 clay structures with a classification of the different subgroups of clays based on their stacking sequences. The difference in term of interlayers distance between the different subgroups of clays is comparable with the real one and depends on the species involved at the interface between the layered units. Reprinted with permission from [2]. Copyright 2018 IOP Publishing.

Furthermore, mixed-layer clays are also available in nature, but for the sake of simplicity, they are not discussed in this document (for details concerning mixed-layer clays, please refers to the following references [74,75]).

Clays belong to the ceramic materials class, characterized by having a crystalline organization with strong bonds (i.e., covalent bonds, and ionic bonds) between atoms forming the layers and secondary bonding (i.e., van der Waals forces, H-bonds) between the layers [2]. This particular architecture indicates a certain level of anisotropy in clays that exhibit cleavage subjected to forces applied parallel respect to the layer direction (thus, acting against the secondary bonding). Furthermore, as previously discussed, clays are characterized by having different chemical species at the interlayer that can somehow exert a different attraction between layers. In fact, depending on the chemical species at the interlayer, clays have different CEC, interlayer thickness (and consequently surface area), and hydration/gel-forming capacity. In detail:

(i) Cation exchange capacity (CEC): The CEC corresponds to the amount of cations (expressed in cmol/kg) that can be exchanged with other cations at the surface of clays. The CEC is influenced by the nature and amount of cations at the clays interlayer.
(ii) Interlayer thickness: Depending on the chemical species forming the interlayer, these generate different electrostatic forces (and consequently different degree of attraction) between the different sheets forming the layer structures of clays. These electrostatic forces influence the interlayer thickness.
(iii) Hydration/gel-forming (or swelling) capacity: Mechanisms at the basis of hydration are mostly two: (a) electrical properties of both clay's inorganic surface and aqueous medium affecting the water molecules orientation at the clay's surface, and (b) osmosis. Furthermore, both hydration and swelling properties are strongly affected by the nature and the quantity of exchangeable cations present at the interlayer, and these

values can be predicted considering the hydration energy of the different ions. In fact, the swelling capability follows the order: Mg > Ca > Li > Na > K [2].

Typical values of CEC, interlayer thickness, specific surface area, and swelling capacity of the different clay subgroups are summarized in Table 1 [2,76].

Table 1. Properties of the different subgroups of clays [2,7,76].

Clays Subgroups	CEC (cmol/kg)	Interlayer Thickness (Å)	Specific Surface Area (m^2/g)	Swelling Capacity
Kaolinite-Serpentine	3–15	7	5–40	None
Pyrophyllite-Talc	<1	9	5–40	None
Smectite	70–100	10–11	40–800	High
Vermiculite	100–150	12–15	500–700	High
Mica	10–40	10–11	50–200	Low
Chlorites	10–40	12–15	10–60	None

3. Advanced Applications: The Second (Technological) Life of Clays

As previously introduced, the peculiar properties of clays are the basis of the evolution (or better, the revolution) around this class of materials, going from being the simplest and oldest material handled and modeled by humans toward being still an important protagonist in the more attractive brand new technologies of the new millennium. In order to provide a detailed discussion on the advanced uses of clays in new technologies, the following paragraphs discuss the most catching eye technologies and case studies involving clays.

3.1. Biomedical Applications

Mousa and coworkers [7] point out that clay surface reactivity (due to CEC, ionic species at the interlayers, surface area and sorption capacity) is of paramount relevance for the biomedical interaction toward living tissues and fluids and synthetic prosthetic devices. In particular, clays are typically included in polymeric matrices to form both soft functional biocompatible scaffolds, hydrogels, and physical gels with improved mechanical responses [77], and hard porous scaffolds and sponges mimicking the bone tissue structural organization [78]. In this context, Giannelli et al. [5] produced a hard composite scaffold made by a keratin 3D sponge matrix, with magnetic Mg/Fe-hydrotalcite nanoparticles dispersed as fillers. These magnetic nanoparticles are able to stimulate the cell adhesion and proliferation by applying an external magnetic field. Results showed that the application of an external magnetic stimulus facilitates the cell proliferation, thus confirming the fundamental role of the magnetic clay component in the biomedical device. Furthermore, these systems show enhanced biocompatibility with osteoblasts and fibroblasts cells. Another interesting study is the one by Cui and coworkers [6], where a microporous 3D hydrogel made by modified chitosan and montmorillonite to promote tissue regeneration was developed. In this composite system, clays play a fundamental role in both increasing the mechanical properties of the hydrogel and improving the final interconnected microstructure. Results indicate the clay-induced microporous structure promotes the native cell infiltration, proliferation, and spontaneous differentiation without including growth factor and drugs in the formulation.

Murugesan et al. [8] critically analyzed the most recent literature describing the development of clay-based nanocomposites for biomedical applications. In particular, clay-containing nanocomposites under the shape of thin films are mostly used as drug-delivery patches, biodegradable/biocompatible packaging films, and microfluidic systems. These thin films are mostly produced following the fabrication techniques used in the field of nanocomposites such as solution blending, melt mixing, electrospinning, in situ polymerization [8]. In this process, both clays and polymers are dissolved in a common solvent (eventually, the two dispersions/solutions are separately prepared and added together), and transferred in the desired mold (i.e., solution blending) or deposited (i.e., cast films). Vice versa, the melt intercalation passes through the dispersion of the clays within the molten polymer. Alternatively, the in situ intercalative polymerization requires the dispersion of clays in a monomer solution and subsequently let the monomers polymerize by applying an external stimulus. The advantages of producing thin films are mostly in terms of chemical and mechanical resistance (compared to different morphologies).

Another interesting technology applied to clays is their integration into polymeric matrices to produce fibers, drug-delivery patches, or fiber-based membranes/scaffolds fabricated via electrospinning. In this case, clays are firstly dispersed in the polymer/solution, letting the clay-containing mixture pass through a needle spinneret and finally deposited onto the substrate pushed by the charge difference between the needle and the collector. The fiber morphology is driven by several parameters such as voltage intensity, the working distance between the needle and the collector, and the geometry of the needle [8,79]. In these fiber-based systems, the role of clays is to enhance the mechanical performance and the thermal stability of the composite, also improving the processability. In the study by Hong and coworkers [80], an organically modified montmorillonite prepared by a solution intercalation method in the presence of polyurethane was successfully deposited via electrospinning, producing a nanocomposite with 200%-enhanced mechanical response (both in terms of Young modulus and tensile strength) with respect to the base polymer.

The production of suitable bioinks for 3D printing and biofabrication passes through several printing methods (e.g., inkjet, extrusion, laser-assisted printing) [81,82]. The details concerning 3D printing technologies are provided in the dedicated following paragraph. However, for the sake of completeness, the biomedical applications involving bioprinting of clay-based systems are discussed here. Bioprinting consists of producing scaffolds or tissues via additive manufacturing (i.e., typically extrusion-based printing). These tissues/scaffolds are made by cells included in a host matrix (mostly biopolymers or hydrogels). One of the main concerns related to these very complex formulations (borderline between living organism and unanimated matter) is the rheology of the system that should be perfect (i.e., not too much as it can cause mechanical stresses to the cells, not too low as the final formulation should be printable and maintain the desired shape). In this context, nanoscopic clays (nanoclays) can be a very efficient and effective additive for such type of bioinks since they enhance the mechanical strength and viscosity of the system, maintaining the final shape, showing high processability and enhancing the biological activity [8]. Furthermore, the formulations containing nanoclays can be stimuli-responsive toward electrical, temperature, and pH variation, making these systems very appealing for muscular tissue regeneration [83–85]. A clear example of these stimuli-responsive systems is the study by Guo et al. [86] where an agarose/polyacrylamide-based thermo-responsive hydrogel containing laponite (which is a synthetic clay belonging to the smectite subgroup) has been successfully prepared and tested. In this particular formulation, clays increase the processability, the post-printing stability, and the final mechanical properties with respect to the bare agarose/polyacrylamide hydrogels.

As previously introduced, clays are characterized by a particular particle size, porosity and surface area, and reactivity coupled with their biocompatibility that allow their use in drug-delivery systems, in pharmaceutical applications, and in cancer research, diagnosis and relative therapy [9,87]. In particular, nanoclays are exploitable as nanoscopic carriers for the delivery of anticancer drugs directly to the tumor site. Furthermore, several in vitro and in vivo tests highlighted that nanoclays show bioactivity against different cancers ([9,88] and references therein). This is, for instance, the case of bentonite that inhibits the growth of central nervous system (glioblastoma) cells, while enhancing the growth of lung adenocarcinoma cells [88]. Interestingly, a montomorillonite/palygorskite clay mixture significantly reduces melanoma cell proliferation and viability by in vitro tests, and melanoma growth in an in vivo animal model by inducing a reduction of the tumor size and weight, decreasing tumor cell mitosis, and inducing tumor necrosis [89]. Nanoclays are used (in combination with polymeric and/or magnetic nanoparticles) to transport specific anticancer drugs directly to the tumor site. This is, for instance, the case of the study by

Lin and coworkers [90] where a Na-montmorillonite was loaded with 5-fluorouracil (a drug used in the colorectal cancer therapy) but reaching the maximum loading capacity of only 2 wt.%. To increase the loading capacity of montmorillonite, montmorillonite was amino-modified and grafted with β-cyclodextrin, reaching a higher 5-fluorouracil loading of approximately 28 wt.% [91]. However, Peixoto et al. [9] classified the use of clays in oncologic therapy against different types of cancer depending on the target organ/apparatus, namely lung, colorectal, gastric, breast, pancreatic, brain, skin, thyroid, and bone cancer. For all these cases, clays are used as carriers of bioactive chemicals, either directly as they are or after a surface functionalization (as depicted in Figure 2) [92].

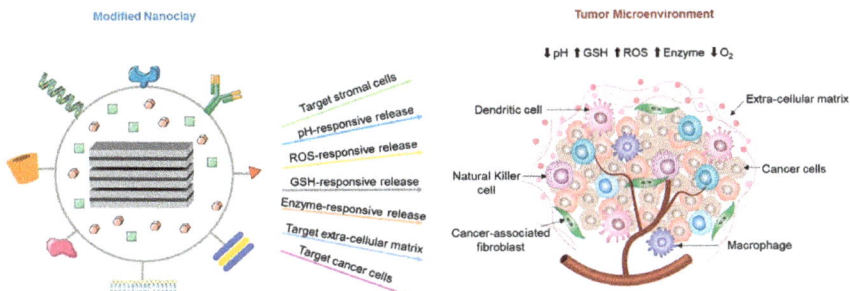

Figure 2. Schematic illustration of strategies used to expand the functions of nanoclays. The modified-nanoclays target the tumor microenvironment (TME), increasing the efficacy of cancer therapy. TME has been shown to have different characteristics from normal tissues, which can be vascular defects, higher expression of given enzymes, elevated glutathione (GSH), hydrogen ion, and reactive oxygen species (ROS) concentrations. Reprinted with permission from [9]. Copyright 2021 Elsevier B.V.

In general, most of the cases investigated used mostly halloysite nanotubes (HNTs), but also montmorillonite, bentonite, laponite, palygorskite, and kaolinite. Figure 3 reports the mechanism involving the pH- and time-responsive release of an anticancer drug loaded by HNTs, delivered through oral administration. As reported by the authors, the role of HNTs (i.e., clays) within this technological approach is very appealing as it minimizes the undesired (early) drug release in the stomach, whereas it maximizes the drug release directly in the intestines, the site affected by the colorectal cancer [9].

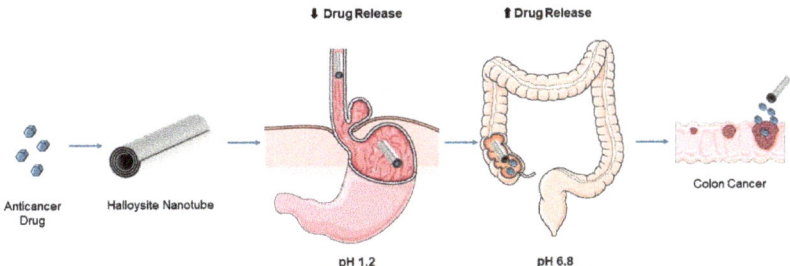

Figure 3. Schematic illustration of a pH-responsive nanoclay system for anticancer drug delivery to the intestines, avoiding early release in the stomach. Reprinted with permission from [9]. Copyright 2021 Elsevier B.V.

Furthermore, these systems are exploitable in diagnostics by simply loading/grafting specific biomarkers or magnetic nanoparticles that can detect the presence of cancer cells in the organism [93,94]. For a detailed study of the diagnostics and imaging of pathological alterations in living systems, please refer to [95].

3.2. Environmental Applications

The continuous depletion of freshwater resources is one of the main concerns affecting the social and economic progress of several underdeveloped areas of Third and Fourth World countries [2]. To overcome this issue, probably the best technological solution is to significantly increase the recyclability of polluted (waste)water into newly clean drinkable water [96,97]. In this context, the scientific literature proposes several alternative approaches for the abatement of anthropogenic (in)organic pollutants from contaminated (waste)water [98–108]. Historically, the use of soil-matter, sand, gravel, and stone as water filtering materials dates back to the Egyptian and ancient Greeks periods [109]. The scientific awareness of the contemporary society pointed out that fundamental parameters influencing the sorption capacity of materials are the surface area and charge, pore volume, and the presence of specific reactive species available at both surface and interface [2,110]. Due to their particular morphological characteristic and related properties, clays represent a very promising class of nanomaterials exploitable in contaminated freshwater and wastewater clean-up processes.

Based on the literature analysis proposed by Uddin [11], clays are effective adsorbents for the removal of heavy metal ions (i.e., As, Cd, Co, Cr, Cu, Hg, Mn, Ni, Pb, Zn) from aqueous solutions, mostly due to their specific surface area and capability of ion exchange (i.e., CEC). In particular, the majority of clays are negative charged with positive cations present at the interface/surface, and available for interacting with metal cations present in a solution following different mechanisms (i.e., mostly ion exchange, but also direct bonding between adsorbent surface and adsorbate species, complexation, and many others). Furthermore, clays can be modified in order to enhance their sorption capacity (e.g., by varying the hydrophilicity/hydrophobicity degree of the surface and by introducing some novel chemical functionalities). In particular, some scientific studies show the importance of both solution pH and thermal treatment (performed on clays) over the adsorption mechanism [11,111–115]. In fact, experimental evidences highlight that the sorption of As(V) over a mixture of different natural clays is maximized at acid conditions [111]. However, in the case of Cr(VI), Cu(II), and Zn(II), an optimum sorption pH at approximately circumneutral conditions [112,113] has been registered. The explanation of this phenomenon is not trivial and depends on both the nature of the clays and heavy metal cations. The literature proposes that pH variations strongly affect the availability of ionic species and the competition for adsorption sites, also influencing the chemical precipitation of metal ions as hydroxyl species [11,114]. Concerning the effect of the thermal pretreatment over clays sorption capacity, even here the trend is still unclear. There is evidence that the sorption capacity is maximized along with the thermal treatment (this is, for instance, the case of Cr(VI) sorption over a 200 °C fired clay) [115]. On the contrary, there is also some evidence that the clay sorption capacity is reduced after firing at higher temperature (and this is the case of the same cation Cr(VI) over a 400 °C fired clay, for details please refer to [11] and references therein). Es-Sahbany and coworkers [15] reported the use of a clay mixture extracted from a Morocco region made by kaolinite, illite, quartz, and traces of vermiculite, showing an overall CEC value of approximately 27 meq/100 g and a pH of 8.8. This mixture was tested for the sorption of heavy metals, namely Co(II), Cu(II), Ni(II), and Pb(II). Experimental results showed the maximum abatement (elimination yield ca. 85%) for all these cations was reached at alkali pH (approximately 8.0–8.5) with a contact time of approximately 90 min [15].

In many cases, modification treatments are proposed as technical solutions to significantly enhance clay sorption performances. Peralta et al. [10] reported the preparation of magnetic clays by introducing magnetic nanoparticles by different mechanisms, namely at the interlayer, within the porous channels, by direct interaction with the siloxane functional groups at the surface/edge of the mineral clay, thus forming magnetic nanocomposites [10,116]. Furthermore, the introduction of a magnet-sensitive fraction is very appealing as it can substantially favor the recovery of the nanocomposites by simply applying an external magnetic stimulus [117–119]. In particular, Peralta and coworkers [10] reported that the typical routes to prepare such magnetic nanocomposites are: coprecipitation, intercalation, and pillaring. Figure 4 reports the main relevant mechanisms [10]. These methods are:

(i) Pillaring: This method consists of introducing a pillar within the structure of the clay by permanently stacking the interlayers, generating a higher porosity [120]. Pillaring is mostly a cationic exchange method in which inorganic species are introduced within the interlayer of clays forming robust oxides strongly bound to the layers of the minerals [17]. In this specific context, the mechanisms proposed are two: either the incorporation of magnetic nanoparticles within the pores of the pillared clays (Figure 4, route A1) or using the magnetic nanoparticles as pillars to expand the interlayer distance of the clays (Figure 4, route A2) [10].

(ii) Coprecipitation: This method consists of the in situ formation of magnetic nanoparticles [121] by performing the synthesis directly in an aqueous dispersion of clays (Figure 4, route B).

(iii) Intercalation: This method consists in the physical insertion of target chemical species within the interlayers/pores of the clays [122]. In this specific context, the mechanisms proposed are two: either the inclusion of magnetic nanoparticles within a previously surfactants-intercalated clay to facilitate the entrance of the magnetic nanoparticles (Figure 4, route C1) or the direct intercalation of surfactant-stabilized magnetic nanoparticles (Figure 4, route C2) [10].

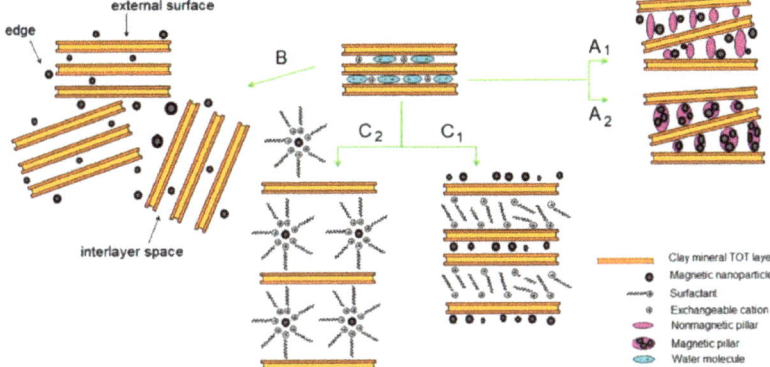

Figure 4. Schematic illustration of typical synthesis routes to obtain magnetic nanoparticles/clay minerals 2:1 type nanocomposites: (A1) pillaring with nonmagnetic pillar; (A2) pillaring with magnetic pillar; (B) coprecipitation route; (C1) intercalation of magnetic nanoparticles into a surfactant intercalated clay; (C2) intercalation of a surfactant-modified magnetic nanoparticles into a clay. Reprinted with permission from [10]. Copyright 2020 MDPI.

In this context, Magdy et al. [123] prepared a kaolinite magnetic nanocomposite via coprecipitation and successfully tested it for the abatement of an anionic dye (Direct Red 23) at neutral pH, reaching the maximum adsorbent capacity of approximately 23 mg/g.

Mukhopadhyay et al. [12], instead, investigated the inorganic modification of two different types of clays, namely smectite and kaolinite, for the adsorption of As(V). In particular, three modified clays were prepared, namely Fe-exchanged smectite, Ti-pillared smectite, and phosphate-bond kaolinite, and results were compared with the bare unmodified clays. Results indicated that these modifications influenced the clays' pH, specific surface area, and CEC. Concerning smectite, both Fe-exchange and Ti-pillaring cause a reduction of the pH (from ca. 8.0 to ca. 4.0–6.0) mainly due to the saturation of the clay sites, a consequent decrement of the CEC since these reactions blocked the availability of sorption sites and increment of the specific surface area (from ca. 200 m^2/g up to 440–480 m^2/g). Concerning kaolinite, instead, phosphate-bonding causes a negligible pH variation, a double increment of the CEC due to ligand adsorption, and a remarkable increment of the specific surface area (from ca. 18 m^2/g up to ca. 90 m^2/g). Results indicate that all these modification significantly better the sorption performances with respect to the bare reference clays [12].

Interestingly, clays might be helpful in advanced oxidation processes (AOPs), which are a very promising category of methods for the removal of organic pollutants present in water by radical-mediated oxidation reactions in most cases light-mediated (i.e., photocatalysis) [124–128]. Heterogeneous AOPs required the use of a semiconductor capable of generating radical species once activated by UV/visible light radiation (i.e., photocatalyst). The most studied semiconductors for this application are TiO$_2$ (mostly anatase phase) [129], ZnO [130] and iron oxides (for photo-Fenton and Fenton processes) [117,131,132]. In particular, clays might be used as either a simple substrate for the delivery of the semiconductors or chemical functionalized to produce a new heterostructure working directly as a photocatalyst. In this context, Akkari et al. [14] developed different ZnO-sepiolite heterostructures and successfully tested them as photocatalysts for the degradation of an aqueous solution contaminated by several anthropogenic pharmaceuticals, taken as model emerging pollutants. Baloyi et al. [17], instead, analyzed the most recent scientific literature describing the use of pillared clays in heterogeneous AOPs for water purification.

Furthermore, concerning the more traditional possibility of using clays as sorbing materials, Azzam et al. [133] reported the treatment of olive mills wastewater (rich in phenols and other organic compounds) by using a mixture containing muscovite and albite, registering a reduction of the chemical oxygen demand (COD below 40%) and phenols concentration (below 80%). Li et al. [16] reported the performances of montmorillonite, kaolinite, and palygorskite in the abatement of a cationic dye (i.e., Rhodamine 6G). Results showed that nonswelling clays (i.e., kaolinite and palygorskite) show a relatively low dye uptake (below 140 mmol/kg) if compared to swelling clay (i.e., montomorillonite, 785 mmol/kg). The rationalization of this behavior is due to the important role of the interlayers (which is available in the case of the swelling clay) in the sorption mechanism.

Sometimes, it is necessary to induce a higher affinity between the inorganic clays and hydrophobic organic species present in the aqueous medium. To do this, researchers investigated the possibility of surface functionalize clays by substituting the interlayer cations with either organocations or covalently-bonded organic moieties, thus generating organoclays [134]. Among clays, smectites, (primarily montmorillonite) have been extensively used to prepare organoclays because of their high CEC, swelling behavior, sorption properties, and large surface area [134]. Other organoclays rely on micas, hectorites, and sepiolites. Organoclays are mostly prepared in solutions by means of either cation exchange reaction or by solid-state reaction. These methods are:

(i) Cation exchange reaction: This method consists of exchanging the interlayers cations with quaternary alkylammonium cations in aqueous solution.
(ii) Solid-state reaction: This method consists of intercalating organic molecules in dried clays (i.e., in absence of solvents).

Depending on the hydrophobic chain length, it is registered an increment of the structural ordering along with the increment of the chain length (as depicted in Figure 5) [135].

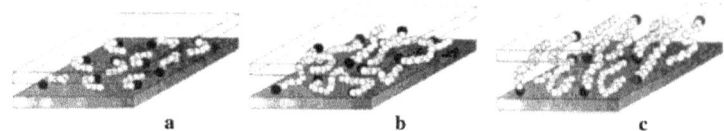

Figure 5. Schematic illustration of chains aggregation models in alkyl-modified organoclays: (**a**) short chains length, lateral monolayer; (**b**) medium chains length, in-plane disorder and interdigitation to form quasi bilayers; (**c**) long chains length: interlayer order increases leading to a liquid–crystalline polymer environment. Reprinted with permission from [135]. Copyright 1994 American Chemical Society.

As reported by Beall et al. [13], organoclays interact against organic contaminants by partitioning them within their hydrophobic layer, which acts as a real organic phase. The efficiency of these systems is driven by the solubility of the target organic pollutant in the water medium and its affinity toward the organic phase in the organoclays channels. One of the main industrial applications of these organoclays is in the recovery of acid emulsified oil well fluids from petroleum offshore platforms [13].

Lastly, further steps forward have been realized in the structural use of clays for the development of advanced ceramic membranes for water purification [19–22]. In this context, it should be noted that the use of a well-consolidated technology such as membrane separation surely guarantees industrial feasibility and speeds up its integration in a productive process. As reported by Abdullayev et al. [19], membranes can be divided into polymeric membranes (cheap, but poor stable) and ceramic ones (expensive, but highly stable). To reduce the costs of ceramic membranes, over the past few years there has been increasing interest in the use of low-cost clays as precursors for the fabrication of cheap (but effective) ceramic membranes. Among these, kaolinite is the most preferred clay used to fabricate ceramic membranes for the relatively low thermal processing and sintering conditions required. Other clays exploited for the fabrication of ceramic membranes are bentonite, sepiolite, and attapulgite [19]. Rashad and coworkers [20] investigated the fabrication of a mullite membrane for microfiltration of oil-in-water emulsion. Results indicated that this system showed an excellent pH stability (from acid to alkaline), high oil rejection, and high regularity in terms of pore size distribution and surface roughness. Elgamouz et al. [21], instead, evaluated the possibility of producing a porous ceramic membrane from a mixture of clays recovered from a particular region of Morocco. Subsequently, the authors modified the clays' porous substrate by further hydrothermal deposition of a templated silicalite coating obtained via sol-gel. Permeability tests against three different gases (namely, N_2, C_3H_8 and SF_6) showed that these membranes have high selectivity with respect to SF_6 relative to N_2, whereas they are hardly selective for C_3H_8. Abubakar et al. [22], instead, exploited a mixture of clays (rich in kaolinite and illite) from a Nigerian mine to fabricate a porous membrane (average pore diameter ca. 5–6 nm). In particular, after sintering at 1300 °C for 2 h, mullite ($3Al_2O_3 \cdot 2SiO_2$, obtained by calcination of kaolinite) and cristobalite (a polymorph of quartz) were formed. The resulting membranes were successfully tested in the separation of U from an aqueous medium, thus simulating the remediation of U-containing wastewaters deriving from fracking, oil exploration, and phosphate mining industries. Foorginezhad et al. [136] produced microfiltration membranes from nanoscopic clays via dry pressing in the presence of natural zeolites and tested these ceramic membranes against cationic and anionic dyes from contaminated water. Due to the negative charge of the clay-membrane, high removal efficiencies are reached for the positively charged species rather than for the negatively ones.

3.3. Other Advanced Applications: Additive Manufacturing and Sol-Gel Processes

Additive manufacturing (AM, also known as either rapid prototyping, or more commonly 3D-printing) is a class of processes extremely useful for obtaining three-dimensional objects starting from a 3D model (by means of a Computer Aided Design, or CAD model) and forming them by depositing layer upon layer [137–140]. Most diffused AM processes are fused deposition modelling (FDM), direct metal laser sintering (DMLS), selective laser melting (SLM), and electron beam melting (EBM) [141–144]. According to the literature, AM processes are exploitable for producing metals [145], polymers [146], ceramics [147], composites and concretes [148,149], carbonaceous materials [137], hydrogels [150], biomaterials [151], engineered tissue and organs [152,153], and food [154]. In this context, clays may also be used in AM to properly manufacture valuable objects.

Chen and coworkers [155] investigated the cement 3D printability by introducing into the formulation ca. 60 wt.% low-grade calcined clay (mainly made by metakaolin), ca. 30 wt.% limestone, and ca. 2 wt.% admixtures (such as plasticizers and/or viscosity modifiers). By increasing the content in clays, it registered a significant reduction of slump, flowability and initial material flow rate. Moreover, results pointed out a buildability improvement (caused by reduced water film thickness), an acceleration of the initial setting and stiffness, together with an increment of the specific surface area. For the sake of comparison, a reduction of the compressive strength due to the dilution effect exerted by the cementitious fraction replacement was also noted. Another interesting study focused on the introduction of calcined clays in cementitious formulation for 3D printing is the one written by Long et al. [156]. The authors registered a significant improvement of several mechanical parameters (i.e., dynamic yield stress, static yield stress), structural recovery, and shape retention during printing of the final mortar containing a 33.33 wt.% in calcined clays. Faksawat and coworkers [157] investigated the possibility of producing composite paste (for future bone-replacement orthopedic implants) made by raw clays and hydroxyapatite (at different ratios from 95:5 to 75:25) by FDM 3D printing. Chikkangoudar et al. [158], instead, evaluated the effects of adding nanoclays in polypropylene filaments for 3D printing, registering an enhancement of the filament dimensional flexibility and a reduction of the deformation of the 3D printed models (counterbalanced by an increment in fragility affecting both filaments and 3D printed models by increasing the nanoclays content). For completeness, case studies discussing the introduction of clays into formulation for 3D printing of biomedical devices are already discussed in the previous paragraph dedicated to the biomedical applications.

Sol-gel process represents a bottom-up procedure to produce ceramic metal oxides under the shape of particles, films, fibers, gels (i.e., xerogels, cryogels, aerogels), and monoliths [159–166]. This technique consists of a series of (acid/alkaline catalyzed) in situ polycondensation reactions involving monomers (i.e., metal oxides precursors), and converting them from a colloidal solution (sol) into an integrated network (gel) [161]. In the previous paragraph, we reported that the surface sites of clays are important sites for further functionalization and grafting. It is also possible to disperse clays in a proper medium in the presence of the selected oxide precursor and directly fabricate a nanocomposite through a sol-gel polymerization mechanism. Meera et al. [35] reported the preparation of silica-(organo-modified) montmorillonite nanocomposites in aqueous medium via a direct sol-gel process. Results showed that the presence of the clay fraction influenced the final morphology of the nanocomposite, and increased the surface hydrophobicity, conferring anti-wetting properties. Furthermore, clays also increased the thermal stability of the nanocomposite (with a shift of ca. 40 °C). In Qian et al. [34], the authors produced a silica-montmorillonite composite registering the formation of mesoporous silica nanostructures covering the clay surface. Results indicate that acid-catalyzed systems show large continuous mesoporous silica covering the clay surfaces, with a substantial increment of the surface area (from ca. 30 m^2/g to ca. 560 m^2/g). In noncatalyzed systems, the morphology is a bit different as silica nanoparticles result being attached on 2D clay platelets, and the increment in the surface area is contained (from ca. 30 m^2/g to ca. 165 m^2/g). The comparison between the two different mechanisms is sketched in Figure 6 [34].

Coming back to the environmental applications, Pronina et al. [167] reported the deposition of different porous titania coatings (sol-gel mediated) onto expanded clays, and their integration into fluidized-bed photocatalytic reactors. These nanocomposites were successfully tested in the photocatalytic abatement of tetracycline antibiotics from aqueous solutions. Results showed a synergy between clay and titania as the degradation mechanism were a combination between adsorption and photocatalytic abatement.

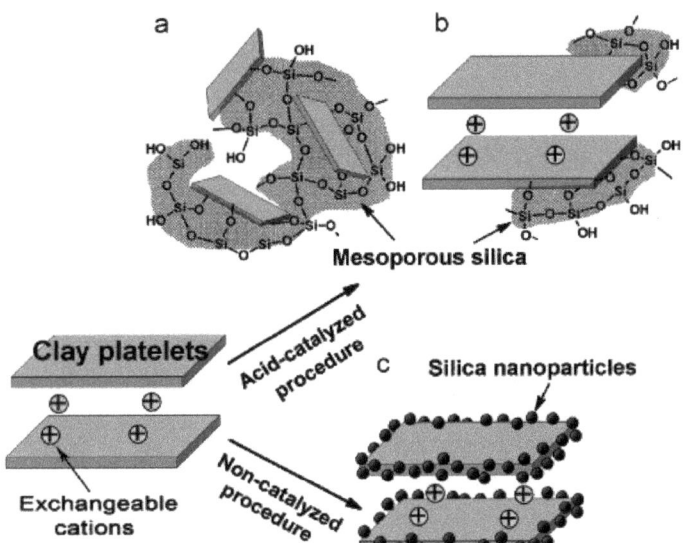

Figure 6. Schematic representation of the proposed structural models for the sol-gel-modified clays in the case of acid-catalyzed procedures: (**a**) a cross-linked structure at high TEOS/clay ratio; (**b**) mesoporous silica attached on clay surface at low TEOS/clay ratio, and in the case of non-catalyzed procedures: (**c**) silica nanoparticles attached on clay surfaces. Reprinted with permission from [34]. Copyright 2008 Elsevier B.V.

4. Conclusions and Future Perspectives

Clays are 2D layered hydrous alumino-silicates extracted from earth. This extremely varied class of materials has been exploited since ancient times for preparing traditional ceramics, such as earthenware, stoneware, porcelains, potteries. It is with the increased scientific awareness of the modern era that some interesting structure-property relationships of clays emerged as a very peculiar characteristic of this class of materials. In particular, what is quite surprising is the continuous interest that even nowadays surrounds clays that result still being appealing for a plethora of novel advanced technological applications, very far from the production of traditional ceramics.

However, the large multitude of different applications in most cases does not facilitate understanding the chemistry behind clays. This is mainly due to the fact that the majority of the review documents found in the scientific literature are primarily focused on the applications, not on the materials. Therefore, with this document, this author hopes to have finally filled this gap. In particular, after providing an economic analysis of the global trading of clays, the text has been organized into two main sections. Part I is dedicated to the classification of clays based on their structural and chemical composition, together with a schematic summary of the main relevant structure-induced properties, which are strongly correlated to the nature and quantity of chemical species at the interlayer, CEC, interlayer thickness, surface area, and hydration/gel-forming capacity among the others. Part II, in contrast, is dedicated to the analysis of the nonconventional applications of clays in technological fields of emerging interest. In particular, several case studies describing the use of clays in biomedicine, environmental remediation, membrane technology, additive manufacturing, and sol-gel processes are presented, and experimental results are critically discussed and correlated with the clay structure-property relationship. In fact, a more correct comprehension of the mechanisms involved is the only way to maximize the valorization of this class of inorganic materials.

Therefore, at the end of this study, we clearly understand that the future of clays is still fluid and very promising. In particular, this author believes that the applications where clays can make the difference are the ones where their biocompatibility can play a pivotal role. For this reason, particular attention has been paid to the application fields such as biomedicine and environmental remediation of contaminated media. However, it should be highlighted that biocompatibility is a major property even considering the end-of-life fate of a given object. This last assumption confirms and encourages once again the growing attention that clays have attracted in the recent years.

Funding: This research received no external funding.

Institutional Review Board Statement: Not applicable.

Informed Consent Statement: Not applicable.

Data Availability Statement: Not applicable.

Acknowledgments: My interest in clays dates back some years, and the inspiration was an afternoon talk with my grandfather Giuseppe "Pino" Salerno (1933–2020) having object the "return to the earth" concept, meaning that the future of humanity passes through sustainability, agriculture, and saving the environment. His life of sacrifice between industry (in youth) and agriculture (his great passion and hobby) was always a model to me. Unfortunately, after a bad lymphoma broke out during the COVID pandemic, he died in April 2020. I think that the resulting manuscript fully respects his philosophy.

Conflicts of Interest: The author declares no conflict of interest.

References

1. Guggenheim, S.; Martin, R.T. Definition of clay and clay mineral: Joint report of the AIPEA and CMS Nomenclature Committees. *Clay Miner.* **1995**, *30*, 257–259. [CrossRef]
2. Nisticò, R. The importance of surfaces and interfaces in clays for water remediation processes. *Surf. Topogr. Metrol. Prop.* **2018**, *6*, 043001. [CrossRef]
3. Zou, Y.-C.; Mogg, L.; Clark, N.; Bacaksiz, C.; Milanovic, S.; Sreepal, V.; Hao, G.-P.; Wang, Y.-C.; Hopkinson, D.G.; Gorbachev, R.; et al. Ion exchange in atomically thin clays and micas. *Nat. Mater.* **2021**, *20*, 1677–1682. [CrossRef]
4. Yang, Y.; Liu, X.; Zhu, Z.; Zhong, Y.; Bando, Y.; Golberg, D.; Yao, J.; Wang, X. The role of geometric sites in 2D materials for energy storage. *Joule* **2018**, *2*, 1075–1094. [CrossRef]
5. Giannelli, M.; Barbalinardo, M.; Riminucci, A.; Belvedere, K.; Boccalon, E.; Sotgiu, G.; Corticelli, F.; Ruani, G.; Zamboni, R.; Aluigi, A.; et al. Magnetic keratic/hydrotalcites sponges as potential scaffolds for tissue regeneration. *Appl. Clay Sci.* **2021**, *297*, 106090. [CrossRef]
6. Cui, Z.-K.; Kim, S.; Baljon, J.J.; Wu, B.M.; Aghaloo, T.; Lee, M. Microporous methacrylated glycol chitosan-montmorillonite nanocomposite hydrogel for bone tissue engineering. *Nat. Commun.* **2019**, *10*, 3523. [CrossRef] [PubMed]
7. Mousa, M.; Evans, N.D.; Oreffo, R.O.C.; Dawson, J.I. Clay nanoparticles for regenerative medicine and biomaterial design: A review of clay bioactivity. *Biomaterials* **2018**, *159*, 204–214. [CrossRef] [PubMed]
8. Murugesan, S.; Scheibel, T. Copolymer/clay nanocomposites for biomedical applications. *Adv. Funct. Mater.* **2020**, *30*, 1908101. [CrossRef]
9. Peixoto, D.; Pereira, I.; Pereira-Silva, M.; Veiga, F.; Hamblin, M.R.; Lvov, Y.; Paiva-Santos, A.C. Emerging role of nanoclays in cancer research, diagnosis, and therapy. *Coord. Chem. Rev.* **2021**, *440*, 213956. [CrossRef]
10. Peralta, M.E.; Ocampo, S.; Funes, I.G.; Onaga Medina, F.; Parolo, M.E.; Carlos, L. Nanomaterials with tailored magnetic properties as adsorbents of organic pollutants from wastewaters. *Inorganics* **2020**, *8*, 24. [CrossRef]
11. Uddin, M.K. A review on the adsorption of heavy metals by clay minerals, with special focus on the past decade. *Chem. Eng. J.* **2017**, *308*, 438–462. [CrossRef]
12. Mukhopadhyay, R.; Manjaiah, K.M.; Datta, S.C.; Yadav, R.K.; Sarkar, B. Inorganically modified clay minerals: Preparation, characterization, and arsenic adsorption in contaminated water and soil. *Appl. Clay Sci.* **2017**, *147*, 1–10. [CrossRef]
13. Beall, G.W. The use of organo-clays in water treatment. *Appl. Clay Sci.* **2003**, *24*, 11–20. [CrossRef]
14. Akkari, M.; Aranda, P.; Belver, C.; Bedia, J.; Haj Amara, A.B.; Ruiz-Hitzky, E. Reprint of ZnO/sepiolite heterostructured materials for solar photocatalytic degradation of pharmaceuticals in wastewater. *Appl. Clay Sci.* **2018**, *160*, 3–8. [CrossRef]
15. Es-Sahbany, H.; Hsissou, R.; El Hachimi, M.L.; Allaoui, M.; Nkhili, S.; Elyoubi, M.S. Investigation of the adsorption of heavy metals (Cu, Co, Ni and Pb) in treatment synthetic wastewater using natural clay as a potential adsorbent (Sale-Marocco). *Mater. Today Proc.* **2021**, *45*, 7290–7298. [CrossRef]
16. Li, Z.; Potter, N.; Rasmussen, J.; Weng, J.; Lv, G. Removal of rhodamine 6G with different types of clay minerals. *Chemosphere* **2018**, *202*, 127–135. [CrossRef] [PubMed]

17. Baloyi, J.; Ntho, T.; Moma, J. Synthesis and application of pillared clay heterogeneous catalysts for wastewater treatment: A review. *RSC Adv.* **2018**, *8*, 5197–5211. [CrossRef]
18. Sun, C.; Zhang, F.; Wang, X.; Cheng, F. Facile preparation of ammonium molybdophosphate/Al-MCM-41 composite material from natural clay and its use in cesium ion adsorption. *Eur. J. Inorg. Chem.* **2015**, *2015*, 2125–2131. [CrossRef]
19. Abdullayev, A.; Bekheet, M.F.; Hanaor, D.A.H.; Gurlo, A. Materials and applications for low-cost ceramic membranes. *Membranes* **2019**, *9*, 105. [CrossRef] [PubMed]
20. Rashad, M.; Logesh, G.; Sabu, U.; Balasubramanian, M. A novel monolithic mullite microfiltration membrane for oil-in-water emulsion separation. *J. Membr. Sci.* **2021**, *620*, 118857. [CrossRef]
21. Elgamouz, A.; Tijani, N. From a naturally occurring-clay mineral to the production of porous ceramic membranes. *Microporous Mesoporous Mater.* **2018**, *271*, 52–58. [CrossRef]
22. Abubakar, M.; Tamin, M.N.; Saleh, M.A.; Uday, M.B.; Ahmad, N. Preparation and characterization of a nigerian mesoporous clay-based membrane for uranium removal from underground water. *Ceram. Int.* **2016**, *42*, 8212–8220. [CrossRef]
23. Yi, H.; Ai, Z.; Zhao, Y.; Zhang, X.; Song, S. Design of 3D-network montmorillonite nanosheet/stearic acid shape-stabilized phase change materials for solar energy storage. *Sol. Energy Mater. Sol. Cells* **2020**, *204*, 110233. [CrossRef]
24. Chen, C.; Ma, Y.; Wang, C. Investigation of electrochemical performance of montmorillonite clay as Li-ion battery electrode. *Sustain. Mater. Technol.* **2019**, *19*, e00086. [CrossRef]
25. Rajapakse, R.M.G.; Murakami, K.; Bandara, H.M.M.; Rajapakse, R.M.M.Y.; Velauthamurti, K.; Wijeratne, S. Preparation and characterization of electronically conducting polypyrrole-montmorillonite nanocomposite and its potential application as a cathode material for oxygen reduction. *Electrochim. Acta* **2010**, *55*, 2490–2497. [CrossRef]
26. Kim, M.H.; Cho, C.H.; Kim, J.S.; Nam, T.U.; Kim, W.-S.; Lee, T.I.; Oh, J.Y. Thermoelectric energy harvesting electronic skin (e-skin) Patch with reconfigurable carbon nanotube clays. *Nano Energy* **2021**, *87*, 106156. [CrossRef]
27. Vaia, R.A.; Price, G.; Ruth, P.N.; Nguyen, H.T.; Lichtenhan, J. Polymer/layered silicate nanocomposites as high performance ablative materials. *Appl. Clay Sci.* **1999**, *15*, 67–92. [CrossRef]
28. Atyia, M.N.; Mahdy, M.G.; Elrahman, M.A. Production and properties of lightweight concrete incorporating lightweight concrete incorporating recycled waste crushed clay bricks. *Constr. Build. Mater.* **2021**, *304*, 124655. [CrossRef]
29. Hassan, A.; Mourad, A.-H.I.; Rashid, Y.; Ismail, N.; Laghari, M.S. Thermal and structural performance of geopolymer concrete containing phase change material encapsulated in expanded clay. *Energy Build.* **2019**, *191*, 72–81. [CrossRef]
30. Madyan, O.A.; Fan, M.; Feo, L.; Hui, D. Physical properties of clay aerogel composites: An overview. *Compos. Part B Eng.* **2016**, *102*, 29–37. [CrossRef]
31. Di Credico, B.; Cobani, E.; Callone, E.; Conzatti, L.; Cristofori, D.; D'Arienzo, M.; Dirè, S.; Giannini, L.; Hanel, T.; Scotti, R.; et al. Size-controlled self-assembly of anisotropic sepiolite fibers in rubber nanocomposites. *Appl. Clay Sci.* **2018**, *152*, 51–64. [CrossRef]
32. Wang, L.; Wang, F.; Huang, B.; Tang, Q. Recent advances in superhydrophobic composites based on clay minerals. *Appl. Clay Sci.* **2020**, *198*, 105793. [CrossRef]
33. Serge, E.J.; Alla, J.P.; Belibi, P.D.B.; Mbadcam, K.J.; Fathima, N.N. Clay/polymer nanocomposites as filler materials for leather. *J. Clean. Prod.* **2019**, *237*, 117837. [CrossRef]
34. Qian, Z.; Hu, G.; Zhang, S.; Yang, M. Preparation and characterization of montmorillonite-silica nanocomposites: A sol-gel approach to modifying clay surfaces. *Phys. B Condens. Matter* **2008**, *403*, 3231–3238. [CrossRef]
35. Meera, K.M.S.; Sankar, R.M.; Murali, A.; Jainsankar, S.N.; Mandal, A.B. Sol-gel network silica/modified montmorillonite clay hybrid nanocomposites for hydrophobic surface coatings. *Colloids Surf. B Biointerfaces* **2012**, *90*, 204–210. [CrossRef] [PubMed]
36. Revelo, C.F.; Colorado, H.A. 3D printing of kaolinite clay ceramics using the Direct Ink Writing (DIW) technique. *Ceram. Int.* **2018**, *44*, 5673–5682. [CrossRef]
37. Chan, S.S.L.; Pennings, R.M.; Edwards, L.; Franks, G.V. 3D printing of clay for decorative architectural applications: Effect of solids volume fraction on rheology and printability. *Addit. Manuf.* **2020**, *35*, 101335. [CrossRef]
38. Njindam, O.R.; Njoya, D.; Mache, J.R.; Mouafon, M.; Messan, A.; Njopwouo, D. Effect of glass powder on the technological properties and microstructure of clay mixture for porcelain stoneware tiles manufacture. *Constr. Build. Mater.* **2018**, *170*, 512–519. [CrossRef]
39. Mahmoudi, S.; Srasra, E.; Zargouni, F. The use of Tunisian Barremian clay in the traditional ceramic industry: Optimization of ceramic properties. *Appl. Clay Sci.* **2008**, *42*, 125–129. [CrossRef]
40. Freyburg, S.; Schwarz, A. Influence of the clay type on the pore structure of structural ceramics. *J. Eur. Ceram. Soc.* **2007**, *27*, 1727–1733. [CrossRef]
41. Randviir, E.P.; Brownson, D.A.C.; Banks, C.E. A decade of graphene research: Production, applications and outlook. *Mater. Today* **2014**, *17*, 426–432. [CrossRef]
42. Santalucia, R.; Vacca, T.; Cesano, F.; Martra, G.; Pellegrino, F.; Scarano, D. Few-layered MoS_2 nanoparticles covering anatase TiO_2 nanosheets: Comparison between ex situ and in situ synthesis approaches. *Appl. Sci.* **2021**, *11*, 143. [CrossRef]
43. Rocha Barreto, I.A.; da Costa, M.L. Viability of Belterra clay, a widespread bauxite cover in the Amazon, as a low-cost raw material for the production of red ceramics. *Appl. Clay Sci.* **2018**, *162*, 252–260. [CrossRef]
44. Lopez-Galindo, A.; Viseras, C.; Cerezo, P. Compositional, technical and safety specifications of clays to be used as pharmaceutical and cosmetic products. *Appl. Clay Sci.* **2007**, *36*, 51–63. [CrossRef]

45. Lonzano-Morales, V.; Gardi, I.; Nir, S.; Undabeytia, T. Removal of pharmaceuticals from water by clay-cationic starch sorbents. *J. Clean. Prod.* **2018**, *190*, 703–711. [CrossRef]
46. Moussi, B.; Hajjaji, W.; Hachani, M.; Hatira, N.; Labrincha, J.A.; Yans, J.; Jamoussi, F. Numidian clay deposits as raw material for ceramics tile manufacturing. *J. Afr. Earth Sci.* **2020**, *164*, 103775. [CrossRef]
47. Gast, J.; Gundolf, K.; Cesinger, B. Doing business in a green way: A systematic review of the ecological sustainability entrepreneurship literature and future research directions. *J. Clean. Prod.* **2017**, *147*, 44–56. [CrossRef]
48. D'Amato, D.; Droste, N.; Allen, B.; Kettunen, M.; Lahtinen, K.; Korhonen, J.; Leskinen, P.; Matthies, B.D.; Toppinen, A. Green, circular, bio economy: A comparative analysis of sustainability avenues. *J. Clean. Prod.* **2017**, *168*, 716–734. [CrossRef]
49. Mies, A.; Gold, S. Mapping the social dimension of the circular economy. *J. Clean. Prod.* **2021**, *321*, 128960. [CrossRef]
50. Chemat, F.; Vian, M.A.; Ravi, H.K. Toward petroleum-free with plant-based chemistry. *Curr. Opin. Green Sustain. Chem.* **2021**, *28*, 100450. [CrossRef]
51. Gholizadeh, M.; Hu, X.; Liu, Q. A mini review of the specialties of the bio-oils produced from pyrolysis of 20 different biomasses. *Renew. Sustain. Energy Rev.* **2019**, *114*, 109313. [CrossRef]
52. Aziz, M.; Darmawan, A.; Juangsa, F.B. Hydrogen production from biomasses and wastes: A technological review. *Int. J. Hydrog. Energy* **2021**, *46*, 33756–33781. [CrossRef]
53. Tabasso, S.; Ginepro, M.; Tomasso, L.; Montoneri, E.; Nisticò, R.; Francavilla, M. Integrated biochemical and chemical processing of municipal bio-waste to obtain bio based products for multiple uses. The case of soil remediation. *J. Clean. Prod.* **2020**, *245*, 119191. [CrossRef]
54. Kerton, F.M.; Liu, Y.; Omari, K.W.; Hawboldt, K. Green chemistry and the ocean-based biorefinery. *Green Chem.* **2013**, *15*, 860–871. [CrossRef]
55. Zhou, Y.; Qin, S.; Verma, S.; Sar, T.; Sarsaiya, S.; Ravindran, B.; Liu, T.; Sindhu, R.; Patel, A.K.; Binod, P.; et al. Production and beneficial impact of biochar for environmental application: A comprehensive review. *Bioresour. Technol.* **2021**, *337*, 125451. [CrossRef]
56. Feng, Q.; Wang, B.; Chen, M.; Wu, P.; Lee, X.; Xing, Y. Invasive plants as potential sustainable feedstock for biochar production and multiple applications: A review. *Resour. Conserv. Recycl.* **2021**, *164*, 105204. [CrossRef]
57. Qiu, M.; Sun, K.; Jin, J.; Gao, B.; Yan, Y.; Han, L.; Wu, F.; Xing, B. Properties of the plant- and manure-derived biochars and their sorption of dibutyl phthalate and phenanthrene. *Sci. Rep.* **2014**, *4*, 5295. [CrossRef]
58. Anceschi, A.; Guerretta, F.; Magnacca, G.; Zanetti, M.; Benzi, P.; Trotta, F.; Caldera, F.; Nisticò, R. Sustainable N-containing biochars obtained at low temperatures as sorbing materials for environmental application: Municipal biowaste-derived substances and nanosponges case studies. *J. Anal. Appl. Pyrolysis* **2018**, *134*, 606–613. [CrossRef]
59. Nisticò, R.; Guerretta, F.; Benzi, P.; Magnacca, G. Chitosan-derived biochars obtained at low pyrolysis temperatures for potential application in electrochemical energy storage devices. *Int. J. Biol. Macromol.* **2020**, *164*, 1825–1831. [CrossRef]
60. Abbasi, A.; Pishvaee, M.S.; Mohseni, S. Third-generation biofuel supply chain: A comprehensive review and future research directions. *J. Clean. Prod.* **2021**, *323*, 129100. [CrossRef]
61. Molino, A.; Larocca, V.; Chianese, S.; Musmarra, D. Biofuels production by biomass gasification: A review. *Energies* **2018**, *11*, 811. [CrossRef]
62. Observatory of Economic Complexity (OEC), Which Countries Exports Clays? Available online: https://oec.world/en/visualize/tree_map/hs92/export/show/all/52508/2019/ (accessed on 10 October 2021).
63. Observatory of Economic Complexity (OEC), Which Countries Imports Clays? Available online: https://oec.world/en/visualize/tree_map/hs92/import/show/all/52508/2019/ (accessed on 10 October 2021).
64. Martin, R.T.; Bailey, S.W.; Eberl, D.D.; Fanning, D.S.; Guggenheim, S.; Kodama, H.; Pevear, D.R.; Środoń, J.; Wicks, F.J. Report of the Clay Minerals Society Nomenclature Committee: Revised classification of clay materials. *Clays Clay Miner.* **1991**, *39*, 333–335. [CrossRef]
65. Salles, F.; Henry, M.; Douilland, J.-M. Determination of the surface energy of kaolinite and serpentine using PACHA formalism—Comparison with immersion experiments. *J. Colloid Interface Sci.* **2006**, *303*, 617–626. [CrossRef]
66. Zhang, J.; Hu, L.; Pant, R.; Yu, Y.; Wei, Z.; Zhang, G. Effects of interlayer interactions on the nanoindentation behavior and hardness of 2:1 phyllosilicates. *Appl. Clay Sci.* **2013**, *80–81*, 267–280. [CrossRef]
67. Bailey, L.; Lekkerkerker, H.N.W.; Maitland, G.C. Smectite clay–inorganic nanoparticle mixed suspensions: Phase behaviour and rheology. *Soft Matter* **2015**, *11*, 222–236. [CrossRef]
68. Rashad, A.M. Vermiculite as a construction material—A short guide for civil engineer. *Constr. Build. Mater.* **2016**, *125*, 53–62. [CrossRef]
69. De Poel, W.; Vaessen, S.L.; Drnec, J.; Engwerda, A.H.J.; Townsend, E.R.; Pintea, S.; de Jong, A.E.F.; Jankowski, M.; Carlà, F.; Felici, R.; et al. Metal ion-exchange on the muscovite mica surface. *Surf. Sci.* **2017**, *665*, 56–61. [CrossRef]
70. Cao, Z.; Liu, G.; Meng, W.; Wang, P.; Yang, C. Origin of different chlorite occurrences and their effects on tight clastic reservoir porosity. *J. Pet. Sci. Eng.* **2018**, *160*, 384–392. [CrossRef]
71. Damasceno Junior, E.; Ferreira de Almeida, J.M.; do Nascimento Silva, I.; Moreira de Assis, M.L.; dos Santos, L.M.; Dias, E.F.; Bezerra Aragao, V.E.; Verissimo, L.M.; Fernandes, N.S.; de Silva, D.R. pH-responsive release system of isoniazid using palygorskite as a nanocarrier. *J. Drug Deliv. Sci. Technol.* **2020**, *55*, 101399. [CrossRef]

72. Garcia-Rivas, J.; Sanchez del Rio, M.; Garcia-Romero, E.; Suarez, M. An insight in the structure of a palygorskite from Palygorskaja: Some questions on the standard model. *Appl. Clay Sci.* **2017**, *148*, 39–47. [CrossRef]
73. Esteban-Cubillo, A.; Pina-Zapardiel, R.; Moya, J.S.; Barba, M.F.; Pecharroman, C. The role of magnesium on the stability of crystalline sepiolite structure. *J. Eur. Ceram. Soc.* **2008**, *28*, 1763–1768. [CrossRef]
74. Hong, H.; Churchman, G.J.; Gu, Y.; Yin, K.; Wang, C. Kaolinite-smectite mixed-layer clays in the Jiujiang red soils and their climate significance. *Geoderma* **2012**, *173–174*, 75–83. [CrossRef]
75. Van Ranst, E.; Kips, P.; Mbogoni, J.; Mees, F.; Dumon, M.; Delvaux, B. Halloysite-smectite mixed-layered clay in fluvio-volcanic soils at the southern foot of Mount Kilimanjaro, Tanzania. *Geoderma* **2020**, *375*, 114527. [CrossRef]
76. Kumari, N.; Mohan., C. Basics of Clay Minerals and Their Characteristic Properties. In *Clay and Clay Minerals*; Morari Do Nascimento, G., Ed.; IntechOpen: London, UK, 2021. Available online: https://www.intechopen.com/online-first/76780 (accessed on 10 October 2021). [CrossRef]
77. Thakur, A.; Jaiswal, M.K.; Peak, C.W.; Carrow, J.K.; Gentry, J.; Dolatshahi-Pirouz, A.; Gaharwar, A.K. Injectable shear thinning nanoengineered hydrogels for stem cell delivery. *Nanoscale* **2016**, *8*, 12362–12372. [CrossRef] [PubMed]
78. Keratvitayanan, P.; Tatullo, M.; Khariton, M.; Joshi, P.; Perniconi, B.; Gaharwar, A.K. Nanoengineered osteoinductive and elastomeric scaffolds for bone tissue enginnering. *ACS Biomater. Sci. Eng.* **2017**, *3*, 590–600. [CrossRef]
79. Schiffman, J.D.; Schauer, C.L. A review: Electrospinning of biopolymer nanofibers and their applications. *Polym. Rev.* **2008**, *48*, 317–352. [CrossRef]
80. Hong, J.-H.; Jeong, E.H.; Lee, H.S.; Baik, D.H.; Seo, S.W.; Youk, J.H. Electrospinning of polyurethane/organically modified montmorillonite nanocomposites. *J. Polym. Sci. Part B Polym. Phys.* **2005**, *43*, 3171–3177. [CrossRef]
81. Gungor-Ozkerim, P.S.; Inci, I.; Zhang, Y.S.; Khademhosseini, A.; Dokmeci, M.R. Bioinks for 3D bioprinting: An overview. *Biomater. Sci.* **2018**, *6*, 915–946. [CrossRef] [PubMed]
82. DeSimone, E.; Schacht, K.; Pellert, A.; Scheibel, T. Recombinant spider silk-based bioinks. *Biofabrication* **2017**, *9*, 044104. [CrossRef] [PubMed]
83. Kang, H.-W.; Lee, S.J.; Ko, I.K.; Kengla, C.; Yoo, J.J.; Atala, A. A 3D bioprinting system to produce human-scale tissue constructs with structural integrity. *Nat. Biotechnol.* **2016**, *34*, 312–319. [CrossRef]
84. El-Husseiny, H.M.; Mady, E.A.; Hamabe, L.; Abugomaa, A.; Shimada, K.; Yoshida, T.; Tanaka, T.; Yokoi, A.; Elbadawy, M.; Tanaka, R. Smart/stimuli-responsive hydrogels: Cutting edge platforms for tissue engineering and other biomedical applications. *Mater. Today Bio* **2022**, *13*, 100186. [CrossRef] [PubMed]
85. Unagolla, J.M.; Jayasuriya, A.C. Hydrogel-based 3D bioprinting: A comprehensive review on cell-laden hydrogels, bioink formulations and future perspectives. *Appl. Mater. Today* **2020**, *18*, 100479. [CrossRef] [PubMed]
86. Guo, J.; Zhang, R.; Zhang, L.; Cao, X. 4D printing of robust hydrogels consisted of agarose nanofibers and polyacrylamide. *ACS Macro Lett.* **2018**, *7*, 442–446. [CrossRef]
87. Lazzara, G.; Cavallaro, G.; Panchal, A.; Fakhrullin, R.; Stavitskaya, A.; Vinokurov, V.; Lvov, Y. An assembly of organic-inorganic composites using halloysite clay nanotube. *Curr. Opin. Colloid Interface Sci.* **2018**, *35*, 42–50. [CrossRef]
88. Cervini-Silva, J.; Ramirez-Apan, M.T.; Kaufhold, S.; Ufer, K.; Palacios, E.; Montoya, A. Role of bentonite clays on cell growth. *Chemosphere* **2016**, *149*, 57–61. [CrossRef] [PubMed]
89. Abduljauwad, S.N.; Ahmed, H.-u.-R.; Moy, V.T. Melanoma treatment via non-specific adhesion of cancer cells using charged nano-clays in pre-clinical studies. *Sci. Rep.* **2021**, *11*, 2737. [CrossRef] [PubMed]
90. Lin, F.H.; Lee, Y.H.; Jian, C.H.; Wong, J.-M.; Shieh, M.-J.; Wang, C.-Y. A study of purified montmorillonite intercalated with 5-fluorouracil as drug carrier. *Biomaterials* **2002**, *9*, 1981–1987. [CrossRef]
91. Yu, M.; Pan, L.; Sun, L.; Li, J.; Shang, J.; Zhang, S.; Liu, D.; Li, W. Supramolecular assemblies constructed from β-cyclodextrin-modified montmorillonite nanosheets as carrier for 5-fluorouracil. *J. Mater. Chem. B* **2015**, *3*, 9043–9052. [CrossRef]
92. Tabasi, H.; Oroojalian, F.; Darroudi, M. Green clay ceramics as potential nanovehicles for drug delivery applications. *Ceram. Int.* **2021**, *47*, 31042–31053. [CrossRef]
93. Zhu, T.; Ma, X.; Chen, R.; Ge, Z.; Xu, J.; Shen, X.; Jia, L.; Zhou, T.; Luo, Y.; Ma, T. Using fluorescently-labeled magnetic nanocomposites as a dual contrast agent for optical and magnetic resonance imaging. *Biomater. Sci.* **2017**, *5*, 1090–1100. [CrossRef] [PubMed]
94. Gianni, E.; Avgoustakis, K.; Papoulis, D. Kaolinite group minerals: Applications in cancer diagnosis and treatment. *Eur. J. Pharm. Biopharm.* **2020**, *154*, 359–376. [CrossRef] [PubMed]
95. Nisticò, R.; Cesano, F.; Garello, F. Magnetic materials and systems: Domain structure visualization and other characterization techniques for the application in the materials science and biomedicine. *Inorganics* **2020**, *8*, 6. [CrossRef]
96. Jury, W.A.; Vaux Jr., H. The role of science in solving the world's emerging water problems. *Proc. Natl. Acad. Sci. USA* **2005**, *102*, 15715–15720. [CrossRef]
97. Maier, J.; Palazzo, J.; Geyer, R.; Steigerwald, D.G. How much potable water is saved by wastewater recycling? Quasi-experimental evidence from California. *Resour. Conserv. Recycl.* **2022**, *176*, 105948. [CrossRef]
98. Jain, M.; Khan, S.A.; Sharma, K.; Jadhao, P.R.; Pant, K.K.; Ziora, Z.M.; Blaskovich, M.A.T. Current perspective of innovative strategies for bioremediation of organic pollutants from wastewater. *Bioresour. Technol.* **2022**, *344*, 126305. [CrossRef] [PubMed]
99. Jain, M.; Khan, S.A.; Pandey, A.; Pant, K.K.; Ziora, Z.M.; Blaskovich, M.A.T. Instructive analysis of engineered carbon materials for potential application in water and wastewater treatment. *Sci. Total Environ.* **2021**, *793*, 148583. [CrossRef] [PubMed]

100. Jaspal, D.; Malviya, A. Composites for wastewater purification: A review. *Chemosphere* **2020**, *246*, 125788. [CrossRef] [PubMed]
101. Rostam, A.B.; Taghizadeh, M. Advanced oxidation processes integrated by membrane reactors and bioreactors for various wastewater treatments: A critical review. *J. Environ. Chem. Eng.* **2020**, *8*, 104566. [CrossRef]
102. Polliotto, V.; Pomilla, F.R.; Maurino, V.; Marcì, G.; Bianco Prevot, A.; Nisticò, R.; Magnacca, G.; Paganini, M.C.; Ponce Robles, L.; Perez, L.; et al. Different approaches for the solar photocatalytic removal of micro-contaminants from aqueous environment: Titania vs. hybrid magnetic iron oxides. *Catal. Today* **2019**, *328*, 164–171. [CrossRef]
103. Bianco Prevot, A.; Baino, F.; Fabbri, D.; Franzoso, F.; Magnacca, G.; Nisticò, R.; Arques, A. Urban biowaste-derived sensitizing materials for caffeine photodegradation. *Environ. Sci. Pollut. Res.* **2017**, *24*, 12599–12607. [CrossRef] [PubMed]
104. Owodunni, A.A.; Ismail, S. Revolutionary technique for sustainable plant-based green coagulants in industrial wastewater treatment-A review. *J. Water Process Eng.* **2021**, *42*, 102096. [CrossRef]
105. Peralta, M.E.; Martire, D.O.; Moreno, M.S.; Parolo, M.E.; Carlos, L. Versatile nanoadsorbents based on magnetic mesostructured silica nanoparticles with tailored surface properties for organic pollutants removal. *J. Environ. Chem. Eng.* **2021**, *9*, 104841. [CrossRef]
106. Nisticò, R.; Tabasso, S.; Magnacca, G.; Jordan, T.; Shalom, M.; Fechler, N. Reactive hypersaline route: One-pot synthesis of porous photoactive nanocomposites. *Langmuir* **2017**, *33*, 5213–5222. [CrossRef] [PubMed]
107. Goh, P.S.; Wong, K.C.; Ismail, A.F. Membrane technology: A versatile tool for saline wastewater treatment and resource recovery. *Desalination* **2022**, *521*, 115377. [CrossRef]
108. Asif, M.B.; Zhang, Z. Ceramic membrane technology for water and wastewater treatment: A critical review of performance, full-scale applications, membrane fouling and prospects. *Chem. Eng. J.* **2021**, *418*, 129481. [CrossRef]
109. Lofrano, G.; Brown, J. Wastewater management through the ages: A history of mankind. *Sci. Total Environ.* **2010**, *408*, 5254–5264. [CrossRef] [PubMed]
110. Plazinski, W.; Rudzinski, W.; Plazinska, A. Theoretical models of sorption kinetics including a surface reaction mechanism: A review. *Adv. Colloid Interface Sci.* **2009**, *152*, 2–13. [CrossRef]
111. Bentahar, Y.; Hurel, C.; Draoui, K.; Khairoun, S.; Marmier, N. Adsorptive properties of Maroccan clays for the removal of arsenic(V) from aqueous solution. *Appl. Clay Sci.* **2016**, *119*, 385–392. [CrossRef]
112. Veli, S.; Alyuz, B. Adsorption of copper and zinc from aqueous solutions by using natural clay. *J. Hazard. Mater.* **2007**, *149*, 226–233. [CrossRef] [PubMed]
113. Bhattacharyya, K.G.; Gupta, S.S. Adsorption of chromium(VI) from water by clays. *Ind. Eng. Chem. Res.* **2006**, *45*, 7232–7240. [CrossRef]
114. Farrah, H.; Pickering, W.F. pH effects in the adsorption of heavy metal ions by clays. *Chem. Geol.* **1979**, *25*, 317–326. [CrossRef]
115. Priyantha, N.; Bandaranayaka, A. Interaction of Cr(VI) species with thermally treated brick clay. *Environ. Sci. Pollut. Res.* **2011**, *18*, 75–81. [CrossRef]
116. Chen, L.; Zhou, C.H.; Fiore, S.; Tong, D.S.; Zhang, H.; Li, C.S.; Ji, S.F.; Yu, W.H. Functional magnetic nanoparticle/clay mineral nanocomposites: Preparation, magnetism and versatile applications. *Appl. Clay Sci.* **2016**, *127–128*, 143–163. [CrossRef]
117. Leonel, A.G.; Mansur, A.A.P.; Mansur, H.S. Advanced functional nanostructures based on magnetic iron oxide nanomaterials for water remediation: A review. *Water Res.* **2021**, *190*, 116693. [CrossRef]
118. Panda, S.K.; Aggarwal, I.; Kumar, H.; Prased, L.; Kumar, A.; Sharma, A.; Vo, D.-V.N.; Thuan, D.V.; Mishra, V. Magnetite nanoparticles as sorbents for dye removal: A review. *Environ. Chem. Lett.* **2021**, *19*, 2487–2525. [CrossRef]
119. Nisticò, R. Magnetic materials and water treatments for a sustainable future. *Res. Chem. Intermed.* **2017**, *43*, 6911–6949. [CrossRef]
120. Najafi, H.; Farajfaed, S.; Zolgharnian, S.; Mirak, S.H.M.; Asasian-Kolur, N.; Sharifian, S. A comprehensive study in modified-pillared clays as ad adsorbent in wastewater treatment processes. *Process Saf. Environ. Prot.* **2021**, *147*, 8–36. [CrossRef]
121. Nisticò, R. A synthetic guide toward the tailored production of magnetic iron oxide nanoparticles. *Bol. Soc. Esp. Cerám. V.* **2021**, *60*, 29–40. [CrossRef]
122. Chiu, C.-W.; Huang, T.-K.; Wang, Y.-C.; Alamani, B.G.; Lin, J.-J. Intercalation strategies in clay/polymer hybrids. *Prog. Polym. Sci.* **2014**, *39*, 443–485. [CrossRef]
123. Magdy, A.; Fouad, Y.O.; Abdel-Aziz, M.H.; Konsowa, A.H. Synthesis and characterization of Fe_3O_4/kaolin magnetic nanocomposite and its application in wastewater treatment. *J. Ind. Eng. Chem.* **2017**, *56*, 299–311. [CrossRef]
124. Coha, M.; Farinelli, G.; Tiraferri, A.; Minella, M.; Vione, D. Advanced oxidation processes in the removal of organic substances form produced water: Potential, configurations, and research needs. *Chem. Eng. J.* **2021**, *414*, 128668. [CrossRef]
125. Brillas, E. A review on the photoelectro-Fenton process as efficient electrochemical advanced oxidation for wastewater remediation. Treatment with UV light, sunlight, and coupling with conventional and other photo-assisted advanced technologies. *Chemosphere* **2020**, *250*, 126198. [CrossRef] [PubMed]
126. Mansouri, L.; Tizaoui, C.; Geissen, S.-U.; Bousselmi, L. A comparative study on ozone, hydrogen peroxide and UV based advanced oxidation processes for efficient removal of diethyl phthalate in water. *J. Hazard. Mater.* **2019**, *363*, 401–411. [CrossRef] [PubMed]
127. Rekhate, C.V.; Srivastava, J.K. Recent advances in ozone-based advanced oxidation processes for treatment of wastewater—A review. *Chem. Eng. J. Adv.* **2020**, *3*, 100031. [CrossRef]
128. Kurian, M. Advanced oxidation processes and nanomaterials—A review. *Clean. Eng. Technol.* **2021**, *2*, 100090. [CrossRef]

129. Andreozzi, R.; Caprio, V.; Insola, A.; Marotta, R. Advanced oxidation processes (AOP) for water purification and recovery. *Catal. Today* **1999**, *53*, 51–59. [CrossRef]
130. Sanakousar, F.M.; Vidyasagar, C.C.; Jimenez-Perez, V.M.; Prakash, K. Recent progress on visible-light-driven metal and non-metal doped ZnO nanostructures for photocatalytic degradation of organic pollutants. *Mater. Sci. Semicond. Process.* **2022**, *140*, 106390. [CrossRef]
131. Bianco Prevot, A.; Arques, A.; Carlos, L.; Laurenti, E.; Magnacca, G.; Nisticò, R. Innovative sustainable materials for the photoinduced remediation of polluted water. In *Sustainable Water and Wastewater Processes*; Galanakis, C.M., Agrafioti, E., Eds.; Elsevier Inc.: Amsterdam, The Netherlands, 2019; Chapter 7; pp. 203–238, ISBN 978-0-12-816170-8. [CrossRef]
132. Singh, P.; Sharma, K.; Hasija, V.; Sharma, V.; Sharma, S.; Raizada, P.; Singh, M.; Saini, A.K.; Hosseini-Bandegharaei, A.; Thakur, V.K. Systematic review on applicability of magnetic iron oxides-integrated photocatalysts for degradation of organic pollutants in water. *Mater. Today Chem.* **2019**, *14*, 100186. [CrossRef]
133. Azzam, M.O.J. Olive mills wastewater treatment using mixed adsorbents of volcanic tuff, natural clay and charcoal. *J. Environ. Chem. Eng.* **2018**, *6*, 2126–2136. [CrossRef]
134. De Paiva, L.B.; Morales, A.R.; Valenzuela Diaz, F.R. Organoclays: Properties, preparation and applications. *Appl. Clay Sci.* **2008**, *42*, 8–24. [CrossRef]
135. Vaia, R.A.; Teukolsky, R.K.; Giannelis, E.P. Interlayer structure and molecular environment of alkylammonium layered silicates. *Chem. Mater.* **1994**, *6*, 1017–1022. [CrossRef]
136. Foorginezhad, S.; Zerafat, M.M. Microfiltration of cationic dyes using nano-clay membranes. *Ceram. Int.* **2017**, *43*, 15146–15159. [CrossRef]
137. Blyweert, P.; Nicolas, V.; Fierro, V.; Celzard, A. 3D printing of carbon-based materials: A review. *Carbon* **2021**, *183*, 449–485. [CrossRef]
138. Ngo, T.D.; Kashani, A.; Imbalzano, G.; Nguyen, K.T.Q.; Hui, D. Additive manufacturing (3D printing): A review of materials, methods, applications and challenges. *Compos. Part B Eng.* **2018**, *143*, 172–196. [CrossRef]
139. Singh, T.; Kumar, S.; Sehgal, S. 3D printing of engineering materials: A state of the art review. *Mater. Today Proc.* **2020**, *28*, 1927–1931. [CrossRef]
140. Zhang, B.; Goel, A.; Ghalsasi, O.; Anand, S. CAD-based design and pre-processing tools for additive manufacturing. *J. Manuf. Syst.* **2019**, *52*, 227–241. [CrossRef]
141. Cano-Vicent, A.; Tambuwala, M.M.; Hassan, S.S.; Barh, D.; Aljabali, A.A.A.; Birkett, M.; Arjunan, A.; Serrano-Aroca, A. Fused deposition modelling: Current status, methodology, applications and future prospects. *Addit. Manuf.* **2021**, *47*, 102378. [CrossRef]
142. Anand, M.; Das, A.K. Issues in fabrication of 3D components through DMLS Technique: A review. *Opt. Laser Technol.* **2021**, *139*, 106914. [CrossRef]
143. Nouri, A.; Shirvan, A.R.; Li, Y.; Wen, C. Additive manufacturing of metallic and polymeric load-bearing biomaterials using laser powder bed fusion: A review. *J. Mater. Sci. Technol.* **2021**, *94*, 196–215. [CrossRef]
144. Dowling, L.; Kennedy, J.; O'Shaughnessy, S.; Trimble, D. A review of critical repeatability and reproducibility issues in powder bed fusion. *Mater. Des.* **2020**, *186*, 108346. [CrossRef]
145. Aboulkhair, N.T.; Simonelli, M.; Perry, L.; Ashcroft, I.; Tuck, C.; Hague, R. 3D printing of aluminium alloys: Additive manufacturing of aluminium alloys using selective laser melting. *Prog. Mater. Sci.* **2019**, *106*, 100578. [CrossRef]
146. Gonzalez-Henriquez, C.M.; Sarabia-Vallejos, M.A.; Rodriguez-Hernandez, J. Polymers for additive manufacturing and 4D-printing: Materials, methodologies, and biomedical applications. *Prog. Polym. Sci.* **2019**, *94*, 57–116. [CrossRef]
147. Chen, Z.; Li, Z.; Li, J.; Liu, C.; Lao, C.; Fu, Y.; Liu, C.; Li, Y.; Wang, P.; He, Y. 3D printing of ceramics: A review. *J. Eur. Ceram. Soc.* **2019**, *39*, 661–687. [CrossRef]
148. Mustapha, K.B.; Metwalli, K.M. A review of fused deposition modelling for 3D printing of smart polymeric materials and composites. *Eur. Polym. J.* **2021**, *156*, 110591. [CrossRef]
149. Xiao, J.; Ji, G.; Zhang, Y.; Ma, G.; Mechtcherine, V.; Pan, J.; Wang, L.; Ding, T.; Duan, Z.; Du, S. Large-scale 3D printing concrete technology: Current status and future opportunities. *Cem. Concr. Compos.* **2021**, *122*, 104115. [CrossRef]
150. Rajabi, M.; McConnell, M.; Cabral, J.; Ali, M.A. Chitosan hydrogels in 3D printing for biomedical applications. *Carbohydr. Polym.* **2021**, *260*, 117768. [CrossRef]
151. Oladapo, B.I.; Zahedi, S.A.; Ismail, S.O.; Omigbodun, F.T. 3D printing of PEEK and its composite to increase biointerfaces as a biomedical material—A review. *Colloids Surf. B Biointerfaces* **2021**, *203*, 111726. [CrossRef]
152. Gao, C.; Lu, C.; Jian, Z.; Zhang, T.; Chen, Z.; Zhu, Q.; Tai, Z.; Liu, Y. 3D bioprinting for fabricating artificial skin tissue. *Colloids Surf. B Biointerfaces* **2021**, *208*, 112041. [CrossRef]
153. Hann, S.Y.; Cui, H.; Esworthy, T.; Miao, S.; Zhou, X.; Lee, S.-J.; Fisher, J.P.; Zhang, L.G. Recent advances in 3D printing: Vascular network for tissue and organ regeneration. *Transl. Res.* **2019**, *211*, 46–63. [CrossRef]
154. Wilms, P.; Daffner, K.; Kern, C.; Gras, S.L.; Schutyser, M.A.I.; Kohlus, R. Formulation engineering of food systems for 3D-printing applications—A review. *Food Res. Int.* **2021**, *148*, 110585. [CrossRef]
155. Chen, Y.; He, S.; Zhang, Y.; Wan, Z.; Copuroglu, O.; Schlangen, E. 3D printing of calcined clay-limestone-based cementitious materials. *Cem. Concr. Res.* **2021**, *149*, 106553. [CrossRef]
156. Long, W.-J.; Lin, C.; Tao, J.L.; Ye, T.-H.; Fang, Y. Printability and particle packing of 3D-printable limestone calcined clay cement composites. *Constr. Build. Mater.* **2021**, *282*, 122647. [CrossRef]

157. Faksawat, K.; Limsuwan, P.; Naemchanthara, K. 3D printing technique of specific bone shape based on raw clay using hydroxyapatite as an additive material. *Appl. Clay Sci.* **2021**, *214*, 106269. [CrossRef]
158. Chikkangoudar, R.N.; Sachidananda, T.G.; Pattar, N. Influence of 3D printing parameters on the dimensional stability of polypropylene/clay printed parts using laser scanning technique. *Mater. Today Proc.* **2021**, *44*, 4118–4123. [CrossRef]
159. Brinker, C.J.; Scherer, G.W. *Sol-Gel Science: The Physics and Chemistry of Sol-Gel Processing*; Academic Press: San Diego, CA, USA, 1990; ISBN 9780080571034.
160. Hench, L.L.; West, J.K. The sol-gel process. *Chem. Rev.* **1990**, *90*, 33–72. [CrossRef]
161. Nisticò, R.; Scalarone, D.; Magnacca, G. Sol-gel chemistry, templating and spin-coating deposition: A combined approach to control in a simple way the porosity of inorganic thin films/coatings. *Microporous Mesoporous Mater.* **2017**, *248*, 18–29. [CrossRef]
162. Ciriminna, R.; Fidalgo, A.; Pandarus, V.; Beland, F.; Ilharco, L.M.; Pagliaro, M. The sol-gel route to advanced silica-based materials and recent applications. *Chem. Rev.* **2013**, *113*, 6592–6620. [CrossRef]
163. Pierre, A.C.; Pajonk, G.M. Chemistry of aerogels and their applications. *Chem. Rev.* **2002**, *102*, 4243–4266. [CrossRef]
164. Lofgreen, J.E.; Ozin, G.A. Controlling morphology and porosity to improve performance of molecularly imprinted sol-gel silica. *Chem. Soc. Rev.* **2014**, *43*, 911–933. [CrossRef]
165. Zhang, Q.; Wang, W.; Goebl, J.; Yin, Y. Self-templated synthesis of hollow nanostructures. *Nano Today* **2009**, *4*, 494–507. [CrossRef]
166. Nisticò, R.; Magnacca, G.; Antonietti, M.; Fechler, N. "Salted silica": Sol-gel chemistry of silica under hypersaline conditions. *Z. Anorg. Allg. Chem.* **2014**, *640*, 582–587. [CrossRef]
167. Pronina, N.; Klauson, D.; Moiseev, A.; Deubener, J.; Krichevskaya, M. Titanium dioxide sol-gel coated expanded clay granules for use in photocatalytic fluidized-bed reactor. *Appl. Catal. B Environ.* **2015**, *178*, 117–123. [CrossRef]

MDPI
St. Alban-Anlage 66
4052 Basel
Switzerland
Tel. +41 61 683 77 34
Fax +41 61 302 89 18
www.mdpi.com

Inorganics Editorial Office
E-mail: inorganics@mdpi.com
www.mdpi.com/journal/inorganics

www.ingramcontent.com/pod-product-compliance
Lightning Source LLC
LaVergne TN
LVHW070400100526
838202LV00014B/1358